AF595001

Machine Learning Modeling for Spatial-Temporal Prediction of Geohazard

Machine Learning Modeling for Spatial-Temporal Prediction of Geohazard

Editors

Junwei Ma
Jie Dou

Basel • Beijing • Wuhan • Barcelona • Belgrade • Novi Sad • Cluj • Manchester

Editors

Junwei Ma
China University of Geosciences
Wuhan, China

Jie Dou
China University of Geosciences
Wuhan, China

Editorial Office
MDPI
St. Alban-Anlage 66
4052 Basel, Switzerland

This is a reprint of articles from the Special Issue published online in the open access journal *Sensors* (ISSN 1424-8220) (available at: https://www.mdpi.com/journal/sensors/special_issues/spatial_temporal_prediction_geohazard).

For citation purposes, cite each article independently as indicated on the article page online and as indicated below:

Lastname, A.A.; Lastname, B.B. Article Title. *Journal Name* **Year**, *Volume Number*, Page Range.

ISBN 978-3-0365-9786-7 (Hbk)
ISBN 978-3-0365-9787-4 (PDF)
doi.org/10.3390/books978-3-0365-9787-4

Contents

About the Editors

Junwei Ma

Junwei Ma is currently an Associate Professor at Badong National Observation and Research Station of Geohazards (BNORSG), China University of Geosciences. He is also a visiting scholar of Purdue University. He received his B.S. and Ph.D. degrees in geological engineering from China University of Geosciences in 2011 and 2016, respectively. His research concerns the engineering properties of soil and rock, landslide stability and reliability analysis, and the application of big data, data mining, and machine learning in geohazards. He has published more than 30 papers in SCI journals. Four of his papers have been deemed ESI Highly Cited Papers, including two ESI Hot Papers. He won the Scientific and Technological Progress Award of Hubei Province. He has also contributed to three Chinese books.

Jie Dou

Jie Dou is a Professor at China University of Geosciences, China. He obtained his Ph.D. from the University of Tokyo and furthered his expertise through collaborations with institutions such as the University of Tokyo and the Public Works Research Institute. His research primarily focuses on geohazards using artificial intelligence (AI) and risk mitigation. Jie Dou is a recipient of the Japan Society for the Promotion of Science (JSPS) fellowship. He has authored over 100 peer-reviewed articles. Dou Jie has served as an Associate Editor of Frontiers in Earth Science and holds positions on the editorial boards of several international journals, including the Journal of Mountains Science, Geocarto International, Geomatics, Natural Hazards, and Risk. Additionally, he is a distinguished reviewer for more than 45 ISI-listed international journals. His commitment extends to being a steering committee member for various commissions/working groups of international academic societies, such as the World Landslide Forum 5 and 6, BIGS2021, BIGS2023, XIV IAEG, etc.

Editorial

Machine Learning Modeling for Spatial-Temporal Prediction of Geohazard

Junwei Ma [1,2,*] and Jie Dou [1,2]

[1] Badong National Observation and Research Station of Geohazards, China University of Geosciences, Wuhan 430074, China; doujie@cug.edu.cn
[2] Three Gorges Research Center for Geohazards of the Ministry of Education, China University of Geosciences, Wuhan 430074, China
* Correspondence: majw@cug.edu.cn

1. Introduction

Geohazards, such as landslides, rock avalanches, debris flow, ground fissures, and ground subsidence, pose significant threats to people's lives and property [1]. Recently, machine learning (ML) has become the predominant approach in geohazard modeling [2–13], offering advantages, like an excellent generalization ability and accurately describing complex and nonlinear behaviors. However, the utilization of advanced algorithms in deep learning remains poorly understood in this field [7,8]. Additionally, there are fundamental challenges associated with ML modeling, including input variable selection, uncertainty quantification, and hyperparameter tuning [3,5–13].

This Special Issue presents original research exploring new frontiers and challenges in applying ML for the spatial-temporal modeling of geohazards. The topics covered include geohazard modeling, spatial-temporal prediction, ML, deep and reinforcement learning, the metaheuristic optimized ML approach, and physics-based and data-driven hybrid modeling.

2. Overview of Contribution

This Special Issue titled "Machine Learning Modeling for Spatial-Temporal Prediction of Geohazard" comprises eleven high-quality papers, including one systematic review article and ten original research articles conducted by researchers from Canada, China, Iran, Malaysia, Pakistan, and Sweden. These ten research articles can be categorized as follows: the susceptibility analysis of glacier debris flow and landslides (contributions 1–3), the displacement prediction of reservoir landslides (contributions 4–6), slope stability prediction and classification (contributions 7–8), building resilience evaluation (contribution 9), and the prediction of rainfall-induced landslide warning signals (contribution 10). Modern ML techniques, including metaheuristic optimized ML, deep learning, and automated ML, have been applied to the spatial-temporal modeling of geohazards in various regions, such as Kurdistan in Iran, Karakorum Highway in Pakistan, and Chongqing, G318 Linzhi Section, and the Three Gorges Reservoir area in China.

Geohazard susceptibility mapping is the central theme of this Special Issue (contributions 1–3). For instance, the susceptibility mapping of glacier debris flows along the G318 Linzhi Section in China was generated using remote sensing imagery and a convolutional-neural-network-based image segmentation model, DeepLabv3+ (contribution 1). In the context of landslide susceptibility mapping, a deep learning model that combines extreme machine learning, a deep belief network, back propagation, and a genetic algorithm has been proposed and validated in Kamyaran in the Kurdistan Province, Iran (contribution 2). The proposed deep learning models achieved satisfactory performances, with values exceeding 0.90 (contributions 1 and 2), underscoring the effectiveness of deep learning in

Citation: Ma, J.; Dou, J. Machine Learning Modeling for Spatial-Temporal Prediction of Geohazard. *Sensors* **2023**, *23*, 9262. https://doi.org/10.3390/s23229262

Received: 19 October 2023
Accepted: 10 November 2023
Published: 18 November 2023

geohazard susceptibility mapping. In the research conducted by Hussain et al. (contribution 3), landslide susceptibility mapping was compared using random forest, extreme gradient boosting, k-nearest neighbor, and naive Bayes in a case study along Karakorum Highway in Northern Pakistan.

Another significant focus of this Special Issue is the prediction of reservoir landslide displacements. Due to seasonal rainfall and periodic reservoir fluctuations, the deformations of reservoir landslide are characterized by a step-like behavior. Innovative approaches based on the decomposition and ensemble principle have been introduced to predict displacements in the cases of the Shiliushubao (contribution 4), Baijiabao (contribution 5), and Baishuihe landslides (contribution 6). These approaches incorporate mode decomposition, input variable selection, individual prediction, and ensemble prediction to achieve a satisfactory performance, nearly optimizing the goodness of fit. Decomposition techniques, such as complete ensemble empirical mode decomposition (contributions 4 and 5) and variational mode decomposition (contribution 6), are utilized to break down cumulative displacement into trend, periodic, and random components. Methods like edit distance for real sequence (contribution 4), gray relational analysis, and association rule mining (contribution 6) have been proposed for the selection of input variables. For individual prediction, various methods, including metaheuristic optimized support vector regression (contribution 4), back propagation neural networks (contribution 6), and gated recurrent unit deep learning (contribution 5), are employed to predict the decomposed displacements, which are then aggregated into a final ensemble prediction. In particular, Zhang et al. (contribution 4) evaluate the performance of hyperparameter tuning using metaheuristic techniques, such as the bat algorithm, grey wolf optimization, dragonfly algorithm, whale optimization algorithm, grasshopper optimization algorithm, and sparrow search algorithm. The abovementioned works (contributions 4–6) contribute significantly to the field of model decomposition, input variable selection, and hyperparameter tuning.

Slope stability prediction and classification (contributions 7 and 8) represent another prominent theme in this Special Issue. Wu et al. (contribution 7) developed a stability prediction model for slope with predetermined shear planes with Box–Jenkins' modeling approach using a mechanical-informed dataset. For the first time, an automated ML model for slope stability classification has been developed with minimal human intervention by Ma et al. (contribution 8). The AuotML model provides an attractive alternative to traditional ML practice, especially for early-stage researchers with limited expertise in ML.

In the work by Zhang et al. (contribution 9), an ML-based model for assessing the resilience of buildings was developed and evaluated in Banan District, a typical mountainous city in Chongqing, China. Furthermore, Zhang et al. (contribution 10) proposed a hybrid model that combines an attention-based temporal convolutional neural network with entropy weight methods for predicting rainfall-induced landslide warning signals.

Additionally, in a review article entitled "Scientometric Analysis of Artificial Intelligence (AI) for Geohazard Research", Jiang et al. (contribution 11) conducted a scientometric review of artificial intelligence for geohazard research based on thousands of records from the Web of Science core collection. This analysis identified and visualized the most productive researchers, institutions, and emerging research topics using animated maps, and it also provided recommendations for future directions. This scientometric review holds promise in offering a comprehensive and objective overview of publication characteristics and emerging trends for researchers in the field.

3. Conclusions

This Special Issue provides a forum for presenting original research that delves into novel frontiers and confronts challenges in utilizing ML for geohazard susceptibility mapping, geohazard prediction, slope stability prediction, building resilience evaluation, and landslide early warning systems. Within these domains, advanced ML techniques, including deep learning, metaheuristic optimized ML, ensemble learning, and AutoML, have been introduced. We anticipate that these innovative techniques and approaches will be valuable contributions that are warmly received by both researchers and practitioners in the field.

Funding: This work was supported by the National Natural Science Foundation of China (Grant Nos. 42177147 and 71874165), the Key Research and Development Project of Hubei Provincial Technology Innovation Plan (No. 2023BCB117), and the Fundamental Research Funds for the Central Universities, China University of Geosciences (Wuhan) (CUG2642022006).

Acknowledgments: The Guest Editors express their sincere gratitude to all the authors, as well as MDPI and the Managing Editor for their invaluable advice and support for this Sensors Special Issue "Machine Learning Modeling for Spatial-Temporal Prediction of Geohazard".

Conflicts of Interest: The authors declare no conflict of interest.

List of Contributions

1. Chen, J.; Gao, H.; Han, L.; Yu, R.; Mei, G. Susceptibility Analysis of Glacier Debris Flow Based on Remote Sensing Imagery and Deep Learning: A Case Study along the G318 Linzhi Section. *Sensors* **2023**, *23*, 6608. https://doi.org/10.3390/s23146608.
2. Ghasemian, B.; Shahabi, H.; Shirzadi, A.; Al-Ansari, N.; Jaafari, A.; Kress, V.R.; Geertsema, M.; Renoud, S.; Ahmad, A. A Robust Deep-Learning Model for Landslide Susceptibility Mapping: A Case Study of Kurdistan Province, Iran. *Sensors* **2022**, *22*, 1573. https://doi.org/10.3390/s22041573.
3. Hussain, M.A.; Chen, Z.; Zheng, Y.; Shoaib, M.; Shah, S.U.; Ali, N.; Afzal, Z. Landslide Susceptibility Mapping Using Machine Learning Algorithm Validated by Persistent Scatterer In-SAR Technique. *Sensors* **2022**, *22*, 3119. https://doi.org/10.3390/s22093119.
4. Zhang, J.; Tang, H.; Tannant, D.D.; Lin, C.; Xia, D.; Wang, Y.; Wang, Q. A Novel Model for Landslide Displacement Prediction Based on EDR Selection and Multi-Swarm Intelligence Optimization Algorithm. *Sensors* **2021**, *21*, 8352. https://doi.org/10.3390/s21248352.
5. Yang, B.; Xiao, T.; Wang, L.; Huang, W. Using Complementary Ensemble Empirical Mode Decomposition and Gated Recurrent Unit to Predict Landslide Displacements in Dam Reservoir. *Sensors* **2022**, *22*, 1320. https://doi.org/10.3390/s22041320.
6. Miao, F.; Xie, X.; Wu, Y.; Zhao, F. Data Mining and Deep Learning for Predicting the Displacement of "Step-like" Landslides. *Sensors* **2022**, *22*, 481. https://doi.org/10.3390/s22020481.
7. Wu, T.; Yu, H.; Jiang, N.; Zhou, C.; Luo, X. Slope with Predetermined Shear Plane Stability Predictions under Cyclic Loading with Innovative Time Series Analysis by Mechanical Learning Approach. *Sensors* **2022**, *22*, 2647. https://doi.org/10.3390/s22072647.
8. Ma, J.; Jiang, S.; Liu, Z.; Ren, Z.; Lei, D.; Tan, C.; Guo, H. Machine Learning Models for Slope Stability Classification of Circular Mode Failure: An Updated Database and Automated Machine Learning (AutoML) Approach. *Sensors* **2022**, *22*, 9166. https://doi.org/10.3390/s22239166.
9. Zhang, C.; Wen, H.; Liao, M.; Lin, Y.; Wu, Y.; Zhang, H. Study on Machine Learning Models for Building Resilience Evaluation in Mountainous Area: A Case Study of Banan District, Chongqing, China. *Sensors* **2022**, *22*, 1163. https://doi.org/10.3390/s22031163.
10. Zhang, D.; Wei, K.; Yao, Y.; Yang, J.; Zheng, G.; Li, Q. Capture and Prediction of Rainfall-Induced Landslide Warning Signals Using an Attention-Based Temporal Convolutional Neural Network and Entropy Weight Methods. *Sensors* **2022**, *22*, 6240. https://doi.org/10.3390/s22166240.
11. Jiang, S.; Ma, J.; Liu, Z.; Guo, H. Scientometric Analysis of Artificial Intelligence (AI) for Geohazard Research. *Sensors* **2022**, *22*, 7814. https://doi.org/10.3390/s22207814.

References

1. Ma, J.W.; Xia, D.; Wang, Y.K.; Niu, X.X.; Jiang, S.; Liu, Z.Y.; Guo, H.X. A Comprehensive Comparison among Metaheuristics (MHs) for Geohazard Modeling using Machine Learning: Insights from A Case Study of Landslide Displacement Prediction. *Eng. Appl. Artif. Intell.* **2022**, *114*, 105150. [CrossRef]
2. Dou, J.; Xiang, Z.; Xu, Q.; Zheng, P.; Wang, X.; Su, A.; Liu, J.; Luo, W. Application and development trend of machine learning in landslide intelligent disaster prevention and mitigation. *Earth Sci.* **2023**, *48*, 1657. [CrossRef]
3. Jiang, S.; Ma, J.; Liu, Z.; Guo, H. Scientometric analysis of artificial intelligence (AI) for geohazard research. *Sensors* **2022**, *22*, 7814. [CrossRef] [PubMed]
4. Zhang, W.; Pradhan, B.; Stuyts, B.; Xu, C. Application of artificial intelligence in geotechnical and geohazard investigations. *Geol. J.* **2023**, *58*, 2187–2194. [CrossRef]
5. Dikshit, A.; Pradhan, B.; Alamri, A.M. Pathways and challenges of the application of artificial intelligence to geohazards modelling. *Gondwana Res.* **2021**, *100*, 290–301. [CrossRef]
6. Phoon, K.-K.; Zhang, W. Future of machine learning in geotechnics. *Georisk Assess. Manag. Risk Eng. Syst. Geohazards* **2022**, *17*, 7–22. [CrossRef]
7. Ma, Z.J.; Mei, G. Deep learning for geological hazards analysis: Data, models, applications, and opportunities. *Earth Sci. Rev.* **2021**, *223*, 103858. [CrossRef]
8. Zhang, W.; Gu, X.; Tang, L.; Yin, Y.; Liu, D.; Zhang, Y. Application of machine learning, deep learning and optimization algorithms in geoengineering and geoscience: Comprehensive review and future challenge. *Gondwana Res.* **2022**, *109*, 1–17. [CrossRef]
9. Ma, Z.J.; Mei, G.; Piccialli, F. Machine learning for landslides prevention: A survey. *Neural Comput. Appl.* **2020**, *33*, 10881–10907. [CrossRef]
10. Bergen, K.J.; Johnson, P.A.; de Hoop, M.V.; Beroza, G.C. Machine learning for data-driven discovery in solid Earth geoscience. *Science* **2019**, *363*, eaau0323. [CrossRef] [PubMed]
11. Pu, Y.; Apel, D.B.; Liu, V.; Mitri, H. Machine learning methods for rockburst prediction-state-of-the-art review. *Int. J. Min. Sci. Technol.* **2019**, *29*, 565–570. [CrossRef]
12. Dramsch, J.S. Chapter One—70 years of machine learning in geoscience in review. In *Advances in Geophysics*; Moseley, B., Krischer, L., Eds.; Elsevier: Amsterdam, The Netherlands, 2020; Volume 61, pp. 1–55.
13. Sun, Z.; Sandoval, L.; Crystal-Ornelas, R.; Mousavi, S.M.; Wang, J.; Lin, C.; Cristea, N.; Tong, D.; Carande, W.H.; Ma, X.; et al. A review of Earth Artificial Intelligence. *Comput. Geosci.* **2022**, *159*, 105034. [CrossRef]

Review

Scientometric Analysis of Artificial Intelligence (AI) for Geohazard Research

Sheng Jiang [1,2], Junwei Ma [1,2,*], Zhiyang Liu [1,2] and Haixiang Guo [3]

1 Badong National Observation and Research Station of Geohazards (BNORSG), China University of Geosciences, Wuhan 430074, China
2 Three Gorges Research Center for Geohazards of the Ministry of Education, China University of Geosciences, Wuhan 430074, China
3 School of Economics and Management, China University of Geosciences, Wuhan 430074, China
* Correspondence: majw@cug.edu.cn; Tel.: +86-158-7243-4086

Abstract: Geohazard prevention and mitigation are highly complex and remain challenges for researchers and practitioners. Artificial intelligence (AI) has become an effective tool for addressing these challenges. Therefore, for decades, an increasing number of researchers have begun to conduct AI research in the field of geohazards leading to rapid growth in the number of related papers. This has made it difficult for researchers and practitioners to grasp information on cutting-edge developments in the field, thus necessitating a comprehensive review and analysis of the current state of development in the field. In this study, a comprehensive scientometric analysis appraising the state-of-the-art research for geohazard was performed based on 9226 scientometric records from the Web of Science core collection database. Multiple types of scientometric techniques, including coauthor analysis, co-citation analysis, and cluster analysis were employed to identify the most productive researchers, institutions, and hot research topics. The results show that research related to the application of AI in the field of geohazards experienced a period of rapid growth after 2000, with major developments in the field occurring in China, the United States, and Italy. The hot research topics in this field are ground motion, deep learning (DL), and landslides. The commonly used AI algorithms include DL, support vector machine (SVM), and decision tree (DT). The obtained visualization on research networks offers valuable insights and an in-depth understanding of the key researchers, institutions, fundamental articles, and salient topics through animated maps. We believe that this scientometric review offers useful reference points for early-stage researchers and provides valuable in-depth information to experienced researchers and practitioners in the field of geohazard research. This scientometric analysis and visualization are promising for reflecting the global picture of AI-based geohazard research comprehensively and possess potential for the visualization of the emerging trends in other research fields.

Keywords: geohazard; artificial intelligence (AI); scientometric; visualization; research cluster

Citation: Jiang, S.; Ma, J.; Liu, Z.; Guo, H. Scientometric Analysis of Artificial Intelligence (AI) for Geohazard Research. *Sensors* **2022**, *22*, 7814. https://doi.org/10.3390/s22207814

Academic Editor: Giulio Iovine

Received: 15 September 2022
Accepted: 12 October 2022
Published: 14 October 2022

1. Introduction

According to the Occupational Safety and Health Administration (OSHA, https://www.ccohs.ca/oshanswers/hsprograms/hazard_risk.html, accessed on 3 October 2022), a hazard is any source of potential damage, harm, or adverse health effects on something or someone under certain conditions at work. Geohazards refer to events caused by geological conditions or processes that pose a threat to human life, property, or the natural environment [1]. According to the Emergency Events Database (EM-DAT, https://public.emdat.be/, accessed on 7 July 2022), a global database of technical and natural disasters, 1877 large-scale geohazards occurred worldwide between 1 January 1990 and 7 July 2022. These disasters killed 2.43 million people, left 25.74 million people homeless, and caused $862 million in damages. Japan and China are the countries with the highest losses due to geohazards, which caused approximately $392 million and $114 million in damages, respectively (Figure 1a). As shown, the number of geohazards increased from 1990 to 2000. Asia and the Americas, which account for 55.5% and 22.9% of the total number of geohazards worldwide, respectively, have suffered the most from geohazards (see Figure 1c).

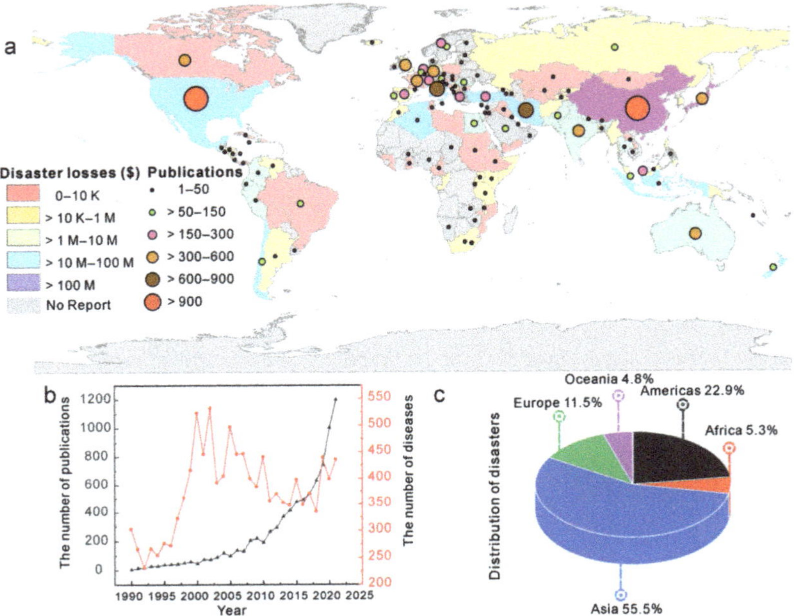

Figure 1. (**a**) The distribution of geohazard loss (Source: https://public.emdat.be/data, accessed on 7 July 2022) and the number of papers. Different colors indicate different degrees of geohazard loss and the size and color of the circles indicate the number of papers published in that country. (**b**) The change in the number of geohazards and the number of publications over time. (**c**) The regional distribution of the number of geohazards.

Great efforts have been made in geohazard prevention and mitigation [2–4]. However, geohazards are characterized as complex and uncertain [5,6]; thus, challenges remain for researchers and practitioners [7]. Recently, artificial intelligence (AI) has become popular among researchers and practitioners and has led to considerable advances in geohazard research. Affected by multiple triggering factors [8,9], the monitoring data of the geohazard are usually characterized with complex and nonlinear relationships. For example, due to seasonal rainfall and periodic reservoir fluctuation, the landslide movements in the Three Gorges Reservoir area are characterized with step-like deformation, which makes the displacement predictions remain as challenges. AI is able to analyze these complex and nonlinear characteristics well by establishing a mapping between the input feature data and the output final results [10]. AI has proven its capability in dealing with high-dimensional and large-scale datasets by providing satisfactory predictions [11]. Moreover, AI, a data-driven approach, relies less on expertise and clear understanding of physical processes [12]. Based on previous review works [13,14], AI is widely used in the geohazard field [15,16] (see Figure 2). For example, Kalantar et al. and Xia et al. [17,18] modeled landslide susceptibility using the support vector machine (SVM) algorithm, logistic regression (LR) algorithm, and artificial neural network (ANN) algorithm. Ghorbanzadeh et al. [19] evaluated the application of deep learning (DL) in landslide identification. Zhang et al. [20] used ML algorithms such as decision tree (DT) and random forest (RF) to map landslide susceptibility. Mousavi et al. [21] proposed a DL model for simultaneous seismic detection and phase selection. Wu et al. [22] used AI algorithms such as SVM for tunnel collapse risk assessment. Choubin et al. [23] adopted AI algorithms such as multivariate discriminant analysis for the prediction of avalanche hazards. Valade et al. [24] implemented intelligent monitoring of global volcanic activity using AI techniques on multisensory satellite-based imagery from Sentinel-1. The rapid development of AI research in geohazards has led to a rapid increase in the number of publications on the subject. This makes it difficult for researchers and practitioners to keep abreast of cutting-edge research information and the overall status of research in this field, which can easily lead to meaningless and repetitive studies. To solve this problem, a scientometric analysis and review of the current state of recent research in this area is necessary.

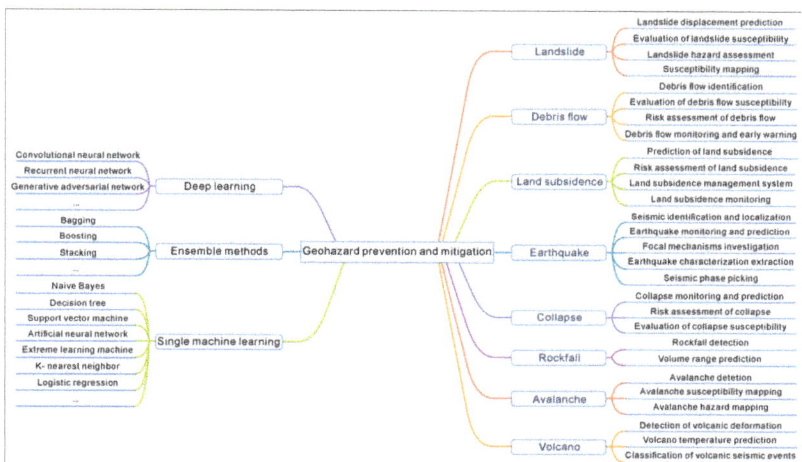

Figure 2. Main AI algorithms and applications in geohazard prevention mitigation.

Several researchers have previously conducted review studies in this field. For example, Dikshit et al. [13] provided a qualitative analysis of the application of AI in geohazards highlighting the direction of development in this field. Huang et al. and Merghadi et al. [25,26] analyzed the application of DL in the field of landslide susceptibility evaluation. Xie et al. [27] provided an overview of the applications and prospects of machine learning (ML) in the field of seismic research. Despite their important contributions to the development of the field, these review studies have some limitations. Most of these review studies are qualitative or are limited to the application of a particular AI to a certain type of geohazard; thus, there is a lack of quantitative and comprehensive review studies of the development of AI in geohazard research. In addition, the current review studies in the field do not include an analysis of the publication characteristics of existing papers, the main authors, institutions, and countries or the studies related to the identification of hot research in the field. Therefore, the current review studies do not provide a comprehensive and objective description of the current state of research of AI in the field of geohazards.

A scientometric analysis, which refers to the quantitative study of science and communication in science [28], is promising for addressing the abovementioned limitations as it can handle large volumes of publications; thus offering a visualization of research networks of key scholars, institutions, fundamental articles, and salient topics. Therefore, scientometric reviews have been applied to various research fields [29–32]. However, so far, no previous reviews have conducted the scientometric analysis of AI-based geohazard research by identification of the salient term and research trend and mapping interconnection.

To fill the gap in quantitative analysis research in geohazard reviews and promote development, quantitative analysis methods are used in this study to analyze and summarize the development of AI in geohazard research from 1990 to 2022. This study contributes to the development of the field of geohazards by objectively presenting the current research status and future directions of AI in this field. The main researchers, institutions, countries, and hot research topics are identified. The advantages and limitations of popular AI algorithms in the field of geohazards are analyzed, and future directions are discussed.

2. Materials and Methods

Scientometric analysis is a method of scientific analysis that shows the logic and connections between documents by mapping, mining, ranking, and analyzing them [33]. Various techniques, such as BibExcel, HistCite, and CiteSpace are available to achieve this goal. CiteSpace (version 5.4. R1 64 bit) [34,35] was chosen in this study because the clarity and interpretability of the resulting visualizations are better than those of other scientometric analysis tools. In the present study, a scientometric analysis of AI for geohazard research was performed based on the following three procedures (see Figure 3).

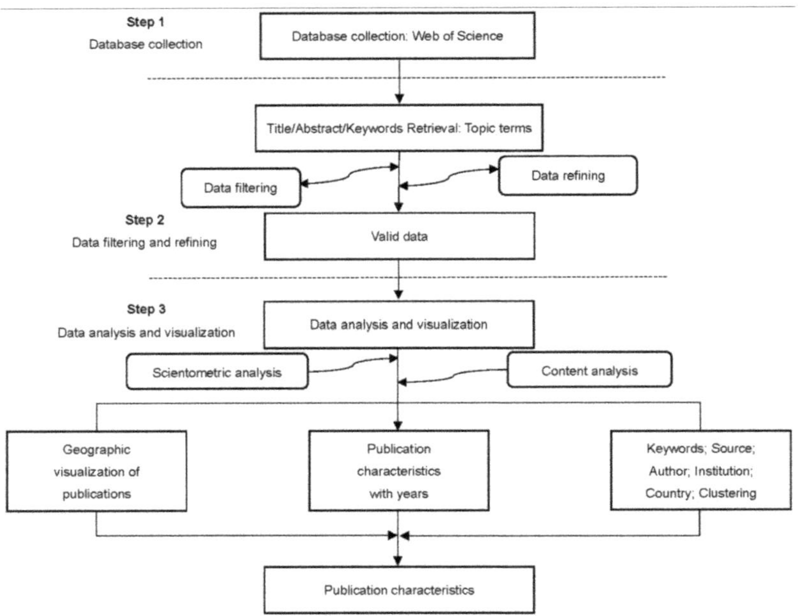

Figure 3. Flow chart for scientometric analysis of AI for geohazard research.

Data collection: Web of Science (http://apps.webofknowledge.com, accessed on 7 March 2022) is a comprehensive database with high-quality citation analysis [36] that is based on high-quality citation data, publication standards, and expert judgment. This database is of higher quality, contains more specialized data than other databases (such as Scopus and Google Scholar [37]) and can support a longer period of citation analysis. Thus, in this study, Web of Science was adopted for data collection. A search was performed for the topic (TS) query in Web of Science using the following formula "TS = ((fuzzy sets or naive Bayes or linear regression or random forests or gradient boosting or reinforcement learning or meta heuristics or AI or artificial intelligence or optimization algorithm or machine learning or deep learning or computational intelligence or decision tree or prediction model) AND (geohazard or landslide or slope or rockfall or collapse or earthquake or debris flow or hazard or tsunami))". Based on the literature search publications in the English language were selected. The year of publication ranged from 1 January 1990 to 1 January 2022, and the subject categories were refined to GeoScience Multidisciplinary and Engineering Geological. A total of 9226 documents were retrieved for scientometric analysis.

Data filtering and refining: Subject terms were identified, subject searches were performed in the Web of Science database ("title, abstract, author keywords, and KeyWords

Plus"), and Boolean operators (OR/AND) were used to expand the search and exclude irrelevant papers. After filtering and refining the search results to determine the time frame, the search results were downloaded and prepared for the next step of the analysis.

Data analysis and visualization: After filtering and refining the papers, the data were visually represented by using the visualization tool CiteSpace. Cluster analysis, an exploratory data mining technique, was adopted for the identification of the salient term and context, research trend, and interconnection. Log-likelihood ratio was used as the clustering index due to advantages of high-quality classification with high intra-class similarity and low inter-class similarity. A cluster overlap indicates that there are relationships between keywords of these different clusters. CiteSpace was used as a tool for performing cluster analysis. A visualization map generated by CiteSpace consists of color-coded nodes and links that describe co-citations or cooccurrences between these nodes. Each representative node, which is made up of a "tree ring" of different colors, denotes one specific item (e.g., country, institution, keyword, author, cited reference, or cited journal). The spectrum of colors denotes the temporal order: oldest in blue and newest in orange. The size of the ring represents the frequency of the corresponding item in a particular year. A red ring present in a particular year denotes a burst, that is, a surge of occurrences or citations in that year. Based on the data visualization, scientometrics and content analysis of the search results in the field are performed and discussed and the results are derived.

3. Results

3.1. Analysis of Publication Characteristics

3.1.1. Publication Distribution Characteristics

The characteristics of the publications on AI for geohazard research over time are shown in Figure 4. As shown, since 1990, the number of papers published in this field has continued to increase. After entering the 21st century, with the rapid development of AI technology, AI technology in the field of geohazards has developed quickly. As shown in Figure 1b, after 2000 there was an overall decreasing trend of geohazards. At the same time the research of AI in the field of geohazards began to grow rapidly, which to some extent reflects the help of AI in geohazard prevention and mitigation.

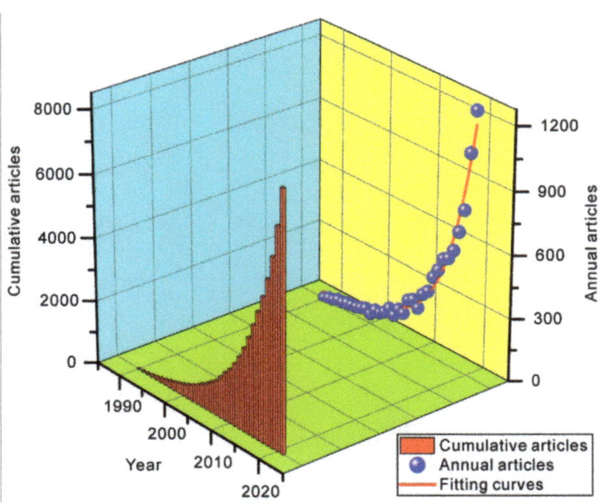

Figure 4. Characteristics of publications of AI for geohazard research by year.

The spatial distribution characteristics of publications are shown in Figure 1a. The circles of different colors and sizes in the figure indicate the total number of papers published in the countries where the circles are located. As shown, China and the United States, with a total of 2320 publications and 1993 publications, respectively, are the two countries with the highest total number of publications in this field. In addition, the total number of national papers is positively related to geohazard losses in that country.

3.1.2. Publication Source Characteristics

To provide researchers with references to authoritative publication sources and to facilitate access to relevant and cutting-edge papers, the top 15 sources of AI for geohazard research are shown in Table 1. The sources of the top 15 papers are all journals and there are no conference papers; these data suggest that researchers prefer to publish their papers in journals rather than academic conferences. In addition, the papers from the top 15 sources account for 48.46% of the total number of papers. Among them, **Natural Hazards** and **Remote Sensing**, with 5.96% and 4.58% of the total number of papers, respectively, are among the top two source journals in this field. Among the top 15 sources, **Geomorphology** (17,656 citations), **Journal of Hydrology** (13,417 citations), and **Catena** (12,810 citations) are the most cited. Those results correspond well with the bibliometric review of Wu et al. [38] in the field of AI. Among the top 15 sources, **Geomorphology** (61.52), **Earthquake Spectra** (50.59), and **Catena** (47.62) have the highest average number of citations. **Geomorphology** and **Catena** not only have a high total number of citations but also a high average number of citations. Therefore, they are considered to be the most active journals in this field.

Table 1. Top 15 source journals according to the number of publications in AI for geohazard research (1990–2022).

No.	Source	Total Papers	Total Citations	Average Citations per Paper	Percentage of Total Papers
1	Natural Hazards (https://www.springer.com/journal/11069, accessed on 3 September 2022)	550	12,286	22.34	5.96%
2	Remote Sensing (https://www.mdpi.com/journal/remotesensing, accessed on 3 September 2022)	423	6376	15.07	4.58%
3	Journal of Hydrology (https://journals.elsevier.com/journal-of-hydrology, accessed on 3 September 2022)	314	13,417	42.73	3.40%
4	Bulletin of Earthquake Engineering (https://www.springer.com/journal/10518/, accessed on 3 September 2022)	306	6068	19.83	3.32%
5	Soil Dynamics and Earthquake Engineering (https://www.sciencedirect.com/journal/soil-dynamics-and-earthquake-engineering, accessed on 3 September 2022)	306	4060	13.27	3.32%
6	Environmental Earth Sciences (https://www.springer.com/journal/12665, accessed on 3 September 2022)	298	7798	26.17	3.23%
7	Engineering Geology (https://www.sciencedirect.com/journal/engineering-geology, accessed on 3 September 2022)	290	11,417	39.37	3.14%
8	Geomorphology (www.elsevie r.com/locate/geomorph, accessed on 3 September 2022)	287	17,656	61.52	3.11%
9	Catena (www.elsevier.com/locate/catena, accessed on 3 September 2022)	269	12,810	47.62	2.92%

Table 1. *Cont.*

No.	Source	Total Papers	Total Citations	Average Citations per Paper	Percentage of Total Papers
10	Natural Hazards and Earth System Sciences (https://www.natural-hazards-and-earth-system-sciences.net/, accessed on 3 September 2022)	264	6611	25.04	2.86%
11	Landslides (https://www.springer.com/journal/10346, accessed on 3 September 2022)	262	8970	34.24	2.84%
12	Arabian Journal of Geosciences (https://www.springer.com/journal/12517, accessed on 3 September 2022)	243	3298	13.57	2.63%
13	Earthquake Engineering & Structural Dynamics (https://onlinelibrary.wiley.com/journal/10969845, accessed on 3 September 2022)	233	7756	33.29	2.53%
14	Earthquake Spectra (https://journals.sagepub.com/home/eqs, accessed on 3 September 2022)	213	10,776	50.59	2.31%
15	Geophysical Research Letters (https://agupubs.onlinelibrary.wiley.com/journal/19448007, accessed on 3 September 2022)	213	5269	24.74	2.31%

3.1.3. Publication Keyword Characteristics

Figure 5 shows the 10 keywords with the strongest citation burst of AI research citations in the field of geohazards, representing the main interests of researchers in the field. As shown, researchers have been interested in researching AI techniques in geohazards since 2012 when fuzzy logic was a popular research topic in the field. New research hotspots have gradually emerged. In 2016, the analytical hierarchy process appeared. In 2017, SVM became the third most cited keyword in this citation burst. Subsequently, LR, DL, and many other ML algorithms began to be widely applied in geohazards and became hot topics in the field. Furthermore, the different sizes of circles in Figure 5 indicate different occurrence frequencies; the larger the circle is, the higher the occurrence frequency. Among them, LR appeared 671 times and was the most popular keyword in the field. Moreover, an increasing number of AI algorithms have been applied to the field of geohazards by researchers.

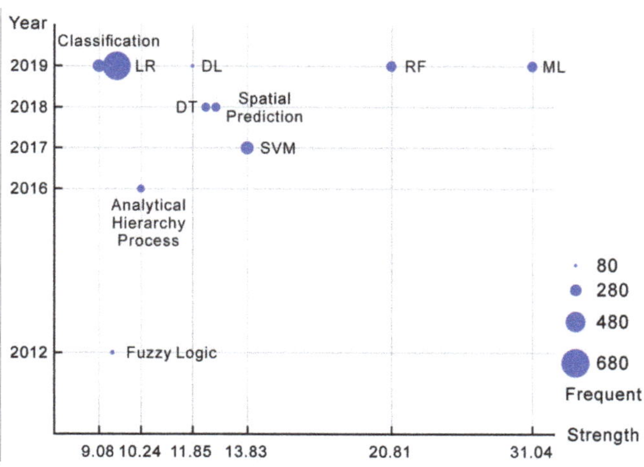

Figure 5. Citation burst of AI in geohazard research during 1990–2022 (logarithmic scale).

3.2. Analysis of Authors, Institutions, and Countries

3.2.1. Most Productive Authors in AI Research in the Field of Geohazards

The affiliations of the top 15 most productive authors in terms of AI research in the field of geohazards and their H-index, total number of papers, and total number of citations are shown in Table 2. As shown in this table, each of these authors published 26 or more papers. These authors have published a cumulative total of 732 papers, which accounts for 7.93% of the papers published by researchers worldwide. Six of these researchers are affiliated with Asian institutions, with three researchers from China and three from Iran. Therefore, AI in the field of geohazards is considered to be developing rapidly in Asia, with China and Iran being the main participating countries in this field of research.

Table 2. Top 15 most productive authors for AI research in the field of geohazards (1990–2022).

No.	Name	Institution, Country	Total Papers	Total Citations	Average Citations per Paper	H-Index	Related Citations Impact
1	Pradhan, Biswajeet	University of Technology Sydney, Australia	136	13,146	96.66	94	1.39
2	Dieu Tien Bui	University of South-Eastern Norway, Norway	70	5604	80.06	68	1.16
3	Pourghasemi, Hamid Reza	Shiraz University, Iran	65	6332	97.42	66	1.41
4	Chen, Wei	Xi'an University of Science and Technology, China	63	4104	65.14	53	0.94
5	Lee, Saro	Korea Institute of Geoscience and Mineral Resources, KIGAM, Korea	63	3880	61.59	64	0.89
6	Hong, Haoyuan	Universität Vienna, Austria	52	3637	69.94	45	1.01
7	Binh Thai Pham	University of Transport Technology, Vietnam	45	2851	63.36	26	0.91
8	Arabameri, Alireza	Tarbiat Modares University, Iran	32	684	21.38	29	0.31
9	Bradley, Brendon A.	University of Canterbury, New Zealand	32	583	18.22	36	0.26
10	Xu, Chong	Institute of Geology, China Earthquake Administration, China	31	1693	54.61	29	0.79
11	Shahabi, Himan	University of Kurdistan, Iran	30	2538	84.60	52	1.22
12	Xu, Qiang	Chengdu University of Technology, China	30	659	21.97	43	0.32
13	Rahmati, Omid	Agricultural Research, Education and Extension Organization (AREEO), Vietnam	29	2231	76.93	36	1.11
14	Prakash, Indra	Bhaskaracharya Institute for Space Applications and Geoinformatics, India	28	1551	55.39	37	0.80
15	Blaschke, Thomas	Universität Salzburg, Austria	26	1235	47.50	50	0.69

Note: The H-index in the table header means that the author has H papers cited H times.

The most productive authors are Pradhan, Biswajeet (136); Dieu Tien Bui (70); and Pourghasemi, Hamid Reza (65); who are also the most cited authors. Six of the top 15 authors have a higher number of average citations (i.e., higher than 69.30) than the rest. They are Pourghasemi, Hamid Reza (97.42); Pradhan, Biswajeet (96.66); Shahabi, Himan (84.60); Dieu Tien Bui (80.06); Rahmati, Omid (76.93); and Hong, Haoyuan (69.94). Four authors have an H-index above 60: Pradhan, Biswajeet (94); Dieu Tien Bui (68); Pourghasemi, Hamid Reza (66); and Lee, Saro (64). Six authors have a relative citation impact greater than 1 relative to the other top 15 authors. They are Pourghasemi, Hamid Reza (1.41); Pradhan, Biswajeet (1.39); Shahabi, Himan (1.22); Dieu Tien Bui (1.16); Rahmati,

Omid (1.11); and Hong, Haoyuan (1.01). According to these data, Pradhan, Biswajeet; Dieu Tien Bui; and Pourghasemi, Hamid Reza have better performances under all parameters. Therefore, they are considered to be strong influential researchers in the field.

3.2.2. Most Productive Institutions in Terms of AI Research in the Field of Geohazards

Among 9226 scientometric records of AI studies in the field of geohazards, 841 institutions are identified. Table 3 shows the data related to the top 15 most productive institutions. At each of these institutions, 83 or more papers have been published. These papers (1806 total) account for 19.58% of the cumulative number of papers in the field. The institutions with the most published papers are the Chinese Academy of Sciences (384), China University of Geosciences (158), and U.S. Geological Survey (156). The Chinese Academy of Sciences has not only published the most papers but also has the highest total citations and is one of the most active research institutions in AI research in the field of geohazards. Those results correspond well with the bibliometric review of Ho and Wang [39] in the field of AI.

Table 3. Top 15 most productive institutions in terms of AI research in the field of geohazards (1990–2022).

No.	Institution	Total Papers	Total Citations	Average Citations per Paper	Relate Citations Impact
1	Chinese Academy of Sciences (https://english.cas.cn, accessed on 3 September 2022)	384	9088	23.67	0.83
2	China University of Geosciences (https://en.cug.edu.cn, accessed on 3 September 2022)	158	3303	20.91	0.73
3	U.S. Geological Survey (https://www.usgs.gov, accessed on 3 September 2022)	156	8928	57.23	2.00
4	University of Chinese Academy of Sciences (https://english.ucas.ac.cn, accessed on 3 September 2022)	115	1264	10.99	0.38
5	Chengdu University of Technology (http://www.cdut.edu.cn. accessed on 3 September 2022)	101	1549	15.34	0.54
6	Tongji University (https://en.tongji.edu.cn, accessed on 3 September 2022)	100	1749	17.49	0.61
7	University of California, Berkeley (https://www.berkeley.edu, accessed on 3 September 2022)	97	4938	50.91	1.78
8	University of Technology Sydney (https://www.uts.edu.au, accessed on 3 September 2022)	93	2583	27.77	0.97
9	Duy Tan University (https://duytan.edu.vn, accessed on 3 September 2022)	91	3278	36.02	1.26
10	University of Tehran (https://ut.ac.ir/en. accessed on 3 September 2022)	88	2027	23.03	0.81
11	Islamic Azad University (https://iau.ae, accessed on 3 September 2022)	86	3074	35.74	1.25
12	China Earthquake Administration (https://www.cea.gov.cn, accessed on 3 September 2022)	85	2431	28.60	1.00
13	Sejong University (https://en.sejong.ac.kr/eng/index.do, accessed on 3 September 2022)	85	3640	42.82	1.50
14	Tarbiat Modares University (https://en.modares.ac.ir, accessed on 3 September 2022)	84	2309	27.49	0.96
15	Kyoto University (https://www.kyoto-u.ac.jp/en, accessed on 3 September 2022)	83	1238	14.92	0.52

The average citations per paper for all papers related to AI research in the field of geohazards from the top 15 institutions is 28.46%. Six institutions have higher average citations per paper than the others. These institutions include the U.S. Geological Survey

(57.23); University of California, Berkeley (50.91); Sejong University (42.82); Duy Tan University (36.02); Islamic Azad University (35.74); and China Earthquake Administration (28.60). From 1990 to 2022, the relative citation impact of the top 15 most productive institutions relative to the total global research output of AI in geohazards was 1.00. Five institutions exceeded this relative citation impact, including the U.S. Geological Survey (2.00); University of California, Berkeley (1.78); Sejong University (1.5); Duy Tan University (1.26); and Islamic Azad University (1.25).

According to average citations per paper and relative citation impact, the U.S. Geological Survey; University of California, Berkeley; and Sejong University are considered to be the most active institutions in this field.

3.2.3. Top Countries in Terms of AI Research in the Field of Geohazards

Information on publications by country and region is closely related to publication characteristics but reflects different information (see Table 4). This table shows that China is the country with the most publications in this field during 1990–2022, with 2349 historical publications. The United States and Italy rank second and third with a total of 1993 and 894 publications, respectively. In addition, the United States has the highest total citations with 61,656 historical citations, followed by China (41,179) and Italy (27,388). Malaysia has the highest average citations per paper with an average of 75.71, followed by Norway and Vietnam with an average of 56.19 and 41.37 citations per paper, respectively. In the area of intercountry cooperation, the United States is the most influential country in this field and has the highest number of collaborations among 10 countries, including China, Italy, and the United Kingdom. In addition, the United States and China are the closest collaborators, with 279 collaborations.

Table 4. Top 20 countries or regions in terms of number of publications.

Country	Total Papers	Total Citations	Average Citations per Paper	Closest Collaborating Country	Number of Total Collaborators
China	2349	41,179	17.53	United States	279
United States	1993	61,656	30.94	China	279
Italy	894	27,388	30.64	United States	72
Iran	629	19,302	30.69	Vietnam	85
England	572	18,371	32.12	United States	111
Japan	505	11,036	21.85	China	83
India	504	9430	18.71	Vietnam	63
Australia	443	11,376	25.68	China	100
Germany	410	11,086	27.04	United States	51
Canada	409	14,586	35.66	United States	78
France	386	10,326	26.75	United States	51
South Korea	323	10,793	33.41	Australia	65
Turkey	289	9311	32.22	United States	34
Spain	277	7582	27.37	Italy	47
Switzerland	240	7625	31.77	United States	42
Netherlands	239	9496	39.73	United States	35
Vietnam	182	7529	41.37	Iran	85
Greece	174	4913	28.24	Italy	26
Malaysia	171	12,946	75.71	Iran	52
Norway	162	9103	56.19	Vietnam	45

Figure 6 shows the cooperation of major countries and regions in this field. The figure clearly shows that China, the United States, Italy, and Iran are the countries with the most AI studies in geohazard research. They are also the countries with the highest number of papers published. Among them, China has the highest number of publications and the second highest number of collaborations with other countries after the United States. The United States is the second most published country and has the most collaborations with other countries. In addition, China and the United States are the countries that cooperate most closely with each other. Italy has published more papers than Iran but has collaborated less with other countries than Iran. These results indicate that China and the United States are the two most representative countries in AI research in the field geohazards. Those results correspond well with the previous bibliometric review of Ho and Wang and Wu et al. [38,39]. Additionally, Italy and Iran also have a productive role in the AI-based geohazard research.

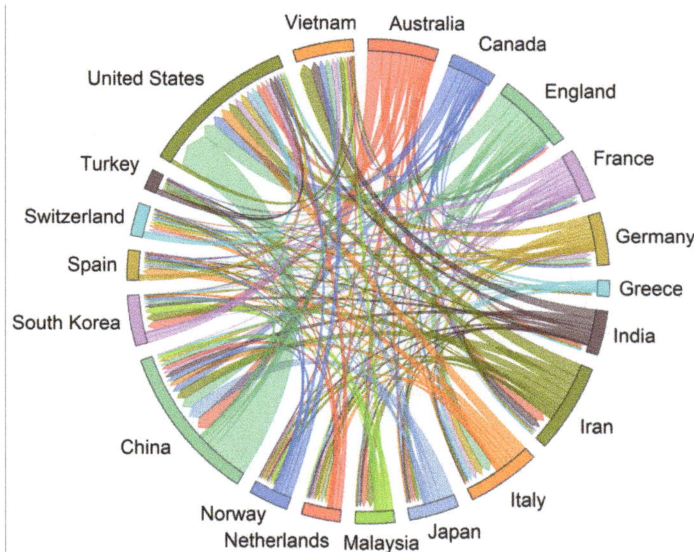

Figure 6. Cooperation between countries and regions.

3.3. *Identification of Salient Research Clusters*

In this study, we selected papers ranked in the top 40% of references each year as the prominent research clusters for identifying the development of AI in geohazard research. By applying the log-likelihood ratio algorithm, 10 prominent research clusters were identified based on the keywords of the top-cited references (see Table 5). The identified clusters for #0 to #9 are ground motion, DL, GIS, landslide, impact, segment linkage, prediction, root reinforcement, debris flow, validation, respectively. Figure 7 and Table 5 show the prominent research clusters obtained based on the Web of Science search results. The silhouette value, a measure of the homogeneity of individual clusters, ranges from −1 to 1. The clustering results are considered convincing only when the silhouette value is greater than 0.5. As shown in Table 5, the silhouette values determined in this study ranged from 0.81 to 1, which indicates that the clustering results are convincing and that the members of each cluster have good consistency. For brevity, only the first five clusters (#0 to #4) were analyzed in this study and are discussed below.

Table 5. Research clusters of AI in geohazard research between 1990 and 2022.

Cluster ID	Size	Silhouette	Cluster Label	Representative Document
#0	70	0.81	Ground motion	Boore & Atkinson [40]
#1	64	0.91	Deep learning	Bui et al. [41]
#2	61	0.97	GIS	Guzzetti et al. [42]
#3	54	0.91	Landslide	Pradhan [43]
#4	47	0.92	Impact	Kim et al. [44]

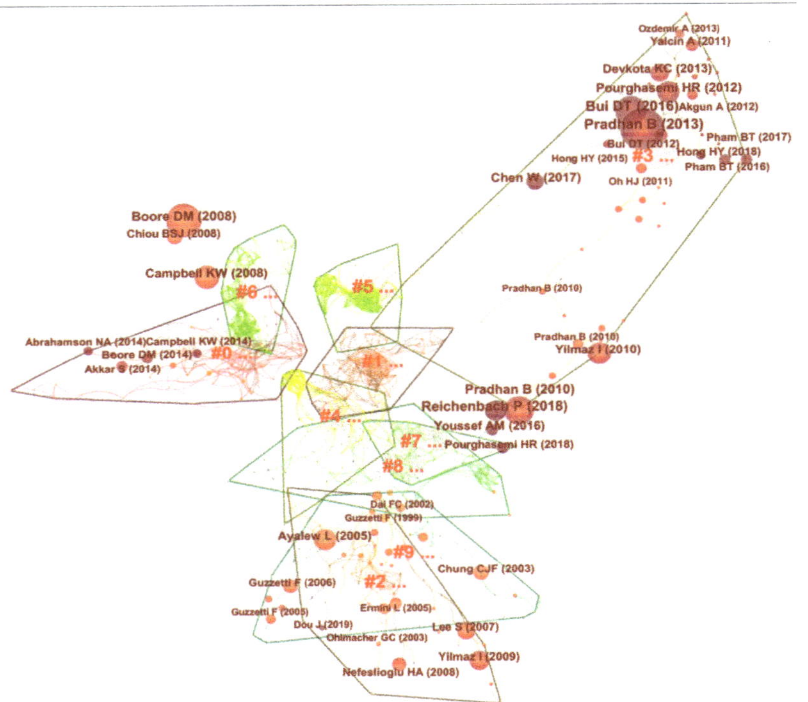

Figure 7. Research cluster network for AI in geohazard research.

Ground motion: The largest cluster (#0) is labeled ground motion, with a total of 70 members and a silhouette value of 0.81. A representative paper is that by Boore and Atkinson [40]. Ground motion usually refers to the surface movement of an area caused by an earthquake or explosion that results from waves generated by the sudden sliding of a fault or the sudden appearance of pressure from an explosion source and propagates along the surface of the Earth.

Deep learning: The second largest cluster (#1) is labeled DL, with 64 members and a silhouette value of 0.91. A representative paper is that by Bui et al. [41]. DL, a major branch of ML, is an algorithm for learning representations of information based on ANNs [45]. With the development of DL, its powerful nonlinear data processing capability has received increasing attention from geohazard researchers. DL is very powerful in geohazard processing and is effective in information extraction. It has since been introduced into geohazard analysis and prevention [14], including landslide and mudflow detection, seismic data interpolation and noise reduction.

GIS: The third largest cluster (#2) is labeled GIS, with 61 members and a silhouette value of 0.97. A representative paper is that by Guzzetti et al. [42]. GIS, or geographic information systems, is a comprehensive discipline of geography, cartography, and computer technology and is now widely used in the field of geohazards. GIS technology has contributed to quantitative studies of geohazard risk assessment and mapping. It has made important contributions in delineating geohazard susceptibility and sensitivity maps, land planning and utilization, and disaster loss reduction [46].

Landslide: The fourth largest cluster (#3) is labeled landslide, with a total of 54 members and a silhouette value of 0.91. Landslides are one of the most common geohazards, causing large economic losses and safety threats to people every year. Representative papers include those by Pradhan [43], Tien Bui et al. [47], and Pourghasemi et al. [48]. These papers consider specific applications of AI in landslide hazards. AI is commonly used in landslide hazards such as displacement prediction and susceptibility mapping.

Impact: The fifth largest cluster (#4) is labeled impact, with 47 members and a profile value of 0.92. The impact label includes both the impact factors and the impacts caused by geohazards. In representative papers by Kim et al. and Ma et al. [44,49], the impact factors of geohazards were studied. Claessens et al. [50] studied the impact of geohazards.

3.4. Top Algorithms and Future Trends in AI Research of Geohazards

In addition to the keyword characteristic analysis of publications, the 10 keywords with the strongest keyword citation burst are all related to AI technology. Seven of these keywords are AI algorithms. This also indicates that AI algorithms have become an important method in geohazard research. Therefore, it is necessary to summarize and analyze the AI algorithms commonly used in geohazard research. Table 6 shows a brief summary and some advantages and limitations of some common AI algorithms in the field of geohazards. Among them, NB, DT, and SVM are the classic ML algorithms. These common single ML algorithms had seen citation outbreaks one after another in 2017 and are widely used by researchers of geohazards. Recently, DL methods, including autoencoders and convolutional and recurrent neural networks, have been widely used by researchers because of their greater processing power of raw natural data [51] and higher accuracy of qualitative hazard prediction [16] than traditional ML methods. DL is the second largest cluster in the cluster analysis and a silhouette of 0.91 is a convincing result. Therefore, it can be considered that DL is one of the trends of AI in geohazards.

Table 6. Summary of popular AI algorithms in the field of geohazards.

		Naive Bayes
	Summary	NB classifiers are simple probabilistic classifiers based on the Bayes theorem and the strong (naive) independence assumption between features [52].
1	**Advantages**	• NB is simple, efficient, and reliable [53]; • The NB classifier does not require a complex iterative parameter estimation scheme and is easy to construct [54]; • The NB classifier handles correlated noise and irrelevant attributes and has very good robustness [55].
	Limitations	• NB determines the posterior probability by using the prior probability and data to determine the classification, so there is a certain error rate in the classification decision [56]; • NB classifiers require strong independence between attributes [57].
		Decision Tree
	Summary	The DT is a basic classification and regression method. The DT model has a tree-like structure and represents the process of classifying instances based on features in a classification problem [58].
2	**Advantages**	• DT can solve complex problems [59]; • It provides expressive representation for learning discrete functions; • The time complexity of the decision tree is small [60].
	Limitations	• A large amount of storage is required to store all classifier results [61]; • It is difficult to understand the reasoning process when multiple classifiers are involved in the decision [58]; • It is easy to overfit the data [62].

Table 6. *Cont.*

		Support Vector Machine
3	**Summary**	SVM is a supervised learning machine proposed by Vapnik et al. [63,64]. It is a powerful tool for solving pattern classification problems and regression problems [65] and has been used in various fields [66–68].
	Advantages	• SVM can achieve better results with fewer samples [25]; • SVM is insensitive to dimensionality and outliers [69]; • SVM is very robust and accurate [70].
	Limitations	• The SVM algorithm has high time complexity and memory training complexity [64]; • The SVM principle is complex and computationally expensive [71].
		Artificial Neural Networks
4	**Summary**	ANNs are algorithmic models inspired by biological neural networks. They are massively parallel systems with a large number of interconnected simple processors [72].
	Advantages	• ANNs have significant advantages in data classification and regression [73]; • ANNs are not constrained by predefined mathematical relationships between variables [74]; • ANNs have a powerful ability to handle complex nonlinear problems.
	Limitations	• The model is considered a black box, making it difficult to understand the internal mechanism [74]; • ANNs have a high demand for computing resources; • ANNs can easily fall into the local minima and sometimes it is difficult to adjust the structure [75].
		Extreme Learning Machine
5	**Summary**	The ELM is a single-layer feedforward neural network that overcomes the difficulty of parameter initialization. It is one of the most widely used algorithms for predicting time series data [76].
	Advantages	• The theoretical basis of an ELM is relatively simple [76]; • ELM can achieve global minimum optimization and has powerful generalization [77]; • ELM computes much faster than other feedforward neural networks.
	Limitations	• An ELM with a fixed number of hidden layer nodes reduces model prediction accuracy [78]; • When the training dataset is relatively small ELM has more erroneous results [76].
		K-Nearest Neighbor
6	**Summary**	KNN is a nonparametric method that is considered one of the top 10 data mining algorithms because of its simplicity, efficiency, and implementation power for classification [79].
	Advantages	• The structure of the KNN algorithm is relatively simple and has good portability [80]; • KNN algorithms are powerful in classification with good effectiveness and implementation [79].
	Limitations	• Outliers in KNN algorithms can have a large adverse effect on the results [79]; • The problems of similarity measurement of two data points and K-value selection in the KNN algorithm still need to be solved [81].
		Logistics Regression
7	**Summary**	LR analysis is a statistical technique for analyzing the relationship between an independent variable and two dependent variables (dichotomous variables) and is widely used in various fields [82,83].
	Advantages	• A regression relationship is formed between the dependent variable and one or more independent variables [84]; • LR algorithms are independent of the data distribution [85]; • Continuous explanatory variables can be used [86].
	Limitations	• LR is more sensitive to multiple linear data [87]; • A large sample size is required for goodness-of-fit measurements [59].

Table 6. *Cont.*

	Ensemble Methods	
8	**Summary**	EMs refer to the combination of individual AI models into one model that has higher accuracy and stronger generalization ability than the individual AI models [88,89].
	Advantages	• The results of the integrated approach are more accurate than those predicted by individual models [90–92]; • EMs can avoid overfitting and local optima [93]; • EMs have better generalization ability than a single AI algorithm [88,94].
	Limitations	• EMs tend to ignore local clustering diversity [95]; • The accuracy of EMs is determined by the choice of the base model [88]; • Much maintenance is required.
	Deep Learning	
9	**Summary**	DL is a branch of machine learning based on ANN [96]. It has excellent performance in processing a large amount of high level data and has a wide range of applications in various fields [97–99].
	Advantages	• DL has powerful feature learning and expression capabilities [100]; • DL is highly efficient in processing high-dimensional data [96]; • DL has more frameworks to use and is more compatible.
	Limitations	• DL requires much data and computing power and the computational cost is high [101,102]; • DL requires high hardware requirements due to its high computing power.

4. Future Directions

AI has been extensively applied to geohazard research, yielding tremendous success. Based on the scientometric analysis of the literature to date, we recommend the following aspects should be addressed for AI-based geohazard research.

4.1. Establishment of Benchmark Database

AI modeling is driven by data [103]; therefore, the quantity and quality of the data may directly affect the performance of AI [14]. However, some fundamental constraints remain for data acquisition and preparation. Firstly, the high cost of monitoring equipment limits the coverage of field monitoring and limits researchers' access to high-quality field data. Another impediment is the lack of large and generalized geohazard datasets. Although tens of thousands of papers have been published for AI-based geohazard research, it is difficult to extract and utilize openly available, curated, and labeled training data. Generally, researchers from different institutions often use different datasets and research methods for their studies, with the terminology and data completeness in the papers varying tremendously. This has led to strong calls from researchers for the establishment of a benchmark database, data sharing, and standardization of data reporting [104] which will be an important boost for the development of AI-based geohazard researches. Some researchers have already started data sharing work. For example, Ji et al. [105] shared a large landslide dataset (http://study.rsgis.whu.edu.cn/pages/download/, accessed on 4 October 2022) containing landslide images, landslide boundary information, landslide area DEM data, etc. Mousavi et al. [100] contributed a large number of high-quality seismic analysis datasets (https://github.com/smousavi05/STEAD, accessed on 4 October 2022) which contain local seismic waveforms, seismic noise waveforms, and no seismic signals. These publicly available high-quality datasets can be used as benchmark datasets for the evaluation of the performance of different AI algorithms in this field and provide a reproducible evaluation environment. In addition, a standardized baseline database not only provides researchers with high-quality datasets but also eases the work of researchers in data management [106]. Therefore, a standardized geohazard benchmark database is desired by researchers.

4.2. Integration of AI with Physical Processes

AI techniques provide good performances in geohazards such as landslide susceptibility evaluation [20], earthquake identification and phase selection [21], and volcanic activity monitoring [24]. However, researchers and practitioners still face challenges in enhancing the reliability [107]. To improve the reliability of AI, some researchers have attempted to integrate AI and physical processes to embody the powerful data processing capabilities of AI techniques and the reliability of physical processes in an ensemble algorithm [108]. For example, Jiang et al. [108] proposed an algorithm to improve the geoscientific knowledge of AI. Depina et al. [109] used an algorithm for the study of unsaturated groundwater flow using a combination of AI and physical processes. The reliability has been enhanced by adopting data-driven components to improve the unrepresentable parts of physical processes and integration of the evolution of physical processes in AI algorithms.

4.3. Auto ML

A strong mastery of expert knowledge is required for AI-based geohazard research. A general workflow for AI modeling usually consists of data preprocessing, feature engineering, selection of a machine learning model, and optimization of the associated hyperparameters [110,111]. Reducing the requirement of expert knowledge and automating all the processing steps is a common expectation among researchers. Some researchers have offered auto ML platforms that have somewhat overcome the problems of algorithm selection and hyperparameter optimization, reducing the need for expertise in AI algorithms. For example, Auto-sklearn optimized hyperparameter selection using a Bayesian algorithm and automated policy selection using meta-learning and integration structures [112]. Auto-WEKA implements the automatic selection of algorithms and hyperparameter optimization based on Bayesian optimization techniques [113]. These auto ML platforms have proven their capacity in the fields of medicine [114], mechanics [115], and geoscience [116].

4.4. Uncertainty Quantification

AI analyzes geohazard data by building corresponding models to predict the occurrence of geohazards and provide evidence and suggestions for its prevention and mitigation. In this process, the uncertainties existing in the data and models may bias the analysis results. Data uncertainty is generated due to class overlap and noise in the training data and is non-approachable due to limitations in how the data are collected. Epistemic uncertainty results from errors caused by model inference or model performance [117]. With the widespread use of AI in geohazards, it is becoming more and more crucial to evaluate the validity and reliability of AI systems before using their analysis results.

Currently, accurate uncertainty quantification is the key to enhance the reliability and accuracy of AI analysis results and the future direction of AI in the field of geohazards. A few researchers have started research on uncertainty quantification. The most common approaches can be divided into Bayesian uncertainty quantification that focuses on specifying the training set to approximate the posterior probability distribution, such as Monte Carlo [118] and Markov Chain Monte Carlo [119], and ensemble uncertainty quantification that obtains improved accuracy by combining multiple models [120] such as deep ensemble [121] and Dirichlet Deep Networks [122].

4.5. Interpretable AI

Some AI algorithms cannot provide a reasonable interpretation for their results which makes researchers and practitioners distrust results obtained from AI. This has limited the development of AI-based geohazard research to a large extent and has brought increasing attention to interpretable AI [123]. Based on previous studies, research methods for interpreting AI techniques are maturing [124,125], terminology and metrics are being harmonized [126,127], and there is some development in the evaluation of interpretable AI and interpretation of AI. Some primary research methods are currently being used to study AI "black boxes"; for example, by decomposing model components into small parts that we can explain [128] and by visualizing the weights of different models to improve the interpretability of DL for seismic monitoring and phase selection [21]. Future works should include overcoming the obstacles to development caused by the uncertainty of quantitative AI interpretation methods, causal interpretation, feature dependence, and other problems [129].

5. Conclusions

AI has been extensively applied to geohazard research and yielding tremendous success. The present study performed a scientometric-assisted review for AI-based geohazard research by visualization of the research status quo and identification of the salient term and context, research trend, and mapping interconnection based on 9226 scientometric records. The analysis of the research publication trend indicates that AI has obtained continuous development in geohazard research over the past 30 years and entered a period of rapid growth beginning in 2000. An analysis of publication source characteristics has revealed that **Natural Hazards** and **Remote Sensing** are the top two source journals. **Geomorphology** and **Catena** are considered to be the most active journals in this field. The analysis of keyword features revealed that ML is a popular research method in this field. **Pradhan, Biswajeet**; **Dieu Tien Bui**, and **Pourghasemi, Hamid Reza** are among the three most productive researchers in this field. Three organizations including the **U.S. Geological Survey**; **University of California, Berkeley**; and **Sejong University** are considered to be the most productive institutions in this field. China, the United States, and Italy are the countries with the highest number of publications and the highest number of total citations among all countries. Identification of salient research clusters indicates that ground motion, **DL**, **GIS**, and landslides are current research hotspots.

Future studies on AI-based geohazard research themes may focus on the establishment of benchmark database, integration of AI with physical processes, Auto ML, uncertainty quantification and interpretable AI.

This scientometric review offers useful reference points for early-stage researchers and provides valuable in-depth information to experienced researchers and practitioners in the field of geohazard research. This scientometric analysis and visualization are promising for comprehensively reflecting the global picture of AI-based geohazard research and are potential for visualization the emerging trends in other research fields.

Author Contributions: Conceptualization, J.M. and S.J.; methodology, J.M. and S.J.; validation, J.M., Z.L. and H.G.; formal analysis, S.J.; data curation, J.M. and Z.L.; writing—original draft preparation, S.J.; writing—review and editing, S.J., J.M., Z.L. and H.G.; visualization, S.J. and Z.L.; supervision J.M. All authors have read and agreed to the published version of the manuscript.

Funding: This research was funded by the Major Program of the National Natural Science Foundation of China (Grant No. 42090055), the National Natural Science Foundation of China (Grant Nos. 42177147 and 71874165), and the Fundamental Research Funds for the Central Universities, China University of Geosciences (Wuhan) (CUG2642022006).

Institutional Review Board Statement: Not applicable.

Informed Consent Statement: Not applicable.

Data Availability Statement: Not applicable.

Conflicts of Interest: The authors declare no conflict of interest.

Abbreviations

AI: artificial intelligence; ANN: artificial neural network; DL: deep learning; DT: decision tree; EM-DAT: Emergency Events Database; EM: ensemble method; ELM: extreme learning machine; KNN: k-nearest neighbor; LR: logistic regression; ML: machine learning; NB: naive Bayes; RF: random forest; SVM: support vector machine.

References

1. Liu, Z.; Li, W.; Zhang, L.; Li, J. Fine Geological Modeling of Complex Fault Block Reservoir Based on Deep Learning. *Wireless Commun. Mobile Comput.* **2022**, *2022*, 9670311. [CrossRef]
2. Zhang, K.; Wang, L.; Dai, Z.; Huang, B.; Zhang, Z. Evolution trend of the Huangyanwo rock mass under the action of reservoir water fluctuation. *Nat. Hazard.* **2022**, *113*, 1583–1600. [CrossRef]
3. Shi, W.; Zhang, M.; Ke, H.; Fang, X.; Zhan, Z.; Chen, S. Landslide Recognition by Deep Convolutional Neural Network and Change Detection. *IEEE Trans. Geosci. Remote Sens.* **2021**, *59*, 4654–4672. [CrossRef]
4. Miao, F.; Zhao, F.; Wu, Y.; Li, L.; Xue, Y.; Meng, J. A novel seepage device and ring-shear test on slip zone soils of landslide in the Three Gorges Reservoir area. *Eng. Geol.* **2022**, *307*, 106779. [CrossRef]
5. Guo, W.; Zuo, X.; Yu, J.; Zhou, B. Method for Mid-Long-Term Prediction of Landslides Movements Based on Optimized Apriori Algorithm. *Appl. Sci.* **2019**, *9*, 3819. [CrossRef]
6. Miao, F.; Wu, Y.; Török, Á.; Li, L.; Xue, Y. Centrifugal model test on a riverine landslide in the Three Gorges Reservoir induced by rainfall and water level fluctuation. *Geosci. Front.* **2022**, *13*, 101378. [CrossRef]
7. Wang, L.; Zhang, Z.; Huang, B.; Hu, M.; Zhang, C. Triggering mechanism and possible evolution process of the ancient Qingshi landslide in the Three Gorges Reservoir. *Geomat. Nat. Hazards Risk* **2021**, *12*, 3160–3174. [CrossRef]
8. Wu, S.; Hu, X.; Zheng, W.; He, C.; Zhang, G.; Zhang, H.; Wang, X. Effects of reservoir water level fluctuations and rainfall on a landslide by two-way ANOVA and K-means clustering. *Bull. Eng. Geol. Environ.* **2021**, *80*, 5405–5421. [CrossRef]
9. Miao, F.; Wu, Y.; Xie, Y.; Li, Y. Prediction of landslide displacement with step-like behavior based on multialgorithm optimization and a support vector regression model. *Landslides* **2018**, *15*, 475–488. [CrossRef]
10. Ma, J.; Niu, X.; Tang, H.; Wang, Y.; Wen, T.; Zhang, J. Displacement prediction of a complex landslide in the Three Gorges Reservoir Area (China) using a hybrid computational intelligence approach. *Complexity* **2020**, *2020*, 2624547. [CrossRef]
11. L'Heureux, A.; Grolinger, K.; Elyamany, H.F.; Capretz, M.A.M. Machine Learning with Big Data: Challenges and Approaches. *IEEE Access* **2017**, *5*, 7776–7797. [CrossRef]
12. Niu, X.; Ma, J.; Wang, Y.; Zhang, J.; Chen, H.; Tang, H. A Novel Decomposition-Ensemble Learning Model Based on Ensemble Empirical Mode Decomposition and Recurrent Neural Network for Landslide Displacement Prediction. *Appl. Sci.* **2021**, *11*, 4684. [CrossRef]
13. Dikshit, A.; Pradhan, B.; Alamri, A.M. Pathways and challenges of the application of artificial intelligence to geohazards modelling. *Gondwana Res.* **2021**, *100*, 290–301. [CrossRef]
14. Ma, Z.; Mei, G. Deep learning for geological hazards analysis: Data, models, applications, and opportunities. *Earth Sci. Rev.* **2021**, *223*, 103858. [CrossRef]
15. Pradhan, B.; Lee, S. Landslide risk analysis using artificial neural network model focussing on different training sites. *Int. J. Phys. Sci.* **2009**, *4*, 1–15.
16. Ma, J.W.; Xia, D.; Wang, Y.K.; Niu, X.X.; Jiang, S.; Liu, Z.Y.; Guo, H.X. A comprehensive comparison among metaheuristics (MHs) for geohazard modeling using machine learning: Insights from a case study of landslide displacement prediction. *Eng. Appl. Artif. Intell.* **2022**, *114*, 105150. [CrossRef]
17. Kalantar, B.; Pradhan, B.; Naghibi, S.A.; Motevalli, A.; Mansor, S. Assessment of the effects of training data selection on the landslide susceptibility mapping: A comparison between support vector machine (SVM), logistic regression (LR) and artificial neural networks (ANN). *Geomat. Nat. Hazards Risk* **2017**, *9*, 49–69. [CrossRef]
18. Xia, D.; Tang, H.M.; Sun, S.X.; Tang, C.Y.; Zhang, B.C. Landslide Susceptibility Mapping Based on the Germinal Center Optimization Algorithm and Support Vector Classification. *Remote Sens.* **2022**, *14*, 2707. [CrossRef]
19. Ghorbanzadeh, O.; Blaschke, T.; Gholamnia, K.; Meena, S.; Tiede, D.; Aryal, J. Evaluation of Different Machine Learning Methods and Deep-Learning Convolutional Neural Networks for Landslide Detection. *Remote Sens.* **2019**, *11*, 196. [CrossRef]
20. Zhang, Y.; Ge, T.; Tian, W.; Liou, Y.-A. Debris Flow Susceptibility Mapping Using Machine-Learning Techniques in Shigatse Area, China. *Remote Sens.* **2019**, *11*, 2801. [CrossRef]
21. Mousavi, S.M.; Ellsworth, W.L.; Zhu, W.; Chuang, L.Y.; Beroza, G.C. Earthquake transformer—An attentive deep-learning model for simultaneous earthquake detection and phase picking. *Nat. Commun.* **2020**, *11*, 3952. [CrossRef]
22. Wu, B.; Qiu, W.; Huang, W.; Meng, G.; Nong, Y.; Huang, J. A Multi-Source Information Fusion Evaluation Method for the Tunneling Collapse Disaster Based on the Artificial Intelligence Deformation Prediction. *Arab. J. Sci. Eng.* **2022**, *47*, 5053–5071. [CrossRef]
23. Choubin, B.; Borji, M.; Mosavi, A.; Sajedi-Hosseini, F.; Singh, V.P.; Shamshirband, S. Snow avalanche hazard prediction using machine learning methods. *J. Hydrol.* **2019**, *577*, 123929. [CrossRef]
24. Valade, S.; Ley, A.; Massimetti, F.; D'Hondt, O.; Laiolo, M.; Coppola, D.; Loibl, D.; Hellwich, O.; Walter, T.R. Towards Global Volcano Monitoring Using Multisensor Sentinel Missions and Artificial Intelligence: The MOUNTS Monitoring System. *Remote Sens.* **2019**, *11*, 1528. [CrossRef]
25. Huang, Y.; Zhao, L. Review on landslide susceptibility mapping using support vector machines. *CATENA* **2018**, *165*, 520–529. [CrossRef]
26. Merghadi, A.; Yunus, A.P.; Dou, J.; Whiteley, J.; ThaiPham, B.; Bui, D.T.; Avtar, R.; Abderrahmane, B. Machine learning methods for landslide susceptibility studies: A comparative overview of algorithm performance. *Earth Sci. Rev.* **2020**, *207*, 103225. [CrossRef]

27. Xie, Y.; Ebad Sichani, M.; Padgett, J.E.; DesRoches, R. The promise of implementing machine learning in earthquake engineering: A state-of-the-art review. *Earthq. Spectra* **2020**, *36*, 1769–1801. [CrossRef]
28. Yalcinkaya, M.; Singh, V. Patterns and trends in Building Information Modeling (BIM) research: A Latent Semantic Analysis. *Autom. Constr.* **2015**, *59*, 68–80. [CrossRef]
29. Martinez, P.; Al-Hussein, M.; Ahmad, R. A scientometric analysis and critical review of computer vision applications for construction. *Autom. Constr.* **2019**, *107*, 102947. [CrossRef]
30. Li, P.; Lu, Y.; Yan, D.; Xiao, J.; Wu, H. Scientometric mapping of smart building research: Towards a framework of human-cyber-physical system (HCPS). *Autom. Constr.* **2021**, *129*, 103776. [CrossRef]
31. Olawumi, T.O.; Chan, D.W.M. A scientometric review of global research on sustainability and sustainable development. *J. Cleaner Prod.* **2018**, *183*, 231–250. [CrossRef]
32. Chatterjee, J.; Dethlefs, N. Scientometric review of artificial intelligence for operations & maintenance of wind turbines: The past, present and future. *Renew. Sustain. Energy Rev.* **2021**, *144*, 111051.
33. Borner, K.; Chen, C.M.; Boyack, K.W. Visualizing knowledge domains. *Annu. Rev. Inf. Sci. Technol.* **2003**, *37*, 179–255. [CrossRef]
34. Chen, C.M. CiteSpace II: Detecting and visualizing emerging trends and transient patterns in scientific literature. *J. Am. Soc. Inf. Sci. Technol.* **2006**, *57*, 359–377. [CrossRef]
35. Chen, C. *CiteSpace: A Practical Guide for Mapping Scientific Literature*; Nova Science Publishers: New York, NY, USA, 2016.
36. Bar-Ilan, J.; Levene, M.; Lin, A. Some measures for comparing citation databases. *J. Informetr.* **2007**, *1*, 26–34. [CrossRef]
37. Falagas, M.E.; Pitsouni, E.I.; Malietzis, G.A.; Pappas, G. Comparison of PubMed, Scopus, Web of Science, and Google Scholar: Strengths and weaknesses. *FASEB J.* **2008**, *22*, 338–342. [CrossRef]
38. Wu, X.; Chen, X.; Zhan, F.B.; Hong, S. Global research trends in landslides during 1991–2014: A bibliometric analysis. *Landslides* **2015**, *12*, 1215–1226. [CrossRef]
39. Ho, Y.-S.; Wang, M.-H. A bibliometric analysis of artificial intelligence publications from 1991 to 2018. *Collnet J. Scientometr. Inf. Manag.* **2020**, *14*, 369–392. [CrossRef]
40. Boore, D.M.; Atkinson, G.M. Ground-Motion Prediction Equations for the Average Horizontal Component of PGA, PGV, and 5%-Damped PSA at Spectral Periods between 0.01 s and 10 s. *Earthq. Spectra* **2008**, *24*, 99–138. [CrossRef]
41. Bui, D.T.; Tsangaratos, P.; Nguyen, V.-T.; Liem, N.V.; Trinh, P.T. Comparing the prediction performance of a Deep Learning Neural Network model with conventional machine learning models in landslide susceptibility assessment. *CATENA* **2020**, *188*, 104426. [CrossRef]
42. Guzzetti, F.; Carrara, A.; Cardinali, M.; Reichenbach, P. Landslide hazard evaluation: A review of current techniques and their application in a multi-scale study, Central Italy. *Geomorphology* **1999**, *31*, 181–216. [CrossRef]
43. Pradhan, B. A comparative study on the predictive ability of the decision tree, support vector machine and neuro-fuzzy models in landslide susceptibility mapping using GIS. *Comput. Geosci.* **2013**, *51*, 350–365. [CrossRef]
44. Kim, H.; Lee, J.-H.; Park, H.-J.; Heo, J.-H. Assessment of temporal probability for rainfall-induced landslides based on nonstationary extreme value analysis. *Eng. Geol.* **2021**, *294*, 106372. [CrossRef]
45. Deng, L. Deep Learning: Methods and Applications. *Found. Trends Signal Processing* **2014**, *7*, 197–387. [CrossRef]
46. Yalcin, A. GIS-based landslide susceptibility mapping using analytical hierarchy process and bivariate statistics in Ardesen (Turkey): Comparisons of results and confirmations. *CATENA* **2008**, *72*, 1–12. [CrossRef]
47. Tien Bui, D.; Tuan, T.A.; Klempe, H.; Pradhan, B.; Revhaug, I. Spatial prediction models for shallow landslide hazards: A comparative assessment of the efficacy of support vector machines, artificial neural networks, kernel logistic regression, and logistic model tree. *Landslides* **2016**, *13*, 361–378. [CrossRef]
48. Pourghasemi, H.R.; Pradhan, B.; Gokceoglu, C. Application of fuzzy logic and analytical hierarchy process (AHP) to landslide susceptibility mapping at Haraz watershed, Iran. *Nat. Hazard.* **2012**, *63*, 965–996. [CrossRef]
49. Ma, J.W.; Tang, H.M.; Hu, X.L.; Bobet, A.; Zhang, M.; Zhu, T.W.; Song, Y.J.; Eldin, M. Identification of causal factors for the Majiagou landslide using modern data mining methods. *Landslides* **2017**, *14*, 311–322. [CrossRef]
50. Claessens, L.; Schoorl, J.M.; Veldkamp, A. Modelling the location of shallow landslides and their effects on landscape dynamics in large watersheds: An application for Northern New Zealand. *Geomorphology* **2007**, *87*, 16–27. [CrossRef]
51. LeCun, Y.; Bengio, Y.; Hinton, G. Deep learning. *Nature* **2015**, *521*, 436–444. [CrossRef]
52. Zhang, H. The optimality of naive Bayes. *Aa* **2004**, *1*, 3.
53. Jiang, L.; Li, C.; Wang, S.; Zhang, L. Deep feature weighting for naive Bayes and its application to text classification. *Eng. Appl. Artif. Intell.* **2016**, *52*, 26–39. [CrossRef]
54. He, Q.; Shahabi, H.; Shirzadi, A.; Li, S.; Chen, W.; Wang, N.; Chai, H.; Bian, H.; Ma, J.; Chen, Y.; et al. Landslide spatial modelling using novel bivariate statistical based Naïve Bayes, RBF Classifier, and RBF Network machine learning algorithms. *Sci. Total Environ.* **2019**, *663*, 1–15. [CrossRef]
55. Das, I.; Stein, A.; Kerle, N.; Dadhwal, V.K. Landslide susceptibility mapping along road corridors in the Indian Himalayas using Bayesian logistic regression models. *Geomorphology* **2012**, *179*, 116–125. [CrossRef]
56. Jiang, L.; Zhang, L.; Li, C.; Wu, J. A Correlation-Based Feature Weighting Filter for Naive Bayes. *IEEE Trans. Knowl. Data Eng.* **2019**, *31*, 201–213. [CrossRef]
57. Tien Bui, D.; Pradhan, B.; Lofman, O.; Revhaug, I. Landslide Susceptibility Assessment in Vietnam Using Support Vector Machines, Decision Tree, and Naïve Bayes Models. *Math. Probl. Eng.* **2012**, *2012*, 974638. [CrossRef]

58. Kotsiantis, S.B. Decision trees: A recent overview. *Artif. Intell. Rev.* **2011**, *39*, 261–283. [CrossRef]
59. Khadse, V.M.; Mahalle, P.N.; Shinde, G.R. Statistical Study of Machine Learning Algorithms Using Parametric and Non-Parametric Tests: A Comparative Analysis and Recommendations. *Int. J. Ambient. Comput. Intell.* **2020**, *11*, 80–105. [CrossRef]
60. Moshkov, M.J. *Time Complexity of Decision Trees;* Transactions on Rough Sets III; Peters, J.F., Skowron, A., Eds.; Springer: Berlin/Heidelberg, Germany, 2005; pp. 244–459.
61. Dietterich, T.G. An experimental comparison of three methods for constructing ensembles of decision trees: Bagging, boosting, and randomization. *Mach. Learn.* **2000**, *40*, 139–157. [CrossRef]
62. Yildiz, O.T.; Alpaydin, E. Omnivariate decision trees. *IEEE Trans. Neural Netw.* **2001**, *12*, 1539–1546. [CrossRef]
63. Tian, Y.; Qi, Z.; Ju, X.; Shi, Y.; Liu, X. Nonparallel Support Vector Machines for Pattern Classification. *IEEE Trans. Cybern.* **2014**, *44*, 1067–1079. [CrossRef] [PubMed]
64. Nalepa, J.; Kawulok, M. Selecting training sets for support vector machines: A review. *Artif. Intell. Rev.* **2018**, *52*, 857–900. [CrossRef]
65. Li, Z.; Weida, Z.; Licheng, J. Wavelet support vector machine. *IEEE Trans. Syst. Man Cybern. Part B* **2004**, *34*, 34–39.
66. Jiang, Q.; Wang, G.; Jin, S.; Li, Y.; Wang, Y. Predicting human microRNA-disease associations based on support vector machine. *Int. J. Data Min. Bioinf.* **2013**, *8*, 282. [CrossRef]
67. Trafalis, T.B.; Ince, H. Support vector machine for regression and applications to financial forecasting. In Proceedings of the IEEE-INNS-ENNS International Joint Conference on Neural Networks. IJCNN 2000. Neural Computing: New Challenges and Perspectives for the New Millennium, Como, Italy, 27 July 2000.
68. Zendehboudi, A.; Baseer, M.A.; Saidur, R. Application of support vector machine models for forecasting solar and wind energy resources: A review. *J. Cleaner Prod.* **2018**, *199*, 272–285. [CrossRef]
69. Wu, X.; Kumar, V.; Ross Quinlan, J.; Ghosh, J.; Yang, Q.; Motoda, H.; McLachlan, G.J.; Ng, A.; Liu, B.; Yu, P.S.; et al. Top 10 algorithms in data mining. *Knowl. Inf. Syst.* **2007**, *14*, 1–37. [CrossRef]
70. Chen, X.; Xu, S.; Li, S.; He, H.; Han, Y.; Qu, X. Identification of architectural elements based on SVM with PCA: A case study of sandy braided river reservoir in the Lamadian Oilfield, Songliao Basin, NE China. *J. Pet. Sci. Eng.* **2021**, *198*, 108247. [CrossRef]
71. Suthaharan, S. Support vector machine. In *Machine Learning Models and Algorithms for Big Data Classification;* Springer: New York, NY, USA, 2016; pp. 207–235.
72. Jain, A.K.; Jianchang, M.; Mohiuddin, K.M. Artificial neural networks: A tutorial. *Computer* **1996**, *29*, 31–44. [CrossRef]
73. Cao, W.; Wang, X.; Ming, Z.; Gao, J. A review on neural networks with random weights. *Neurocomputing* **2018**, *275*, 278–287. [CrossRef]
74. Tu, J.V. Advantages and disadvantages of using artificial neural networks versus logistic regression for predicting medical outcomes. *J. Clin. Epidemiol.* **1996**, *49*, 1225–1231. [CrossRef]
75. Ding, S.; Li, H.; Su, C.; Yu, J.; Jin, F. Evolutionary artificial neural networks: A review. *Artif. Intell. Rev.* **2011**, *39*, 251–260. [CrossRef]
76. Li, H.; Xu, Q.; He, Y.; Deng, J. Prediction of landslide displacement with an ensemble-based extreme learning machine and copula models. *Landslides* **2018**, *15*, 2047–2059. [CrossRef]
77. Huang, G.-B.; Zhu, Q.-Y.; Siew, C.-K. Extreme learning machine: Theory and applications. *Neurocomputing* **2006**, *70*, 489–501. [CrossRef]
78. Zhao, F.; Liu, M.; Wang, K.; Wang, T.; Jiang, X. A soft measurement approach of wastewater treatment process by lion swarm optimizer-based extreme learning machine. *Measurement* **2021**, *179*, 109322. [CrossRef]
79. Gou, J.; Ma, H.; Ou, W.; Zeng, S.; Rao, Y.; Yang, H. A generalized mean distance-based k-nearest neighbor classifier. *Expert Syst. Appl.* **2019**, *115*, 356–372. [CrossRef]
80. Cai, P.; Wang, Y.; Lu, G.; Chen, P.; Ding, C.; Sun, J. A spatiotemporal correlative k-nearest neighbor model for short-term traffic multistep forecasting. *Transp. Res. Part C Emerg. Technol.* **2016**, *62*, 21–34. [CrossRef]
81. Zhang, S.C.; Li, X.L.; Zong, M.; Zhu, X.F.; Cheng, D.B. Learning k for kNN Classification. *ACM Trans. Intell. Syst. Technol.* **2017**, *8*, 19. [CrossRef]
82. Tripepi, G.; Jager, K.J.; Dekker, F.W.; Zoccali, C. Linear and logistic regression analysis. *Kidney Int.* **2008**, *73*, 806–810. [CrossRef]
83. Mehrolia, S.; Alagarsamy, S.; Solaikutty, V.M. Customers response to online food delivery services during COVID-19 outbreak using binary logistic regression. *Int. J. Consum. Stud.* **2020**, *45*, 396–408. [CrossRef]
84. Muche, R. Logistic regression—A useful tool in rehabilitation research. *Rehabilitation* **2008**, *47*, 56–62. [CrossRef]
85. Bai, S.-B.; Wang, J.; Lü, G.-N.; Zhou, P.-G.; Hou, S.-S.; Xu, S.-N. GIS-based logistic regression for landslide susceptibility mapping of the Zhongxian segment in the Three Gorges area, China. *Geomorphology* **2010**, *115*, 23–31. [CrossRef]
86. Sperandei, S. Understanding logistic regression analysis. *Biochem. Med.* **2014**, *24*, 12–18. [CrossRef] [PubMed]
87. Han, D.; Ma, L.; Yu, C. Financial Prediction: Application of Logistic Regression with Factor Analysis. In Proceedings of the 2008 4th International Conference on Wireless Communications, Networking and Mobile Computing, Dalian, China, 12–14 October 2008.
88. Wang, Z.Y.; Srinivasan, R.S. A review of artificial intelligence based building energy use prediction: Contrasting the capabilities of single and ensemble prediction models. *Renew. Sust. Energ. Rev.* **2017**, *75*, 796–808. [CrossRef]
89. Ma, J.W.; Xia, D.; Guo, H.X.; Wang, Y.K.; Niu, X.X.; Liu, Z.Y.; Jiang, S. Metaheuristic-based support vector regression for landslide displacement prediction: A comparative study. *Landslides* **2022**, *23*, 2489–2511. [CrossRef]

90. Aburomman, A.A.; Ibne Reaz, M.B. A novel SVM-kNN-PSO ensemble method for intrusion detection system. *Appl. Soft Comput.* **2016**, *38*, 360–372. [CrossRef]
91. Ma, J.W.; Liu, X.; Niu, X.X.; Wang, Y.N.; Wen, T.; Zhang, J.R.; Zou, Z.X. Forecasting of Landslide Displacement Using a Probability-Scheme Combination Ensemble Prediction Technique. *Int. J. Environ. Res. Public Health* **2020**, *17*, 4788. [CrossRef]
92. Zhang, J.R.; Tang, H.M.; Wen, T.; Ma, J.W.; Tan, Q.W.; Xia, D.; Liu, X.; Zhang, Y.Q. A Hybrid Landslide Displacement Prediction Method Based on CEEMD and DTW-ACO-SVR-Cases Studied in the Three Gorges Reservoir Area. *Sensors* **2020**, *20*, 4287. [CrossRef]
93. Sagi, O.; Rokach, L. Ensemble learning: A survey. *WIREs Data Min. Knowl. Discov.* **2018**, *8*, e1249. [CrossRef]
94. Zhang, J.R.; Tang, H.M.; Tannant, D.D.; Lin, C.Y.; Xia, D.; Liu, X.; Zhang, Y.Q.; Ma, J.W. Combined forecasting model with CEEMD-LCSS reconstruction and the ABC-SVR method for landslide displacement prediction. *J. Cleaner Prod.* **2021**, *293*, 18. [CrossRef]
95. Huang, D.; Wang, C.-D.; Lai, J.-H. Locally Weighted Ensemble Clustering. *IEEE Trans. Cybern.* **2018**, *48*, 1460–1473. [CrossRef]
96. Janiesch, C.; Zschech, P.; Heinrich, K. Machine learning and deep learning. *Electron. Mark.* **2021**, *31*, 685–695. [CrossRef]
97. Min, S.; Lee, B.; Yoon, S. Deep learning in bioinformatics. *Brief. Bioinform.* **2016**, *18*, 851–869. [CrossRef]
98. Shen, D.; Wu, G.; Suk, H.-I. Deep Learning in Medical Image Analysis. *Annu. Rev. Biomed. Eng.* **2017**, *19*, 221–248. [CrossRef]
99. Mater, A.C.; Coote, M.L. Deep Learning in Chemistry. *J. Chem. Inf. Model.* **2019**, *59*, 2545–2559. [CrossRef]
100. Zhang, W.; Li, H.; Li, Y.; Liu, H.; Chen, Y.; Ding, X. Application of deep learning algorithms in geotechnical engineering: A short critical review. *Artif. Intell. Rev.* **2021**, *54*, 5633–5673. [CrossRef]
101. Spring, R.; Shrivastava, A. Scalable and Sustainable Deep Learning via Randomized Hashing. In Proceedings of the 23rd ACM SIGKDD International Conference on Knowledge Discovery and Data Mining, New York, NY, USA, 13–17 August 2017.
102. Yuan, X.; Ou, C.; Wang, Y.; Yang, C.; Gui, W. A Layer-Wise Data Augmentation Strategy for Deep Learning Networks and Its Soft Sensor Application in an Industrial Hydrocracking Process. *IEEE Trans. Neural Netw. Learn. Syst.* **2021**, *32*, 3296–3305. [CrossRef]
103. Ghahramani, Z. Probabilistic machine learning and artificial intelligence. *Nature* **2015**, *521*, 452–459. [CrossRef]
104. Gewin, V. Data sharing: An open mind on open data. *Nature* **2016**, *529*, 117–119. [CrossRef]
105. Ji, S.; Yu, D.; Shen, C.; Li, W.; Xu, Q. Landslide detection from an open satellite imagery and digital elevation model dataset using attention boosted convolutional neural networks. *Landslides* **2020**, *17*, 1337–1352. [CrossRef]
106. Reichstein, M.; Camps-Valls, G.; Stevens, B.; Jung, M.; Denzler, J.; Carvalhais, N.; Prabhat. Deep learning and process understanding for data-driven Earth system science. *Nature* **2019**, *566*, 195–204. [CrossRef]
107. Ebert-Uphoff, I.; Samarasinghe, S.; Barnes, E. Thoughtfully using artificial intelligence in Earth science. *Eos* **2019**, *100*, 10.1029. [CrossRef]
108. Jiang, S.; Zheng, Y.; Solomatine, D. Improving AI System Awareness of Geoscience Knowledge: Symbiotic Integration of Physical Approaches and Deep Learning. *Geophys. Res. Lett.* **2020**, *47*, e2020GL088229. [CrossRef]
109. Depina, I.; Jain, S.; Mar Valsson, S.; Gotovac, H. Application of physics-informed neural networks to inverse problems in unsaturated groundwater flow. *Georisk Assess. Manage. Risk Eng. Syst. Geohazards* **2021**, *16*, 21–36. [CrossRef]
110. Hutter, F.; Kotthoff, L.; Vanschoren, J. *Automated Machine Learning: Methods, Systems, Challenges*; Springer Nature: New York, NY, USA, 2019.
111. Sun, Z.; Sandoval, L.; Crystal-Ornelas, R.; Mousavi, S.M.; Wang, J.; Lin, C.; Cristea, N.; Tong, D.; Carande, W.H.; Ma, X.; et al. A review of Earth Artificial Intelligence. *Comput. Geosci.* **2022**, *159*, 105034. [CrossRef]
112. Feurer, M.; Eggensperger, K.; Falkner, S.; Lindauer, M.; Hutter, F. Auto-sklearn 2.0: Hands-free automl via meta-learning. *arXiv* **2020**, arXiv:2007.04074.
113. Kotthoff, L.; Thornton, C.; Hoos, H.H.; Hutter, F.; Leyton-Brown, K. Auto-WEKA: Automatic model selection and hyperparameter optimization in WEKA. In *Automated Machine Learning*; Springer: Cham, Switzerland, 2019; pp. 81–95.
114. Kim, I.K.; Lee, K.; Park, J.H.; Baek, J.; Lee, W.K. Classification of pachychoroid disease on ultrawide-field indocyanine green angiography using auto-machine learning platform. *Br. J. Ophthalmol.* **2021**, *105*, 856–861. [CrossRef]
115. Sader, S.; Husti, I.; Daróczi, M. Enhancing failure mode and effects analysis using auto machine learning: A case study of the agricultural machinery industry. *Processes* **2020**, *8*, 224. [CrossRef]
116. Hill, E.; Pearce, M.A.; Stromberg, J.M. Improving automated geological logging of drill holes by incorporating multiscale spatial methods. *Math. Geosci.* **2021**, *53*, 21–53. [CrossRef]
117. Hüllermeier, E.; Waegeman, W. Aleatoric and epistemic uncertainty in machine learning: An introduction to concepts and methods. *Mach. Learn.* **2021**, *110*, 457–506. [CrossRef]
118. Neal, R.M. *Bayesian Learning for Neural Networks*; Springer Science & Business Media: Berlin/Heidelberg, Germany, 2012; Volume 118.
119. Kupinski, M.A.; Hoppin, J.W.; Clarkson, E.; Barrett, H.H. Ideal-observer computation in medical imaging with use of Markov-chain Monte Carlo techniques. *JOSA A* **2003**, *20*, 430–438. [CrossRef]
120. Abdar, M.; Pourpanah, F.; Hussain, S.; Rezazadegan, D.; Liu, L.; Ghavamzadeh, M.; Fieguth, P.; Cao, X.; Khosravi, A.; Acharya, U.R.; et al. A review of uncertainty quantification in deep learning: Techniques, applications and challenges. *Inf. Fusion* **2021**, *76*, 243–297. [CrossRef]
121. Hu, R.; Huang, Q.; Chang, S.; Wang, H.; He, J. The MBPEP: A deep ensemble pruning algorithm providing high quality uncertainty prediction. *Appl. Intell.* **2019**, *49*, 2942–2955. [CrossRef]

122. Tsiligkaridis, T. Information Aware max-norm Dirichlet networks for predictive uncertainty estimation. *Neural Netw.* **2021**, *135*, 105–114. [CrossRef]
123. Adadi, A.; Berrada, M. Peeking Inside the Black-Box: A Survey on Explainable Artificial Intelligence (XAI). *IEEE Access* **2018**, *6*, 52138–52160. [CrossRef]
124. Carvalho, D.V.; Pereira, E.M.; Cardoso, J.S. Machine Learning Interpretability: A Survey on Methods and Metrics. *Electronics* **2019**, *8*, 832. [CrossRef]
125. Guidotti, R.; Monreale, A.; Ruggieri, S.; Turini, F.; Giannotti, F.; Pedreschi, D. A Survey of Methods for Explaining Black Box Models. *ACM Comput. Surv.* **2019**, *51*, 1–42. [CrossRef]
126. Hoffman, R.R.; Mueller, S.T.; Klein, G.; Litman, J. Metrics for explainable AI: Challenges and prospects. *arXiv* **2018**, arXiv:1812.04608.
127. Mohseni, S.; Zarei, N.; Ragan, E.D. A Multidisciplinary Survey and Framework for Design and Evaluation of Explainable AI Systems. *ACM Trans. Interact. Intell. Syst.* **2021**, *11*, 1–45. [CrossRef]
128. Molnar, C.; Casalicchio, G.; Bischl, B. *Interpretable Machine Learning—A Brief History, State-of-the-Art and Challenges*; ECML PKDD 2020 Workshops; Koprinska, I., Kamp, M., Appice, A., Loglisci, C., Antonie, L., Zimmermann, A., Guidotti, R., Özgöbek, Ö., Ribeiro, R.P., Gavaldà, R., et al., Eds.; Springer International Publishing: Cham, Switzerland, 2020; pp. 417–431.
129. Molnar, C.; König, G.; Herbinger, J.; Freiesleben, T.; Dandl, S.; Scholbeck, C.A.; Casalicchio, G.; Grosse-Wentrup, M.; Bischl, B. Pitfalls to avoid when interpreting machine learning models. *arXiv* **2020**, arXiv:2007.04131v1.

Article

Susceptibility Analysis of Glacier Debris Flow Based on Remote Sensing Imagery and Deep Learning: A Case Study along the G318 Linzhi Section

Jiaqing Chen [1], Hong Gao [1], Le Han [1], Ruilin Yu [1] and Gang Mei [1,2,*]

1 School of Engineering and Technology, China University of Geosciences (Beijing), Beijing 100083, China; 1002202205@email.cugb.edu.cn (J.C.); 1012201106@email.cugb.edu.cn (H.G.); 1002202203@email.cugb.edu.cn (L.H.); 1002202222@email.cugb.edu.cn (R.Y.)

2 Engineering and Technology Innovation Center for Risk Prevention and Control of Major Project Geosafety, Ministry of Natural Resources, Beijing 100083, China

* Correspondence: gang.mei@cugb.edu.cn

Abstract: Glacial debris flow is a common natural disaster, and its frequency has been increasing in recent years due to the continuous retreat of glaciers caused by global warming. To reduce the damage caused by glacial debris flows to human and physical properties, glacier susceptibility assessment analysis is needed. Most research efforts consider the effect of existing glacier area and ignore the effect of glacier ablation volume change. In this paper, we consider the impact of glacier ablation volume change to investigate the susceptibility of glacial debris flow. The susceptibility to mudslide was evaluated by taking the glacial mudslide-prone ditch of G318 Linzhi section of Sichuan-Tibet Highway as the research object. First, by using a simple band ratio method with manual correction, we produced a glacial mudslide remote sensing image dataset, and second, we proposed a deep-learning-based approach using a weight-optimized glacial mudslide semantic segmentation model for accurately and automatically mapping the boundaries of complex glacial mudslide-covered remote sensing images. Then, we calculated the ablation volume by the change in glacier elevation and ablation area from 2015 to 2020. Finally, glacial debris flow susceptibility was evaluated based on the entropy weight method and Topsis method with glacial melt volume in different watersheds as the main factor. The research results of this paper show that most of the evaluation indices of the model are above 90%, indicating that the model is reasonable for glacier boundary extraction, and remote sensing images and deep learning techniques can effectively assess the glacial debris flow susceptibility and provide support for future glacial debris flow disaster prevention.

Keywords: geological hazards; glacial debris flow; remote sensing; deep learning

Citation: Chen, J.; Gao, H.; Han, L.; Yu, R.; Mei, G. Susceptibility Analysis of Glacier Debris Flow Based on Remote Sensing Imagery and Deep Learning: A Case Study along the G318 Linzhi Section. *Sensors* **2023**, *23*, 6608. https://doi.org/10.3390/s23146608

Academic Editor: Junwei Ma and Jie Dou

Received: 18 June 2023
Revised: 18 July 2023
Accepted: 21 July 2023
Published: 22 July 2023

1. Introduction

Glacier debris flow is a kind of mountain natural disaster caused by ice melting, which is characterized by suddenness, large scale, and hazard. In the context of global warming, glacier retreat is accelerating, and the frequency and scale of glacier debris flows are increasing [1,2]. Therefore, accurate assessment and prediction of glacial debris flow susceptibility is important to protect people's lives and properties and to maintain the ecological environment.

Remote sensing images and deep learning techniques have an important role in the analysis of glacier debris flow susceptibility. Remote sensing images can provide large-scale, high-resolution surface information, including topography, vegetation, and hydrology [3], which provides basic data for the formation and occurrence of glacier debris flows. Deep learning techniques can quickly and accurately identify potential glacier debris flow hazard areas by feature extraction and classification of remotely sensed images [4], providing an effective means for glacier debris flow prediction and early warning. Therefore, glacier debris flow susceptibility analysis based on remote sensing images and deep learning has important research significance and application value [5].

Glacier debris flow susceptibility analysis is important research work that can help people better understand the formation mechanism and influencing factors of glacier debris flows, so as to take effective prevention and control measures. In recent years, with the continuous development of remote sensing technology and deep learning algorithms, the analysis of glacier debris flow susceptibility based on remote sensing images and deep learning has also gained wide attention. The occurrence of glacial debris flow has obvious indicators of catastrophic changes, such as increased density of hanging glacier crevasses, enhanced glacier velocity, and rapid increase in glacial lake area.

The influencing factors of its glacial debris flow susceptibility analysis mainly include the following aspects: (1) Topographic factors: Topography is one of the important factors in the formation of glacial debris flow, including topographic height difference, slope, and slope direction [6]. In remote sensing images, topographic information can be obtained by data such as digital elevation model (DEM). (2) Climatic factors: Climatic factors also have an important influence on the formation and development of glacier debris flows, including rainfall, temperature, and humidity. Remote sensing images can obtain meteorological data, such as rainfall. (3) Geological factors: Geological factors are also one of the important factors in the formation of glacial debris flow, including lithology, faults, and earthquakes. Remote sensing images can obtain geological information, such as lithology, faults, etc. (4) Vegetation factor: Vegetation cover also has some influence on the formation and development of glacial debris flow. Remote sensing images can obtain vegetation information, such as vegetation cover, etc. [7]. Glacier debris flow susceptibility analysis based on remote sensing images and deep learning can obtain glacier debris flow susceptibility information by feature extraction and classification of remote sensing images. Commonly used deep learning algorithms include convolutional neural networks (CNN), recurrent neural networks (RNN), etc. The prediction results of glacier debris flow susceptibility can be obtained by training and testing the remotely sensed images. In conclusion, the analysis of glacier debris flow susceptibility based on remote sensing images and deep learning can provide an important scientific basis for glacier debris flow prevention and control [8].

In recent years, there has been a new research trend aimed at glacier debris flow susceptibility assessment through remote sensing images and deep learning techniques. For example, Ji et al. used deep learning and remote sensing image analysis to establish a mudslide susceptibility assessment model based on topographic and geomorphological features, and validated it in the Bijie City area of northwestern Guizhou, showing that the model can more accurately assess mudslide susceptibility in the area [9]. Ref. [10] used high-resolution remote sensing image data for glacier prediction. As for the method of remote sensing image susceptibility analysis, Lin et al. considered the influence of the change in glacier ablation volume and conducted a mudslide susceptibility analysis using the G217 glacier mudslide-prone trench on the Dukku Highway in Xinjiang. This study showed that accurate prediction of glacier mudslide susceptibility could be achieved using high-resolution remote sensing image data and machine learning algorithms [11].

However, these methods have some limitations, such as low accuracy in conducting glacier debris flow susceptibility assessment. Therefore, we need more accurate models for glacier debris flow susceptibility assessment.

In recent years, deep learning techniques have been widely used in remote sensing image processing [12]. Among them, the DeepLabv3+ model is a deep convolutional-neural-network-based method with high segmentation accuracy and fast operation speed. This model improves the accuracy of image segmentation by optimizing the traditional convolutional neural network through the null convolution and decoder modules. Using this model, we can semantically segment glacier debris flows by labelling them as "Ice" category and other terrain as "Background" category. This method not only ensures high accuracy segmentation results, but also fast evaluation, avoiding the errors in traditional methods. In addition, the method can actively reduce the harm caused by glacier debris flow to humans and the natural environment.

We chose the G318—Linzhi section of the Sichuan-Tibet Highway, which has been affected by global warming in recent years, and the glacial melt in the Linzhi area has accelerated, resulting in frequent glacier debris flow disasters and serious hazards. According to incomplete statistics, which are only from April 2006 to September 2007, the mud-slide disaster in the Linzhi area endangered the safety of 264 villagers in 17 villages, resulting in one death and seven people missing. A comprehensive analysis of mudslide disasters and environmental factors in the Linzhi region shows that the current glacier debris flow disasters in the Linzhi region are at a high incidence and seriously affect the local economic development. Therefore, it is important to study the development pattern of mudslide under climate change in the Linzhi region for monitoring, early warning and prevention of mudslide disasters in the Linzhi region and the whole of southeast Tibet.

Our contributions in this paper can be summarized as follows: (1) the ablation zone of the study area was determined by comparing the glacier boundary in 2015 and the glacier boundary in 2020; (2) the amount of glacier ablation was calculated based on the changes in glacier elevation and ablation area from 2015 to 2020; and (3) an evaluation of the susceptibility of glacial debris flow by using the melting amount of glaciers in different basins.

2. Study Area

The Linzhi section of National Highway 318 is located in Linzhi City, Tibet Autonomous Region of China, with a total length of about 287 km, connecting Linzhi City with Chengdu City in the southwest of Sichuan Province. The starting point of the Linzhi section is located in the town of Motuo within Linzhi City, and the end point is located in the county of Yajiang. The road traverses several natural scenic areas such as the Hengduan Mountains, the Sichuan-Tibet Plateau, and the Yarlung Tsangpo River Grand Canyon.

The Linzhi section is a typical area prone to glacial debris flow. There are many glaciers and rocks piled together, and the geomorphologic conditions are such that glacial debris flow occurs easily. The study area has typical alpine valley and mountain valley landforms developed due to crustal uplift, strong river undercutting, and strong tectonic activity. In the context of high stress, the potential risk of strong earthquakes is high [13]. The Linzhi section has abundant research resources, such as existing satellite remote sensing images and ground monitoring data, to facilitate the implementation of glacier debris flow analysis.

The average elevation of the study area is about 3000 m, the lowest elevation is 115 m in the territory of Murdoch County, the highest elevation is more than 7000 m in Namcha Barwa Peak, which is the zone with the largest vertical landform drop in the world, and the relative elevation difference is generally around 1000–2000 m. The slope of the mountain slope is generally not less than 30°, and the slope of the canyon area is mostly around 80°.

The national highway G318 crosses Gongbu Jiangda County, Bayi District, and Bomi County, respectively, the details of which are shown in Figure 1. Gongbu Jiangda County is located in the transition zone from the valley of southern Tibet to the high mountain valley area of eastern Tibet, bounded by the eastern extension of the Gangdis Mountains in the south and the Tanggula Mountains in the north, with mountains and valleys spreading in an east–west direction. Bayi district in the south for the Gangdis Mountain remnants, the north belongs to the Nianqing Tanggula Mountain branch alpine section. The average elevation of the territory is 3000 m, and the highest peak is Galabaek Peak, which is 7300 m above sea level, while the lowest place is Bayu Village, which is 1600 m above sea level, with a relative height difference of 4700 m. Bomi County is located in the eastern section of Tanggula and the eastern end of the Himalayas, with high north and low south and continuous high mountains, and the central part of the Palongzangbu River Valley and the Egonzangbu River Valley, with the highest elevation of 6648 m and the lowest of 2001.4 m in Bomi County. With the change in topography and geomorphology, the geological effect is also changing. glacier debris flow hazards are generally developed on the margins of modern glaciers and snowpacks, and the topography and excessive relative elevation differences in this region provide favourable spatial conditions for the formation and development of glacier debris flow hazards [13].

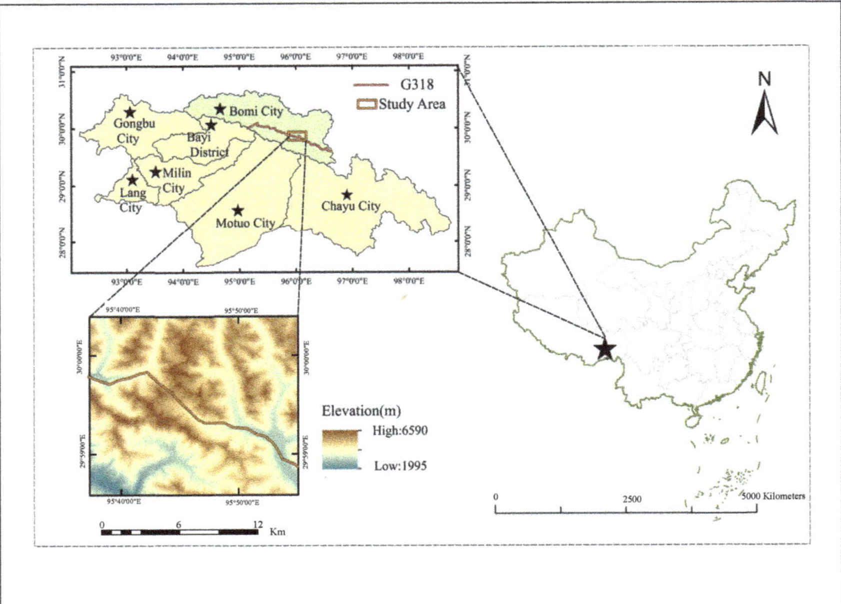

Figure 1. Geographical location of the glacier study area.

While the climate of this study area is obviously influenced by the crustal uplift and more prominently influenced by the topography, the average temperature in the southern part of Dongjiu Township is around +12 °C. The cold-temperate zone becomes more pronounced as we move upstream of the Yarlung Tsangpo River, and Bomi County is known as the centre for modern glaciers because it has high mountains with perennial snow accumulation below 0 °C. In recent years, the glacier area has been retreating, which is mainly influenced by the continuous increase in temperature, and the change in precipitation has little effect on glacier changes [14]. Therefore, we chose Bomi County as our typical study area. The average annual temperature in Kumbumgangda County is +8.7 °C, with a maximum temperature of +31.5 °C and a minimum temperature of −10.4 °C. The average annual rainfall is 640.1 mm, with a maximum annual rainfall of 808.3 mm and a maximum daily rainfall of 45.2 mm. The seasonal distribution of rainfall is uneven, with 80% of the rainfall concentrated between May and September. Bayi district is influenced by the warm and humid airflow of the Indian Ocean, and has a temperate humid monsoon climate with abundant rainfall. The annual average temperature is +8.5 °C, the highest temperature +29 °C, and the lowest temperature −1.8 °C. The average annual rainfall is 654 mm, mainly concentrated in May–September, accounting for about 90% of the annual rainfall. The average annual temperature in Bomi County is +8.5 °C, with a maximum temperature of +31.1 °C and a minimum temperature of −1.8 °C. The average annual rainfall is 977 mm. Since the mid-twentieth century, rising temperatures have led to the melting of many glaciers at unsustainably high rates of melting, resulting in diminishing ice storage [15]. Temperature and rainfall conditions are important factors in triggering the occurrence of glacier debris flow hazards [16,17], and it is important to fully understand and investigate this aspect.

3. Methods

3.1. Overview

In this paper, we investigate the susceptibility of glacier debris flow along the G318 Linzhi section based on remote sensing imagery and deep learning. First, high-quality remotely sensed images are acquired and pre-processed prior to segmentation. Second, the processed images are used to generate a sample set. Third, after the segmentation results are output, post-processing procedures such as boundary extraction, small polygon removal, and edge lubrication are applied to obtain glacier profiles.

Our research is divided into three sections: (1) remote sensing image processing; (2) the glacier ablation volume was calculated by combining elevation data; and (3) eight influencing factors were used to evaluate the susceptibility of glacial debris flow. A flow chart of the study is shown in Figure 2.

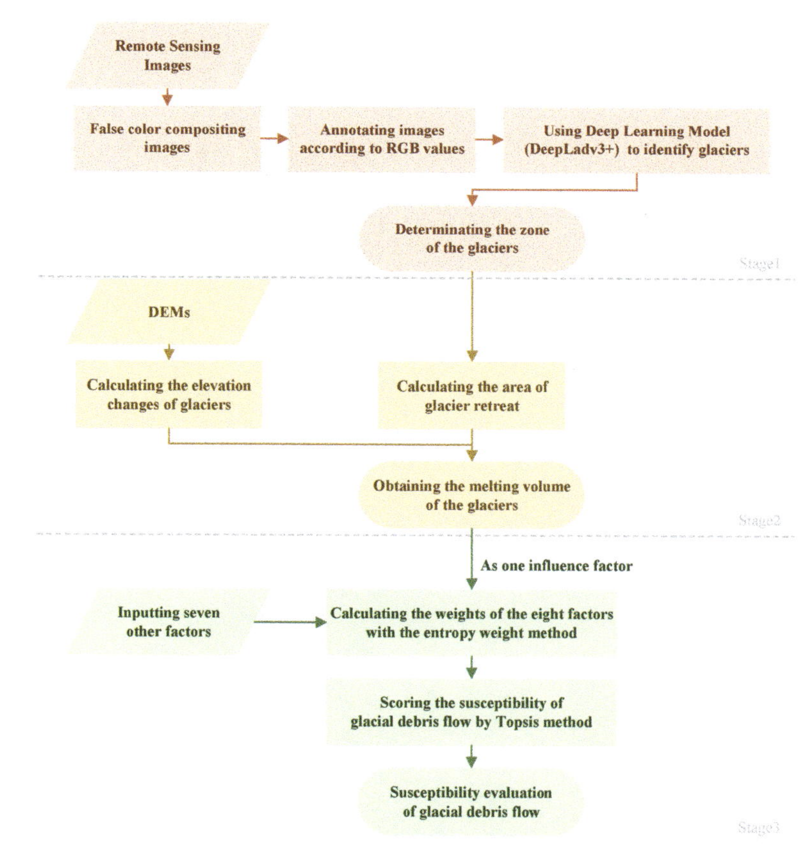

Figure 2. Workflow of the susceptibility analysis of glacier debris flow.

3.2. Step 1: Data Acquisition and Pre-Processing

In this paper, we have obtained freely available Landsat 8 data from the US Geological Survey [18], covering the period of 2015 to 2020. In recent years, remote sensing imagery has been widely used in physical geography and environmental research, especially in areas such as glacier monitoring and geological hazard control. Because of its advantages of high resolution and confidentiality, it often charges fees or provides a limited number of images, making it difficult to achieve long-term monitoring over large areas [19]. Landsat 8, as a high-resolution multispectral satellite, can effectively improve the identification accuracy over large areas and has obvious advantages for glacier identification. Taking into account the climatic conditions, glacier distribution, and change characteristics of the study area, remote sensing images with less cloud shadow in summer were selected. We also selected two global elevation remote sensing data: Copernicus DEM downloaded from the Copernicus Open Access Centre (Available online: https://scihub.copernicus.eu/ (accessed on 12 March 2023)) mapped in 2015, and the latest global 30 m resolution DEM data currently available—NASA DEM was released by NASA on 18 February 2020, and NASA DEM will be the highest resolution, best quality, and widest coverage DEM product in the foreseeable future [20].

Furthermore, in this paper, we generate the corresponding glacier remote sensing dataset with the help of the above-mentioned channels for constructing semantic segmentation models. We selected the Band2 (Blue), Band3 (Green), and Band6 (SWIR 1) bands from the Landsat 8 data and synthesised them as pseudo-colour images. In the synthesised image, the ice appears blue and the bare ground is red, whereby we annotated the obtained remote sensing image with glaciers, as shown in Figure 3.

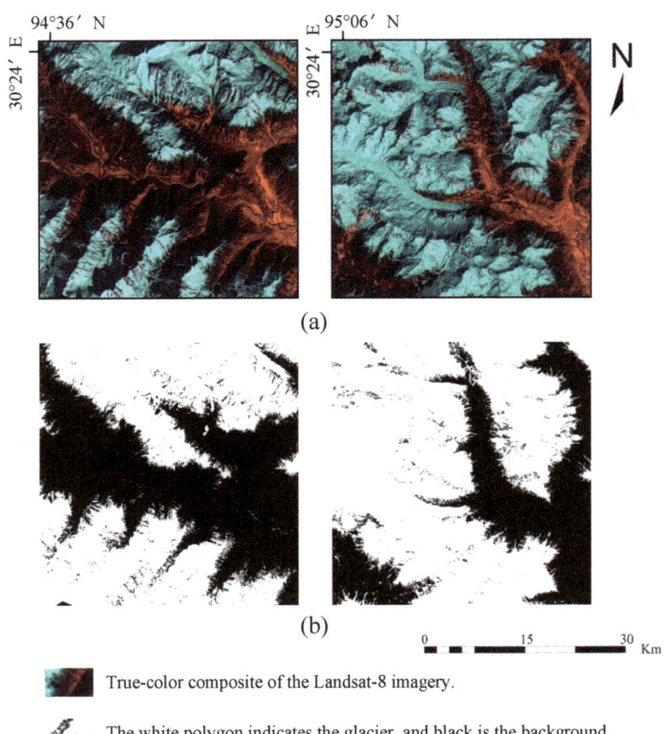

Figure 3. The dataset of semantic segmentation. Examples show the sample labels of glaciers in different images. (**a**) True-colour composite of the Landsat 8 imagery; (**b**) the white polygon indicates the glacier, and black is the background. (Band2 (Blue), Band3 (Green) and Band6 (SWIR 1)).

Due to the influence of external factors such as topographic relief, solar radiation and cloud cover, remote sensing images may suffer from distortions and other problems during the imaging process [21]. Therefore, pre-processing of the raw remote sensing data, including data format conversion, radiometric calibration, atmospheric correction, and geometric correction, is required before classifying the glaciers [22]. Extraction of information, such as elevation and area of the study area, allows for debris flow hazard assessment and predictive analysis, as well as final output of study results and visualisation images, hazard evaluation maps, prediction analysis maps, etc.

3.3. Step 2: Model Architectures and Training

In this section, we detail the model architecture of a deep-learning-based approach to accurately and automatically map complex debris-covered glaciers from remotely sensed images. First, we generate sample sets for training and testing. Afterwards, we perform model training by employing a semantic segmentation model with weight optimisation. The following is an example of the basic process for processing remotely sensed images as Figure 4.

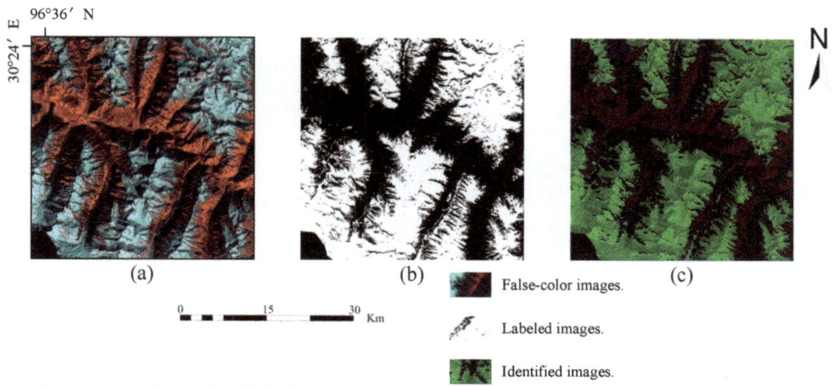

Figure 4. The processing of remote sensing images (**a**) False-colour images; (**b**) labelled images; (**c**) identified images.

DeepLabv3+ is a convolutional-neural-network-based image segmentation model for segmenting objects in images [23]. We use a deep convolutional neural network called Mobilenet as the backbone network for extracting image features. By leveraging the capabilities of DeepLabv3+, the research aims to achieve accurate and reliable segmentation of glacier debris and surrounding terrain from remote sensing imagery. The model's high accuracy can enhance the quality of the susceptibility analysis results. Furthermore, to extend the perceptual field of the convolutional kernel, DeepLabv3+ adds a null convolution layer. Null convolution adds a certain number of voids inside the convolution kernel, allowing larger convolution kernels to process features over large regions without using too many parameters [24]. DeepLabv3+ uses spatial pyramidal pooling for aggregation of multi-scale image features. Specifically, the model samples the feature maps at different scales using separate pooling kernels to capture the features of object regions at different scales. A full convolution decoder is also used for reducing the feature maps extracted in the backbone network to segmented images of the same size as the input image. This decoder obtains results by upsampling and stitching high-resolution features with low-resolution semantic segmentation maps. DeepLabv3+ achieves efficient extraction of multi-scale image features and accurate segmentation of glacier regions by using Mobilenet networks with operations such as null convolution, spatial pyramid pooling and a full convolution decoder [25]. The DeepLabv3+ architecture is shown in Figure 5.

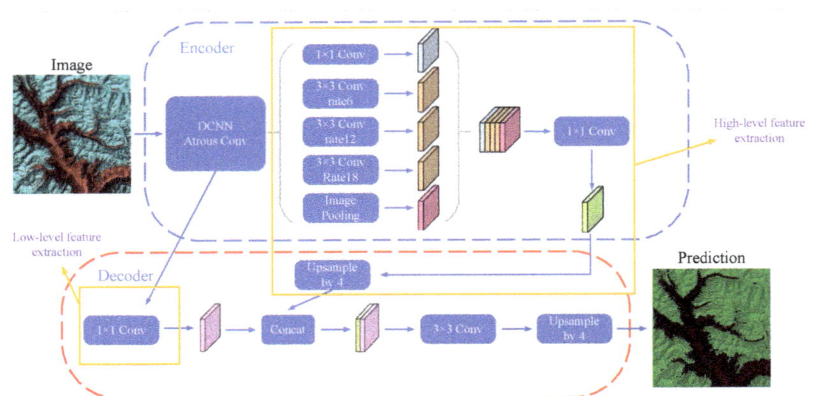

Figure 5. DeepLabv3+ architecture for semantic segmentation (figure adapted from [24]).

3.4. Step 3: Model Architectures and Training

We used the DeepLabv3+ model for glacier debris flow remote sensing image classification, which allows us to obtain information on the susceptibility of glacier debris flow remote sensing images. In evaluating the Deeplabv3+ model, we learned that the accuracy assessment of glacier identification relies heavily on the quality of the sample set. The samples were divided into training and validation sets in a 9:1 ratio for training the model and evaluating its accuracy, respectively. The main evaluation metrics for semantic segmentation are MPA, MIoU, and pixel accuracy, which are calculated based on the confusion matrix. The four basic elements that make up the confusion matrix are true positive (TP), false positive (FP), true negative (TN) and false negative (FN). The performance accuracy of a glacier or non-glacier can be defined based on MPA, MIoU, and pixel accuracy. The following formulas are quoted from [24].

$$MPA = \frac{1}{k+1}\sum_{i=0}^{k} PA_i \tag{1}$$

$$MIoU = \frac{1}{k+1}\sum_{i=0}^{k} \frac{TP}{FN+FP+TP} \tag{2}$$

$$Pixel - Accuracy = \frac{TP+TN}{TP+TN+FP+FN} \tag{3}$$

We use the DeepLabv3+ model to train the annotated remote sensing images and carry out model optimisation to improve the model prediction performance and to evaluate the ease of occurrence. At the same time, there are some limitations and challenges to its application. On the one hand, the model requires a large amount of high quality data for training, which is demanding in terms of data quality and data volume, which may create some limitations in areas of study where there is insufficient data. On the other hand, if the amount of training data is too small or unbalanced, over-fitting can easily occur. In addition to this, the model is relatively complex and requires parameter tuning, which requires a certain level of technical skill and practical experience on the part of the user. When using the DeepLabv3+ model for glacial mudslide susceptibility assessment, these issues need to be considered thoroughly to ensure accurate and reliable results.

3.5. Step 4: Glacial Debris Flow Susceptibility Assessment

In conducting the analysis of glacial debris flow susceptibility, we conducted a comprehensive analysis and judgment of eight factors that affect the occurrence of glacial debris flows, including the volume of physical sources, catchment area, maximum daily rainfall, longitudinal slope drop of the main gully, length of the main gully, glacier volume, total glacial lake area, and vegetation area [26]. Based on our study, we replaced glacier area with the glacier ablation volume in the previous study, and calculated the volume of meltwater by calculating the volume of glacier ablation within the glacial debris flow basin over a five-year period, and combined it with other factors to arrive at a more accurate evaluation method for glacial debris flows.

We first calculated the values of each factor in the glacier debris flow basin in the study area, and then used the entropy weighting method to calculate the weight of each factor on glacier debris flow susceptibility. Finally, we used the Topsis method to score the susceptibility of each glacial debris flow by combining the weights of each factor, and evaluated the susceptibility of glacial debris flows based on the scores.

4. Results and Analysis

4.1. Experimental Environment and Settings

The experimental environment for this paper uses the Python programming language, the windows 11 operating system, a 12th Gen Intel® Core™ i7-12700H CPU, an NVIDIA GeForce RTX 3060 Laptop GPU, the PyTorch framework, and Mobilenet for the backbone network. A larger value is better for parallel computing, while a smaller value affects the GPU's performance. When the batch size is set to 16 or 32, too large a batch may lead to a lack of memory. Finally, considering the convergence speed and random gradient noise and device performance, the batch size was chosen to be 4 for the freezing phase and 2 for the thawing phase. The number of iterations is the number of times the training set is fed into the neural network for training. Usually, when to stop iterating depends on the predictive performance of the model. When the number of iterations is chosen to be 600, overfitting can occur because the number of iterations is too large. It needs to be simplified. When the number of iterations is chosen to be 500, the difference between the test error rate and the training error rate is small and the current number can be considered appropriate. The learning rate is the size of the network weights update in the optimisation algorithm. The maximum learning rate of the model defaults to 0.01 when the learning process is adaptively adjusted according to the current batch_size. The training set is then fed into the DeepLabv3+ network for iteration. As described above, the training parameters were continuously adjusted to finally obtain a weight-optimised semantic segmentation model, which we used to predict the glacier boundary in the study area and to calculate the glacier ablation volume over a five-year period in combination with the elevation data.

4.2. Glacier Boundary Identification Model Training and Evaluation

Before the training, the extracted remote sensing images were divided into 3690 images according to a size of 128*128. If the whole image is trained directly without cropping, it may consume too much computer memory during the training process, resulting in the interruption of the training process. According to 9:1, the data set was divided into the training set and the verification set, which were 3354 and 336 pieces, respectively. In this model training, we used the test set as the validation set and did not divide the test set separately any more, and the remaining training set was used to learn the discriminative information between glaciers and non-glaciers.

We conducted several training sessions of the glacier boundary recognition model, selected several different batch_sizes, obtained several model files, and evaluated their model training performance. Figures 6 and 7 show the relevant evaluation parameter changes for 200 and 500 epochs of model training, respectively.

The analysis of the glacier segmentation results based on different model architectures was performed by continuously adjusting the parameters to obtain weight optimisation. The segmentation results for the test data based on Landsat 8 images are shown in Figure 8 in comparison to the corresponding original false colour images. We can see that the pixel points discriminated as glacier-like are given a green mask during the model prediction process, while the pixel points discriminated as non-glacier-like are left unchanged.

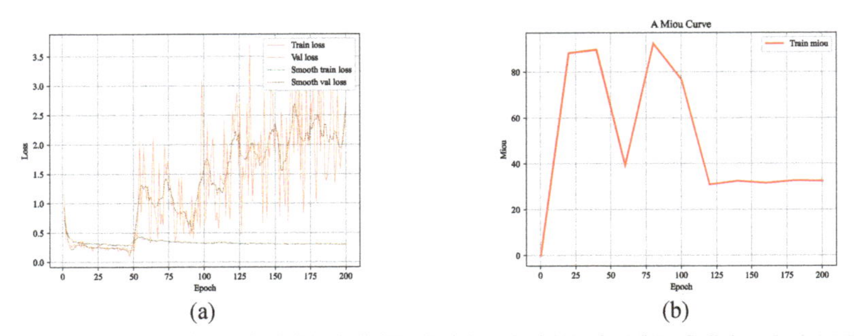

Figure 6. Model parameters after 200 epochs of training: (**a**) loss; (**b**) MIoU.

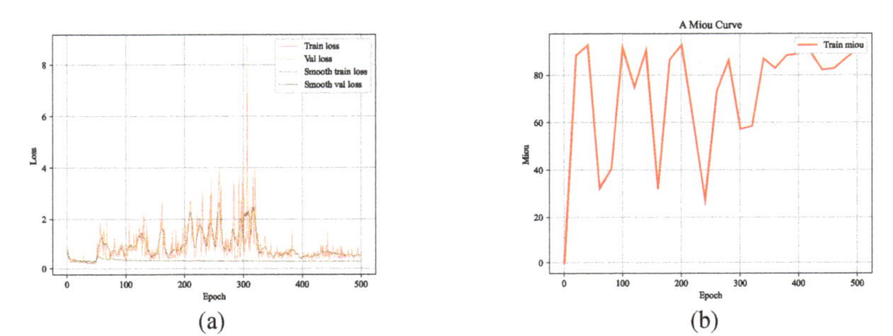

Figure 7. Model parameters after 500 epochs of training: (**a**) loss; (**b**) MIoU.

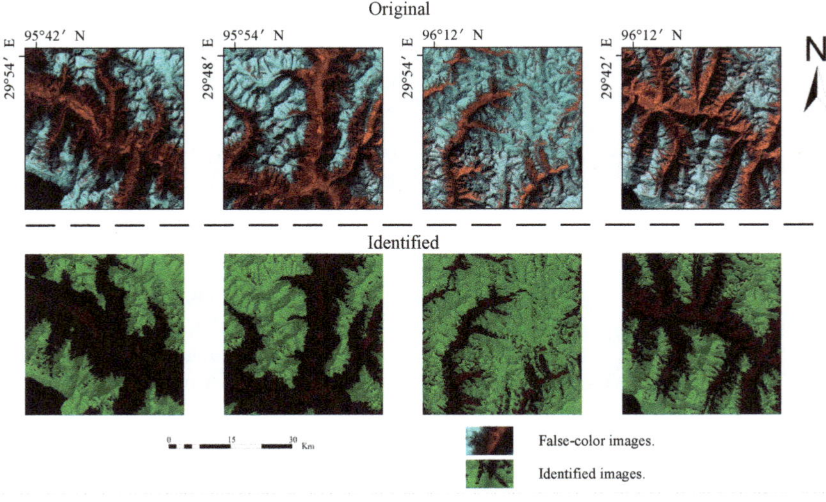

Figure 8. Comparison of the original image and the segmentation model results of glacier boundary part.

The discrimination from these remote sensing images alone is not comprehensive enough for us to quantitatively evaluate the trained glacier boundary recognition model. Therefore, we assessed the quality of the model training through the values of MPA, MIoU, and pixel accuracy to reflect the accuracy of the model discrimination. For the DeepLabv

3+ model performance from the Landsat 8 dataset, the values of MIoU were 90.85%, MPA 96.09%, and pixel accuracy 96.64%, all of which were above 90%, indicating that our trained model can perform reasonable glacier boundary extraction.

In general, the DeepLabv3+ model can extract mudslide susceptibility information from remote sensing images well, but the evaluation needs to consider a variety of indicators and combine with other factors such as topography and rainfall to make a comprehensive analysis and judgment. At the same time, the model results need to be revised and validated by combining the experience of historical mudslide events and actual measurement data when making predictions.

4.3. Assessment Results of the Vulnerability of Glacial Debris Flow

In the analysis of the susceptibility of glacier debris flows, we have addressed material source conditions, slope, precipitation, glacial geological conditions, etc.

We summarised eight influencing factors that have a significant impact on the occurrence of glacier debris flows: volume of physical source (X_1), catchment area (X_2), maximum daily rainfall (X_3), longitudinal slope drop of the main gully (X_4), length of the main gully (X_5), glacier volume (X_6), total glacial lake area (X_7), and vegetation area (X_8). We found a raw data matrix of 132 glacier debris flow gullies in Linzhi city classified by the above influencing factors.

To qualitatively analyse the formation factors of glacier debris flows, we normalised the raw data and used the entropy weighting method to derive the weight of each influencing factor on the susceptibility of glacier debris flows. The weights are shown in the Table 1.

Table 1. Evaluation metrics based on the entropy weighting method (EWM).

Influencing Factors	X_1	X_2	X_3	X_4	X_5	X_6	X_7	X_8
Weights	0.205	0.156	0.006	0.042	0.009	0.219	0.358	0.005

Combining the weights given, we scored the mudslide gullies in the study area using the Topsis method and graded the results as Table 2.

Table 2. Scores for six glacier debris flows based on the Topsis method of analysis.

No.	1	2	3	4	5	6
Score	0.060	0.071	0.055	0.097	0.083	0.048

Figures 9 and 10 shows our evaluation of the susceptibility of six glacier debris flows in the study area and the evaluation of their susceptibility in previous studies.

The direct result of the remote sensing image vulnerability assessment of the G318 Linzhi section of the national highway is an image map with different coloured areas, each colour representing the corresponding mudslide vulnerability level. The red areas represent areas of very high susceptibility, the orange areas represent areas of high susceptibility, the yellow areas represent areas of medium susceptibility, and the green areas represent areas of low susceptibility. These different coloured areas provide a visual indication of the mudslide susceptibility of the area and provide important reference information for the prevention of mudslide disasters. The results of our evaluation are highly accurate when compared with the mudslide susceptibility assessment of the glaciers obtained after the field survey.

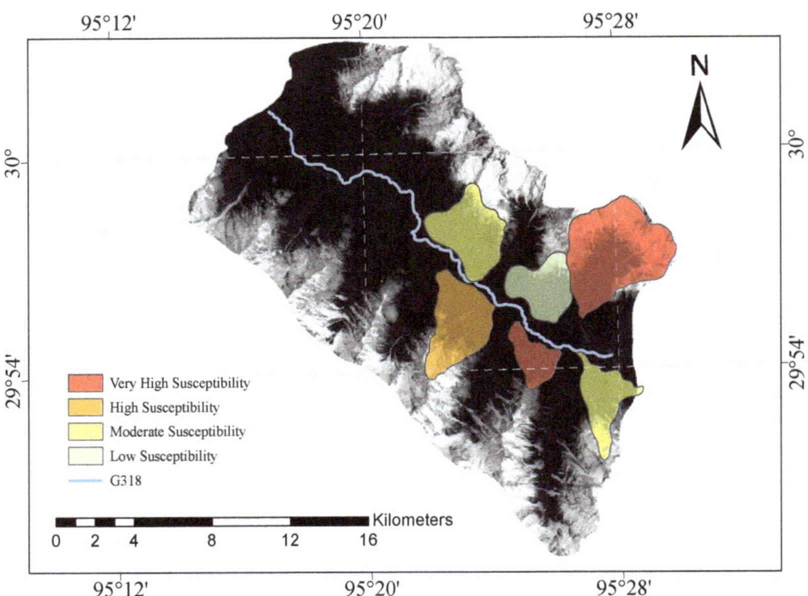

Figure 9. Glacial debris flow susceptibility mapping of the study area.

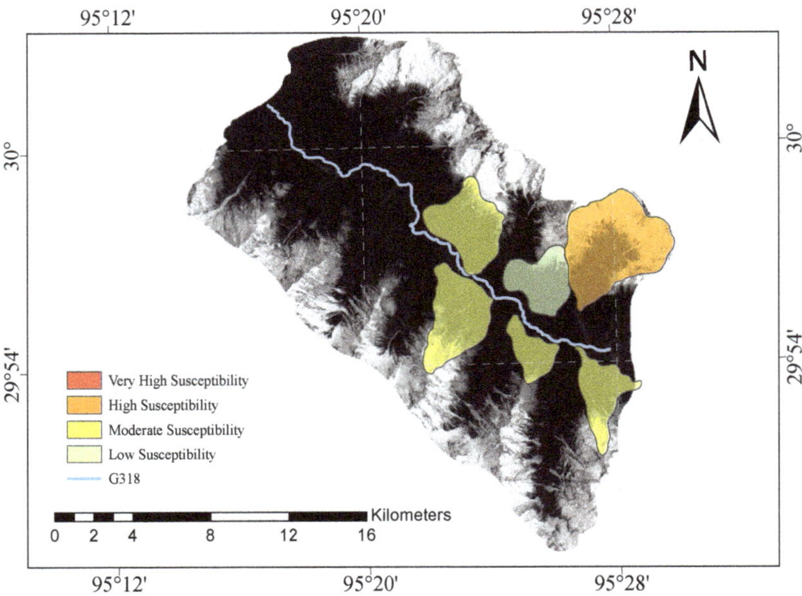

Figure 10. Previous glacial debris flow susceptibility mapping of the study area.

4.4. Different Factors' Influence on the Results

In conducting the glacial debris flow susceptibility analysis, we considered eight factors, namely material source volume, catchment area, maximum daily rainfall, longitudinal slope drop of the main gully, length of the main gully, glacier volume, total glacial lake area, and vegetation area, for comprehensive analysis and judgement.

4.4.1. Volume of Material Source

Loose material is one of the most important precipitating conditions for glacial debris flows, and loose material plays an important role in the estimation of the volume of the material source. The product of the material distribution area and the average thickness is the estimate of the volume of the material source. Based on this method, and combined with remote sensing images, it can be concluded that: 60.1% of the 145 glacier debris flows in the study area have a material source volume greater than 10×10^6 m^3 , 34% are between 1×10^6 and 10×10^6 m^3, and 5.9% are less than 1×10^6 m^3. This shows that the glacier debris flows in this study area have abundant reserves of material sources, providing conditions for the initiation of glacier debris flows, as shown in Figure 11.

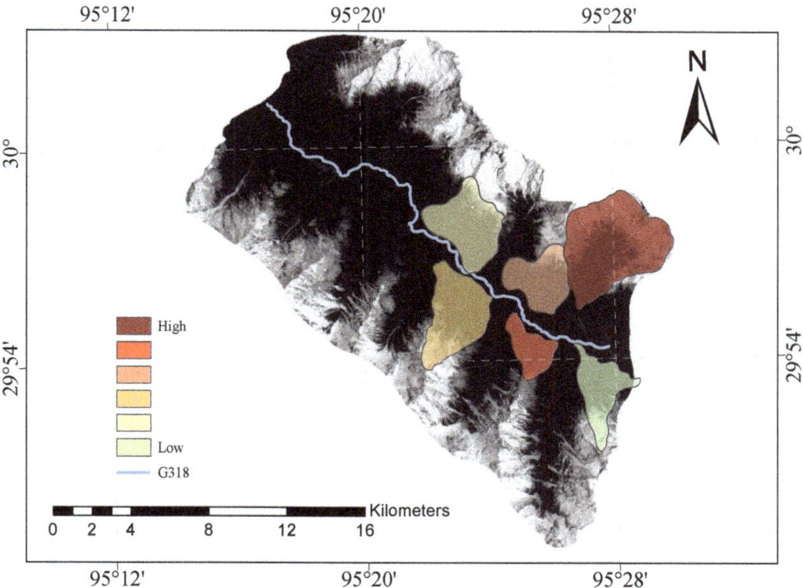

Figure 11. Comparison of the volume of the physical source in the study area.

4.4.2. Catchment Area

The catchment area is the area of water within a valley or watershed that is formed by ground form, precipitation, snow melt, and other factors. Within the catchment area, water, such as precipitation or snow melt and ice melt, collects through the gully and surface to become a river or stream, and eventually flows into the convergence point. The area of the convergence point and its upstream area is called the glacier debris flow catchment area. There are 145 glacier debris flows in the study area, of which the smallest is 0.93 km^2 and the largest is 349 km^2, with 64.7% of the catchments measuring 10 km^2 to 100 km^2. Glacial debris flows have a strong erosion and accumulation effect on the landscape, forming natural catchment areas such as ice buckets and troughs, so the catchment area in this study area is large, as seen in Figure 12.

4.4.3. Maximum Daily Precipitation

The maximum daily precipitation is an important factor in measuring rainfall. The role of rainfall among the many triggering factors for glacier debris flows is obvious. The study area is rich in rainfall and, according to statistics the maximum daily rainfall in the Linzhi section, is very close to the standard for heavy rainfall, which fully meets the requirements for the formation of glacier debris flows. The amount of rainfall influences the confluence of glacier debris flow slopes and the runoff from the gully. Strong rainfall can lead to erosion collapse, while weak rainfall manifests itself as liquefaction of the glacier debris flow slope, as seen in Figure 13.

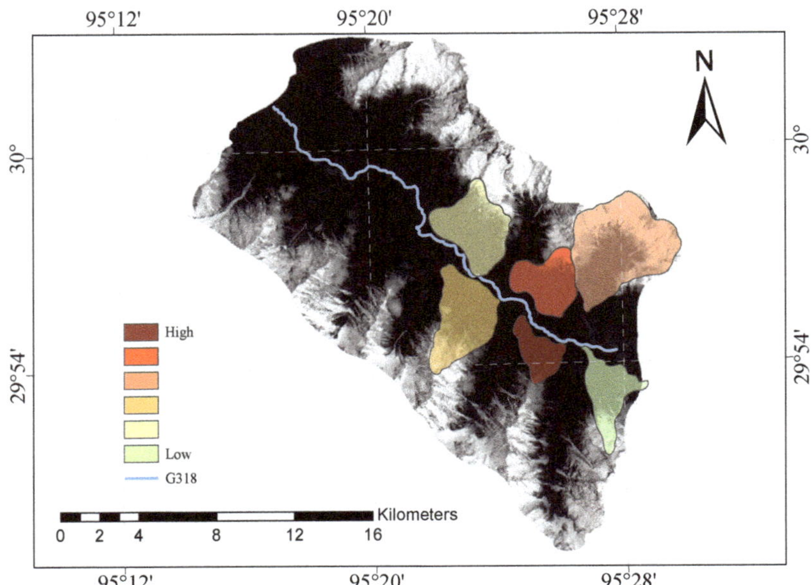

Figure 12. Comparison of catchment areas in the study area.

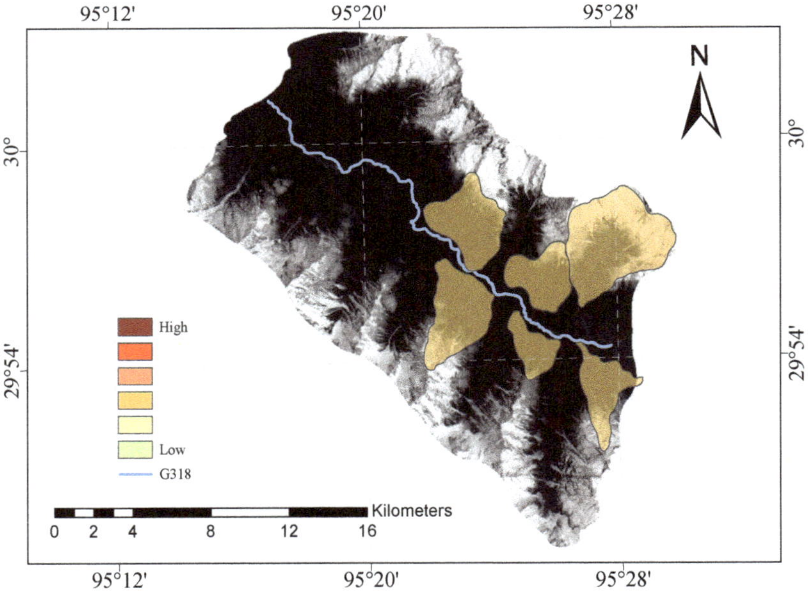

Figure 13. Comparison of maximum daily rainfall in the study area.

4.4.4. Longitudinal Slope Drop of the Main Ditch

The longitudinal slope drop of the main ditch is the change in height difference between the length of the ditch in the direction of the river, and its value is mainly determined by two important basic parameters, the length of the main ditch and the height difference. The main ditch longitudinal slope drop is calculated as follows:

$$W = \frac{H}{L} \tag{4}$$

where W is the longitudinal slope drop of the main gully, H is the relative height difference of the watershed along the main gully, and L is the length of the main gully.

The longitudinal slope drop of the main gully is one of the factors necessary to cause large-scale glacial debris flow hazards, and therefore the analysis of the longitudinal slope drop of the main gully is an integral part of the study of glacial debris flows, as seen in Figure 14.

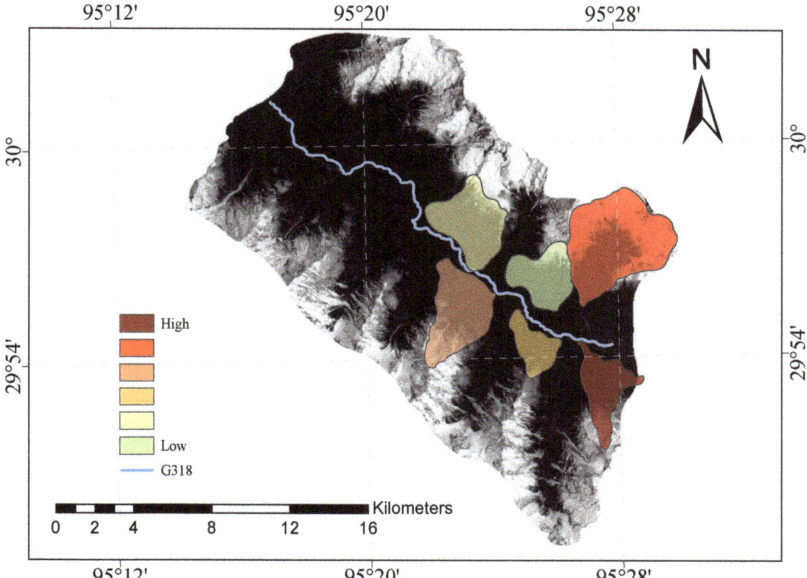

Figure 14. Comparison of the longitudinal slope drop of the main ditch in the study area.

4.4.5. Length of Main Gully

A main gully is a gully with a certain slope, a pronounced gully, and a high water table or surface flow rate, formed by natural factors such as glaciers, rivers, and wind. The length of the main gully is then the length of the main gully within the gully belt or valley. It is closely related to the longitudinal slope drop above, as seen in Figure 15.

4.4.6. Glacier Volume

Modern glaciers are necessary for the formation of glacier debris flows. Glacier volume is proportional to the number of glacier debris flows that occur. Most of the mudslide gullies in the study area contain large volumes of modern glaciers. The freezing and thawing of glaciers produces loose solid material that can trigger glacier debris flows. The analysis of glacial debris flows therefore requires an analysis of the volume of glaciers in the study area. The volume of the glacier and its volume of water is better reflected by combining two-dimensional remote sensing images with elevation than by the area of the glacier. This value has a more accurate impact on the analysis of susceptibility than glacier area, as seen in Figure 16.

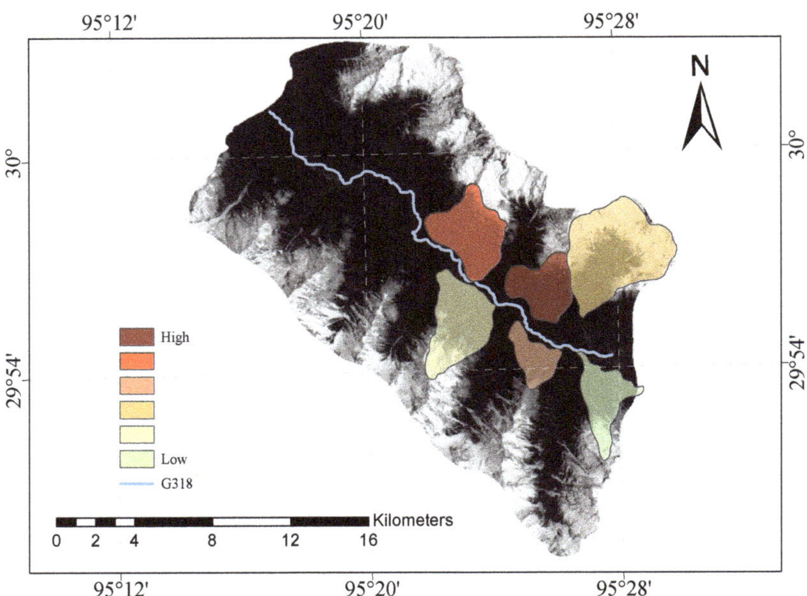

Figure 15. Comparison of the length of the main ditch in the study area.

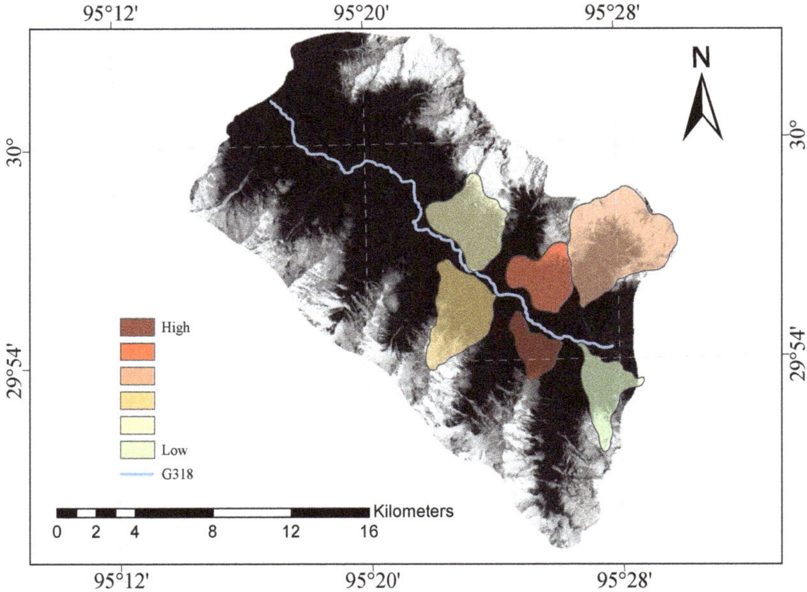

Figure 16. Comparison of glacier volumes in the study area.

4.4.7. Total Glacial Lake Area

A lake formed by glacial melt, flash floods, etc., with ice and glacial water as the main components is a glacial lake. The size and variation in the glacial lake area can reflect the activity and melting rate of the glacier, and is important for the analysis of the susceptibility of glacial debris flow, as seen in Figure 17.

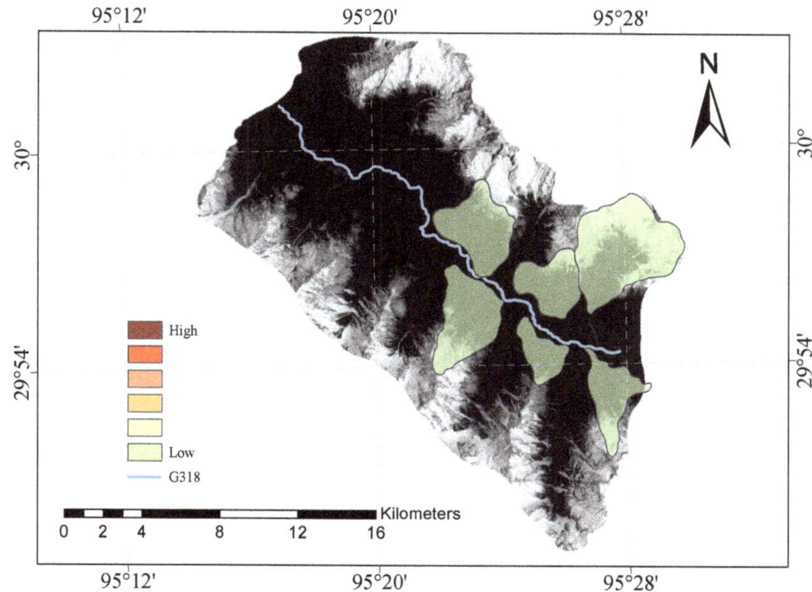

Figure 17. Comparison of the total glacial lake area in the study area.

4.4.8. Vegetation Area

Vegetation plays a suppressive role in the formation of glacier debris flows, with its roots penetrating deep into the soil and interlocking in a web-like pattern, acting similarly to anchors, anchoring the soil against erosion and scouring. The area of vegetation is therefore also an important indicator of the formation and susceptibility of glacier debris flows, and is one of the factors necessary for their analysis, as seen in Figure 18.

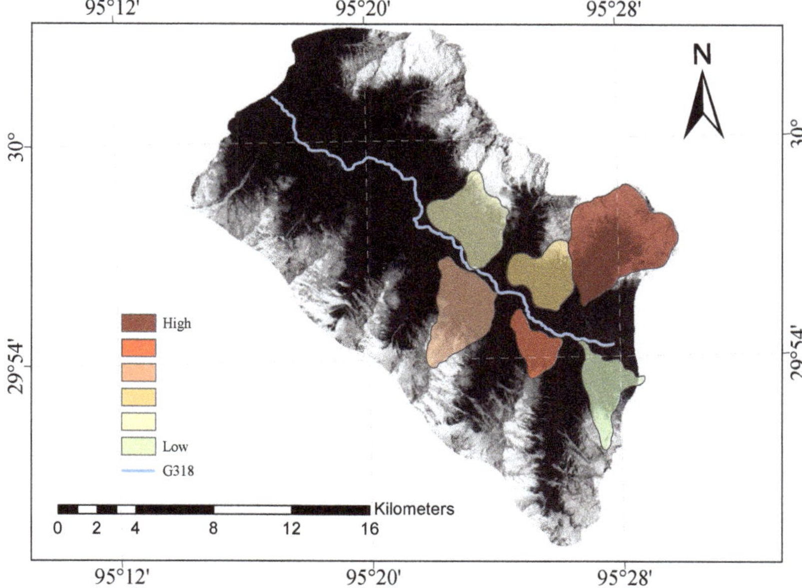

Figure 18. Comparative map of vegetation area in the study area.

In addition, there are a number of other factors that affect the accuracy of glacial debris flow susceptibility assessment. For example, the quality of remote sensing image data, the higher the quality of the data, the more reliable the prediction results; for the coarse and fine classification of feature classes, too fine or too coarse classification of feature classes will affect the model prediction results; for the adjustment of parameter settings in the model, including the adjustment of multi-scale image cropping, learning rate, number of training rounds, etc., will affect the model prediction results; for the characteristics of the sample itself, if the sample classes are unbalanced, it will easily make the mudslide susceptibility class is off from reality; and for the model structure and performance, different types of model structure, size of convolution kernel and other factors will affect the model prediction effect.

Different combinations of the above factors may produce different forms of impact effects, such as less accurate prediction results, failure to meet accuracy requirements, or weaker generalisation ability. Therefore, when analysing the susceptibility of remote sensing images of glacier debris flows in the Linzhi section of National Highway G318, these factors need to be taken into account and optimised and adjusted in the process of model training and prediction in order to improve the prediction effect of the model. At the same time, a comprehensive analysis of the actual terrain and other important factors is also needed to obtain more scientific and reliable prediction results.

5. Discussion

In this paper, we investigate the susceptibility of glacier debris flow along the G318 Linzhi section based on remote sensing imagery and deep learning. The segmentation results demonstrate the effectiveness and accuracy of the method. Its strengths and limitations are discussed below, and our future work to address the drawbacks is noted.

5.1. Advantages

The applications of remote sensing image and deep learning technology are important progress for the field of glacier debris flow analysis. This method provides a more effective and accurate means to study and predict the susceptibility of debris flow on glaciers. In contrast, previous studies may have relied on traditional methods, which were time-consuming and imprecise. The study of the Linzhi section of National Highway 318 is a case study, but its impact extends beyond a specific region. Glacial debris flow is a worldwide phenomenon, and the methods adopted in this study can be applied to other glacial regions around the world. This contributes to a global body of knowledge and provides a valuable tool for assessing and managing glacial debris flow risks in different regions.

The DeepLabv3+ model can significantly improve the accuracy and effectiveness of remote sensing image sensitivity analysis of glacier debris flow in the Linzhi section of the G318 National Highway. The model uses high-resolution image segmentation capability to finely segment high-resolution remote sensing images at the detail level and accurately extract key features of debris flow susceptibility [27]. In addition, the DeepLabv3+ model has stable prediction results, multi-scale input and zero convolution technology can improve the robustness and noise resistance of the model and reduce errors and outliers in the sensitivity analysis of glacial debris flow. The model has strong adaptability and transferability, and can maintain the segmentation effect and obtain good prediction results even in different study areas. In terms of vulnerability prediction, the model provides accurate and intuitive vulnerability map of glacier debris flow, and realizes the visualization of regional debris flow vulnerability information, which is of great significance for preventing and mitigating glacier debris flow disasters.

In addition, replacing the glacier area with the glacier ablation volume can more accurately assess glacier debris flow susceptibility. Sensitivity assessment based on deep learning reduces the high error rate of manual recognition while improving efficiency, especially for large-scale remote sensing images.

In summary, this study introduced research methods, extracted the boundary of glacier debris flow, obtained the vulnerability analysis diagram of glacier debris flow, identified the risk factors, verified the results, provided inspiration for disaster management, and made progress in the study of glacier debris flow. These advances help to understand and mitigate the risks associated with glacier debris flow and enhance the safety and resilience of affected areas.

5.2. Limitations

We know from our research literature that in an evaluation of the changes in elevation and surface velocity of Iran's largest and most dynamic detritus-covered glacier (Alamkouh Glacier) during 2018–2020, The high-resolution images of the UAV were obtained and processed into a digital elevation model (spatial resolution of about 15 cm) and an orthophoto image (spatial resolution of about 8 cm), and the changes in glacier thickness were obtained. However, in our study, we only used satellite images, and the accuracy could not reach the centimetre level [28]. It is extremely difficult to detect glacier surface flow rates in rugged and alpine Himalayan terrain using traditional surface techniques. Karimi, Neamat et al. used the differential band composite method for the first time to estimate the glacier surface velocity in the non-detrital covered area and the detrital covered area of the glacier, respectively. The accuracy is relatively considerable [29].

The assessment of glacier debris flow susceptibility requires a large amount of fine-grained data, and the acquisition of remote sensing images is critical to the accuracy of the assessment results. The remote sensing elevation image data contains the regional topographic information needed for the susceptibility evaluation of glacial debris flow, which is of great significance. However, in practice, it is often difficult to obtain high quality remote sensing images for glacier debris flow vulnerability assessment. First of all, in order to meet the evaluation needs, a large number of remote sensing images need to be obtained continuously over a long time scale, and obtaining these continuous high-quality remote sensing image data is an insurmountable challenge. Second, the lack of high-quality remote sensing imagery in many areas requires an extensive review of relevant sources to obtain more comprehensive topographic information. Finally, the process of remote sensing image pre-processing is also very complicated, and it is usually necessary to perform multiple steps to obtain high-quality remote sensing image data. Therefore, the acquisition and processing of high quality remote sensing image data is an important part of the susceptibility assessment of glacier debris flow, and sufficient attention should be paid in the evaluation work.

The integration of remote sensing image and deep learning technology is an important progress in the field of glacier debris flow analysis, while this study demonstrates the effectiveness of these methods in the specific context of the G318 Linzhi section, it is necessary to evaluate how these techniques compare to existing methods used in different parts of the globe. Exploring the limitations and potential improvements of these techniques helps to gain a more complete understanding of their applicability and effectiveness in different glacial environments.

5.3. Outlook

Glacial debris flow susceptibility analysis using deep learning and remote sensing imagery techniques has a wide range of applications. In order to improve the accuracy of the glacier debris flow susceptibility analysis model in response to this phenomenon, integration of multiple sources of information, including topographic data, land cover data, meteorological data, etc., can be considered. In addition, integrating multiple remote sensing data modalities such as optical images, infrared images, and radar images can help to improve the robustness and prediction accuracy of the model. Once the model has the capability to identify areas of glacier debris flow susceptibility, the scope and number of training sets and the use of global or national remote sensing data from international agencies will need to be expanded to further improve the generalisability and accuracy of the method.

6. Conclusions

In this research, we explore the vulnerability of glacier debris flow along the G318 Linzhi section using remote sensing imagery and deep learning techniques. We have come to the following conclusions: (1) the precursors of glacier debris flows can be monitored and warned in real time using remote sensing technology; (2) glacier retreat and glacial lake formation are important factors in glacier debris flow susceptibility areas and should be given priority consideration; (3) with the help of remote sensing data, slopes, river valleys, cliffs, and water bodies can be effectively identified; and (4) by obtaining glacier morphological parameters, topographic slope and elevation data, and using deep learning techniques to construct complex predictive models, we can predict the susceptibility of glacier debris flows more accurately.

In the future, we will consider how to more accurately monitor changes in glacier edges, and use state-of-the-art deep learning methods to address monitoring changes in glacier edges, observing changes in glaciers, and providing data support for applications such as environmental protection.

Author Contributions: Conceptualization, J.C., G.M., H.G., L.H. and R.Y.; methodology, J.C., G.M., H.G., L.H. and R.Y.; writing—original draft preparation, J.C. and G.M.; writing—review and editing, J.C., G.M., H.G., L.H. and R.Y. All authors have read and agreed to the published version of the manuscript.

Funding: This research was funded by China University of Geosciences (Beijing) Student Innovation and Entrepreneurship Training Programme, Category A Project 202311415016.

Institutional Review Board Statement: Not applicable.

Informed Consent Statement: Not applicable.

Data Availability Statement: Not applicable.

Acknowledgments: The authors would like to thank the editor and the reviewers for their contributions.

Conflicts of Interest: The authors declare no conflict of interest.

Abbreviations

The following abbreviations are used in this manuscript:

DEM	Digital elevation model
CNN	Convolutional neural network
RNN	Recurrent neural network

References

1. Cook, K.L.; Andermann, C.; Gimbert, F.; Adhikari, B.R.; Hovius, N. Glacial lake outburst floods as drivers of fluvial erosion in the Himalaya. *Science* **2018**, *362*, 53–57. [CrossRef]
2. Medeu, A.R.; Popov, N.V.; Blagovechshenskiy, V.P.; Askarova, M.A.; Medeu, A.A.; Ranova, S.U.; Kamalbekova, A.; Bolch, T. Moraine-dammed glacial lakes and threat of glacial debris flows in South-East Kazakhstan. *Earth-Sci. Rev.* **2022**, *229*, 103999. [CrossRef]
3. Lechner, A.M.; Foody, G.M.; Boyd, D.S. Applications in remote sensing to forest ecology and management. *One Earth* **2020**, *2*, 405–412. [CrossRef]
4. Zhang, L.; Zhang, L.; Du, B. Deep learning for remote sensing data: A technical tutorial on the state of the art. *IEEE Geosci. Remote Sens. Mag.* **2016**, *4*, 22–40. [CrossRef]
5. Ma, Z.; Mei, G. Deep learning for geological hazards analysis: Data, models, applications, and opportunities. *Earth-Sci. Rev.* **2021**, *223*, 103858. [CrossRef]
6. Crosta, G.B.; Dal Negro, P.; Frattini, P. Soil slips and debris flows on terraced slopes. *Nat. Hazards Earth Syst. Sci.* **2003**, *3*, 31–42. [CrossRef]
7. Xu, B.; Li, J.; Luo, Z.; Wu, J.; Liu, Y.; Yang, H.; Pei, X. Analyzing the spatiotemporal vegetation dynamics and their responses to climate change along the Ya'an–Linzhi section of the Sichuan–Tibet Railway. *Remote Sens.* **2022**, *14*, 3584. [CrossRef]

8. Wang, S.; Zhuang, J.; Mu, J.; Zheng, J.; Zhan, J.; Wang, J.; Fu, Y. Evaluation of landslide susceptibility of the Ya'an–Linzhi section of the Sichuan–Tibet Railway based on deep learning. *Environ. Earth Sci.* **2022**, *81*, 250. [CrossRef]
9. Ji, S.; Yu, D.; Shen, C.; Li, W.; Xu, Q. Landslide detection from an open satellite imagery and digital elevation model dataset using attention boosted convolutional neural networks. *Landslides* **2020**, *17*, 1337–1352. [CrossRef]
10. Banks, M.E.; McEwen, A.S.; Kargel, J.S.; Baker, V.R.; Strom, R.G.; Mellon, M.T.; Gulick, V.C.; Keszthelyi, L.; Herkenhoff, K.E.; Pelletier, J.D.; et al. High Resolution Imaging Science Experiment (HiRISE) observations of glacial and periglacial morphologies in the circum-Argyre Planitia highlands, Mars. *J. Geophys. Res. Planets* **2008**, *113*. [CrossRef]
11. Lin, R.; Mei, G.; Liu, Z.; Xi, N.; Zhang, X. Susceptibility analysis of glacier debris flow by investigating the changes in glaciers based on remote sensing: A case study. *Sustainability* **2021**, *13*, 7196. [CrossRef]
12. Li, Y.; Zhang, H.; Xue, X.; Jiang, Y.; Shen, Q. Deep learning for remote sensing image classification: A survey. *Wiley Interdiscip. Rev. Data Min. Knowl. Discov.* **2018**, *8*, e1264. [CrossRef]
13. Hu, K.; Zhang, X.; You, Y.; Hu, X.; Liu, W.; Li, Y. Landslides and dammed lakes triggered by the 2017 Ms6. 9 Milin earthquake in the Tsangpo gorge. *Landslides* **2019**, *16*, 993–1001. [CrossRef]
14. Lingzhi, X.; Zhihong, L.; Jinbao, L.; Xiao, Z. Ynamic Variation of Glaciers in Bomi County of Tibet During 1980 2010. *Procedia Environ. Sci.* **2011**, *10*, 1654–1660. [CrossRef]
15. Nie, Y.; Pritchard, H.D.; Liu, Q.; Hennig, T.; Wang, W.; Wang, X.; Liu, S.; Nepal, S.; Samyn, D.; Hewitt, K.; et al. Glacial change and hydrological implications in the Himalaya and Karakoram. *Nat. Rev. Earth Environ.* **2021**, *2*, 91–106. [CrossRef]
16. Chiarle, M.; Iannotti, S.; Mortara, G.; Deline, P. Recent debris flow occurrences associated with glaciers in the Alps. *Glob. Planet. Chang.* **2007**, *56*, 123–136. [CrossRef]
17. Ge, Y.; Cui, P.; Su, F.; Zhang, J.; Chen, X. Case history of the disastrous debris flows of Tianmo Watershed in Bomi County, Tibet, China: Some mitigation suggestions. *J. Mt. Sci.* **2014**, *11*, 1253–1265. [CrossRef]
18. U.S. Geological Survey. Available online: https://earthexplorer.usgs.gov/ (accessed on 12 March 2023).
19. Lin, R.; Mei, G.; Xu, N. Accurate and automatic mapping of complex debris-covered glacier from remote sensing imagery using deep convolutional networks. *Geol. J.* **2022**, *58*, 2254–2267. [CrossRef]
20. Shean, D.E.; Alexandrov, O.; Moratto, Z.M.; Smith, B.E.; Joughin, I.R.; Porter, C.; Morin, P. An automated, open-source pipeline for mass production of digital elevation models (DEMs) from very-high-resolution commercial stereo satellite imagery. *ISPRS J. Photogramm. Remote Sens.* **2016**, *116*, 101–117. [CrossRef]
21. Xu, L.; Chen, Q. Remote-sensing image usability assessment based on ResNet by combining edge and texture maps. *IEEE J. Sel. Top. Appl. Earth Obs. Remote Sens.* **2019**, *12*, 1825–1834. [CrossRef]
22. Foody, G.M. Status of land cover classification accuracy assessment. *Remote Sens. Environ.* **2002**, *80*, 185–201. [CrossRef]
23. Gu, R.; Wang, G.; Song, T.; Huang, R.; Aertsen, M.; Deprest, J.; Ourselin, S.; Vercauteren, T.; Zhang, S. CA-Net: Comprehensive attention convolutional neural networks for explainable medical image segmentation. *IEEE Trans. Med. Imaging* **2020**, *40*, 699–711. [CrossRef] [PubMed]
24. Chen, L.C.; Zhu, Y.; Papandreou, G.; Schroff, F.; Adam, H. Encoder-decoder with atrous separable convolution for semantic image segmentation. In Proceedings of the European Conference on Computer Vision (ECCV), Munich, Germany, 8–14 September 2018; pp. 801–818.
25. Xi, N.; Mei, G.; Liu, Z.; Xu, N. Automatic identification of mining-induced subsidence using deep convolutional networks based on time-series InSAR data: A case study of Huodong mining area in Shanxi Province, China. *Bull. Eng. Geol. Environ.* **2023**, *82*, 78. [CrossRef]
26. Kang, S.; Lee, S.R. Debris flow susceptibility assessment based on an empirical approach in the central region of South Korea. *Geomorphology* **2018**, *308*, 1–12. [CrossRef]
27. Burgueño, A.M.; Aldana-Martín, J.F.; Vázquez-Pendón, M.; Barba-González, C.; Jiménez Gómez, Y.; García Millán, V.; Navas-Delgado, I. Scalable approach for high-resolution land cover: A case study in the Mediterranean Basin. *J. Big Data* **2023**, *10*, 91. [CrossRef]
28. Sam, L.; Bhardwaj, A.; Singh, S.; Kumar, R. Remote sensing flow velocity of debris-covered glaciers using Landsat 8 data. *Prog. Phys. Geogr.* **2016**, *40*, 305–321. [CrossRef]
29. Karimi, N.; Sheshangosht, S.; Roozbahani, R. High-resolution monitoring of debris-covered glacier mass budget and flow velocity using repeated UAV photogrammetry in Iran. *Geomorphology* **2021**, *389*, 107855. [CrossRef]

Article

A Robust Deep-Learning Model for Landslide Susceptibility Mapping: A Case Study of Kurdistan Province, Iran

Bahareh Ghasemian [1], Himan Shahabi [1,*], Ataollah Shirzadi [2], Nadhir Al-Ansari [3], Abolfazl Jaafari [4], Victoria R. Kress [5], Marten Geertsema [6], Somayeh Renoud [7] and Anuar Ahmad [8]

1 Department of Geomorphology, Faculty of Natural Resources, University of Kurdistan, Sanandaj 6617715175, Iran; b.ghassemian@nr.uok.ac.ir
2 Department of Rangeland and Watershed Management, Faculty of Natural Resources, University of Kurdistan, Sanandaj 6617715175, Iran; a.shirzadi@uok.ac.ir
3 Civil, Environmental and Natural Resources Engineering, Lulea University of Technology, 97187 Lulea, Sweden; nadhir.alansari@ltu.se
4 Research Institute of Forests and Rangelands, Agricultural Research, Education and Extension Organization (AREEO), Tehran 1496813111, Iran; jaafari@rifr-ac.ir
5 Department of Ecosystem Science and Management, University of Northern British Columbia, 3333 University Way, Prince George, BC V2N 4Z9, Canada; kressv@unbc.ca
6 Research Geomorphologist, Ministry of Forests, Lands, Natural Resource Operations and Rural Development, 499 George Street, Prince George, BC V2L 1R5, Canada; Marten.Geertsema@gov.bc.ca
7 Data Mining Laboratory, Department of Engineering, College of Farabi, University of Tehran, Tehran 1417935840, Iran; somayehronoud@gmail.com
8 Department of Geoinformation, Faculty of Built Environment and Surveying, Universiti Teknologi Malaysia (UTM), Johor Bahru 81310, Malaysia; anuarahmad@utm.my
* Correspondence: h.shahabi@uok.ac.ir

Abstract: We mapped landslide susceptibility in Kamyaran city of Kurdistan Province, Iran, using a robust deep-learning (DP) model based on a combination of extreme learning machine (ELM), deep belief network (DBN), back propagation (BP), and genetic algorithm (GA). A total of 118 landslide locations were recorded and divided in the training and testing datasets. We selected 25 conditioning factors, and of these, we specified the most important ones by an information gain ratio (IGR) technique. We assessed the performance of the DP model using statistical measures including sensitivity, specificity, accuracy, F1-measure, and area under-the-receiver operating characteristic curve (AUC). Three benchmark algorithms, i.e., support vector machine (SVM), REPTree, and NBTree, were used to check the applicability of the proposed model. The results by IGR concluded that of the 25 conditioning factors, only 16 factors were important for our modeling procedure, and of these, distance to road, road density, lithology and land use were the four most significant factors. Results based on the testing dataset revealed that the DP model had the highest accuracy (0.926) of the compared algorithms, followed by NBTree (0.917), REPTree (0.903), and SVM (0.894). The landslide susceptibility maps prepared from the DP model with AUC = 0.870 performed the best. We consider the DP model a suitable tool for landslide susceptibility mapping.

Keywords: landslide susceptibility; extreme learning machine; deep belief network; genetic algorithm; GIS; Iran

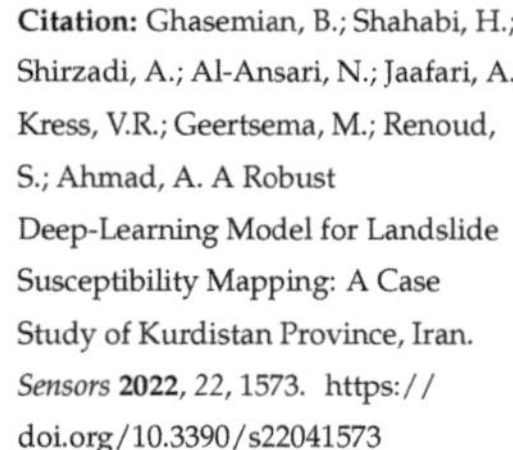

Citation: Ghasemian, B.; Shahabi, H.; Shirzadi, A.; Al-Ansari, N.; Jaafari, A.; Kress, V.R.; Geertsema, M.; Renoud, S.; Ahmad, A. A Robust Deep-Learning Model for Landslide Susceptibility Mapping: A Case Study of Kurdistan Province, Iran. *Sensors* **2022**, *22*, 1573. https://doi.org/10.3390/s22041573

Academic Editors: Junwei Ma and Jie Dou

Received: 24 December 2021
Accepted: 26 January 2022
Published: 17 February 2022

1. Introduction

Landslides occur in a variety of materials and undergo various styles of movement at different rates [1]. Landslides play an important geomorphological role in the evolution of landscapes, impacting the natural (soils, ecosystems, aquatic habitat, etc.) and built (residential areas, roads, pipelines, etc.) environment [2,3]. Landslide hazards are often exacerbated by land use practices such as road building, and deforestation, and may be made worse by increases in precipitation [4]. Therefore, it is important to identify areas that have a high potential for landslides and mitigate landslide damage.

Landslide risk assessment methodologies can be classified into three dominant groups: qualitative, quantitative, and artificial intelligence approaches. Qualitative approaches often rely on air photo and field interpretation and expert judgment (e.g., Schwab and Geertsema [5]). Quantitative methods are based on mathematical rules and expert judgment [6]. Artificial intelligence techniques can use subjective knowledge or pattern recognition techniques to solve a set of mathematical equations. Selection of the most appropriate model is usually based on the type of data available, the scale of the case study and analysis, and the knowledge of the researcher [7].

In recent decades, with the rapid development of geographic information systems (GIS) remote sensing (RS) techniques and improvements in the computing power of artificial intelligence algorithms, machine learning has played an important role in increasing the accuracy and reliability of landslide predictions [8]. Machine learning methods depend on field observations and statistical calculations [9]. Machine learning uses computer algorithms for analyzing and forecasting information by learning training datasets [10]. They have a high ability to detect landslide occurrence behavior using distribution estimation algorithms, they have a data-driven nature, and they utilize high repetition of the modeling process. In several studies, these methods have proven their comparative advantage over bivariate and multivariate statistical models [11,12].

Several machine-learning methods have been applied in landslide susceptibility assessment, such as logistic regression (LR) [13], naive Bayes (NB) [14,15], fuzzy logic (FL) [16], support vector machines (SVM) [17–19], kernel logistic regression (KLR) [20,21], Bayesian logistic regression (BLR) [17,22], artificial neural network (ANN) [23,24], random forest (RF) [25–28], rotation forest [29,30], random subspace (RS) [31], neuro-fuzzy inference system (ANFIS) [32,33], decision tree (DT) [26], classification and regression tree (CART) [34–36] and many other methods [37].

Despite the logical results and high performance of different models, geoscientists are always looking for new methods to more accurately identify landslide-prone areas and produce reliable maps needed for environmental planning. Therefore, presenting a new approach based on artificial intelligence algorithms, deep learning and GIS-RS techniques for landslide modeling is of high necessity in landslide hazards management [38].

One of the major challenges in mountainous areas is the occurrence of landslides, of which the occurrence is naturally inevitable and cannot be completely prevented but can be managed. In this study, we are looking for a technique that can be combined with several methods to achieve an algorithm with higher predictive power than conventional machine-learning algorithms to predict landslide prone areas. Despite the logical results and high performance of different models, geoscientists are always looking for new methods with quantitative criteria to more accurately identify landslide-prone areas and reliable maps needed for environmental planning. Therefore, presenting a new approach based on artificial intelligence algorithms and remote sensing techniques for landslide modeling is of high necessity in landslide management. We applied and developed a deep-learning model based on ELM, DBN, and BP optimized by GA for landslide susceptibility mapping. This model has been used earlier in spatial prediction against floods [39] and also has been applied to predict cancer [40]; however, its ability has not been evaluated to landslide susceptibility mapping so far. The model has been confirmed by some statistical measures and compared with some state-of-the-art benchmark machine-learning algorithms including SVM, NBTree and REPTree. We developed this model in MATLAB 2018a and all landslide susceptibility maps were produced in ArcGIS 10.5. The purpose of this study is to evaluate a robust deep-learning model that will support landslide susceptibility mapping. Here, we build on previous landslide susceptibility modelling for this study area using landslide data belonging to Asadi et al. [41], but with a different set of algorithms.

2. Study Area and Data

2.1. Description of the Study Area

Our study area, around Kamyaran city, is a mountainous area of nearly 150 km^2 in the southwest of the Kurdistan Province, Iran (Figure 1). The elevation ranges from 850 to 2328 m and has a mean annual temperature that varies between 11.3 °C and 17 °C and mean annual precipitation of 528 mm. Geologically, the study area is in the Sirvan drainage basin, located in the structural zone of Sanandaj-Sirjan and Zagros. Bedrock lithologies include outcrops from Cretaceous to Quaternary rocks, the oldest of which include Micrite limestone and dark gray shale. Most of the study areas are covered by Mesozoic and Cretaceous formations, which include Basaltic pillow lava and dark grey shale with intercalations of volcanic rocks. Holocene sediments of the Old Testament include alluvial fans and alluvial barracks. The predominant land covers in the study area are semi-dense forests and dry farming. In addition, dense pasture, semi-dense pasture, low-dense forest and woodlands are other types of land cover/land use in the study area. The area is significantly prone to landslides associated with road developments.

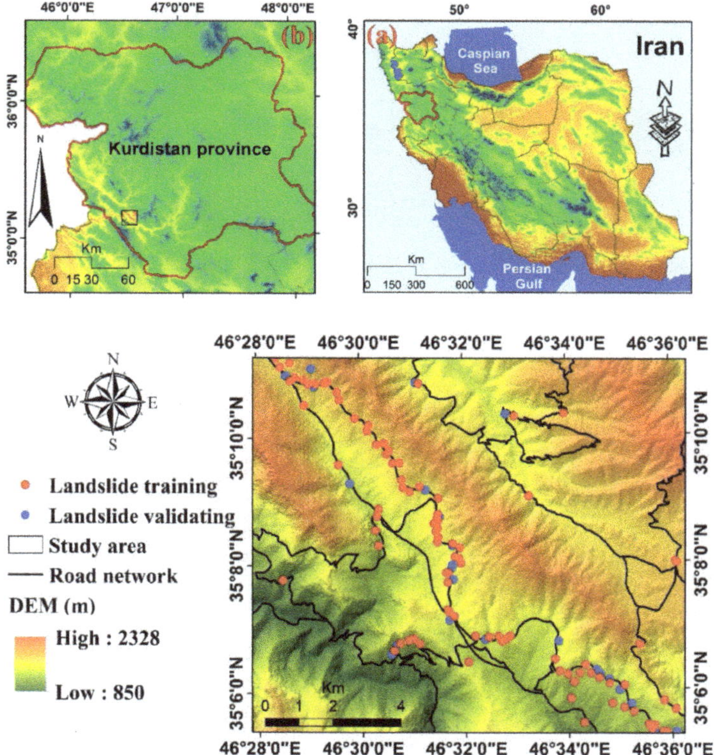

Figure 1. Geographical location of the study area in (a) Iran and (b) Kurdistan province.

2.2. Data

2.2.1. Landslide Inventory Map

It is necessary to prepare a landslide distribution map for landslide modeling because the assumption of the modeling process is that future landslides occur in the same conditions as in the past [42]. That *"the past and the present are key to the future"* is one of the most important principles in earth science. This means landslides that have occurred in the past and present under specific topographic, geological, hydrogeological, and climatic conditions in an area can provide useful information to predict the potential for future landslides in that area [43]. A map showing such information is useful for studying the spatial relationships between landslide distribution and factors affecting landslide occurrence [44]. Galli et al. [45] have mentioned that the quality of a landslide inventory map can lead to reasonable results in landslide modeling. From a total of 118 landslide points detected in the study area, 94 points (~80%) were used as the training dataset, and 24 points (~20%) were considered as the validation dataset. A total of 118 landslide locations used in this study were a part of a total of 175 landslide locations of Asadi et al. [41].

2.2.2. Landslide Conditioning Factors

The selection of the factors affecting the occurrence of landslides is one of the most important steps in landslide susceptibility studies [46]. In this study, we selected 25 conditioning factors that were slope angle, aspect, elevation above sea level, curvature, profile curvature, plan curvature, solar radiation, valley depth (VD), terrain ruggedness index (TRI), vector ruggedness measure (VRM), stream power index (SPI), topographic wetness index (TWI), length slope (LS), topographic position index (TPI), land use, normalized difference vegetation index (NDVI), lithology, soil, distance to fault, distance to river, distance to road, fault density, river density, road density, and rainfall (Figure 2). We used some conditioning factors for this study area that were earlier published by Asadi et al. (2022).

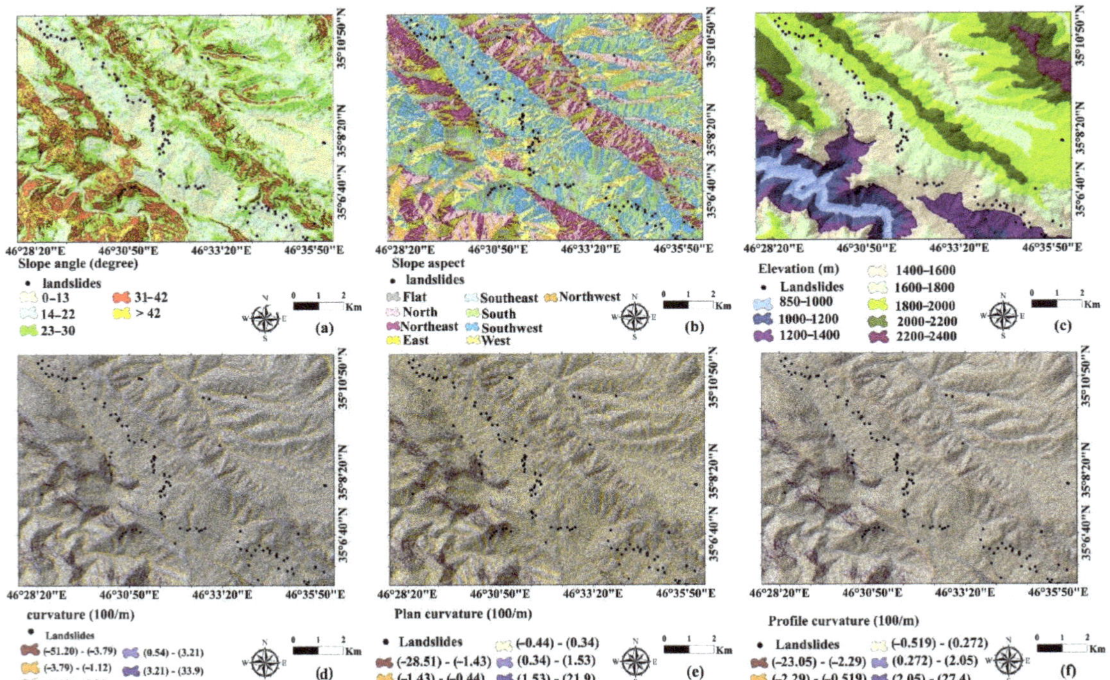

Figure 2. *Cont.*

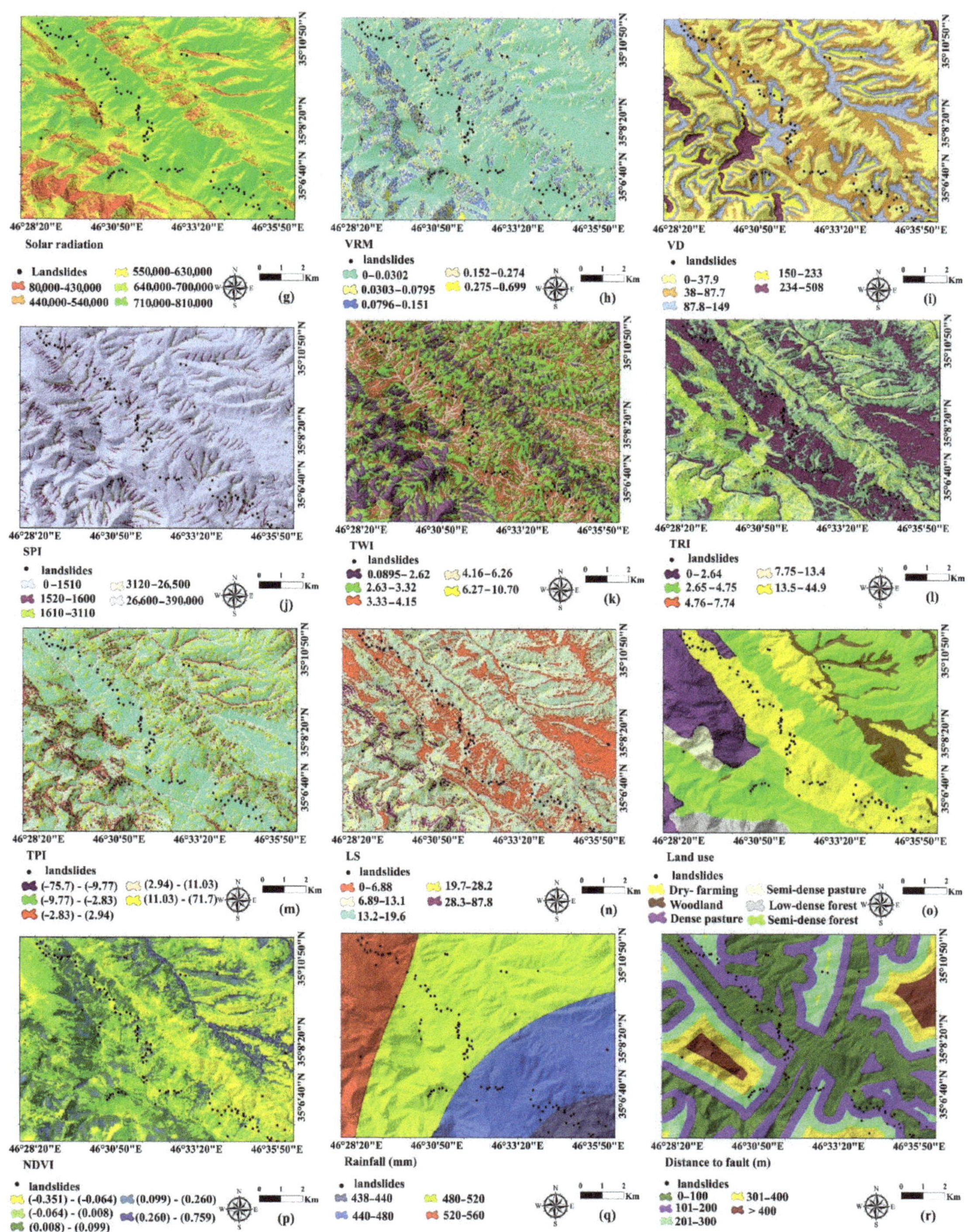

Figure 2. *Cont.*

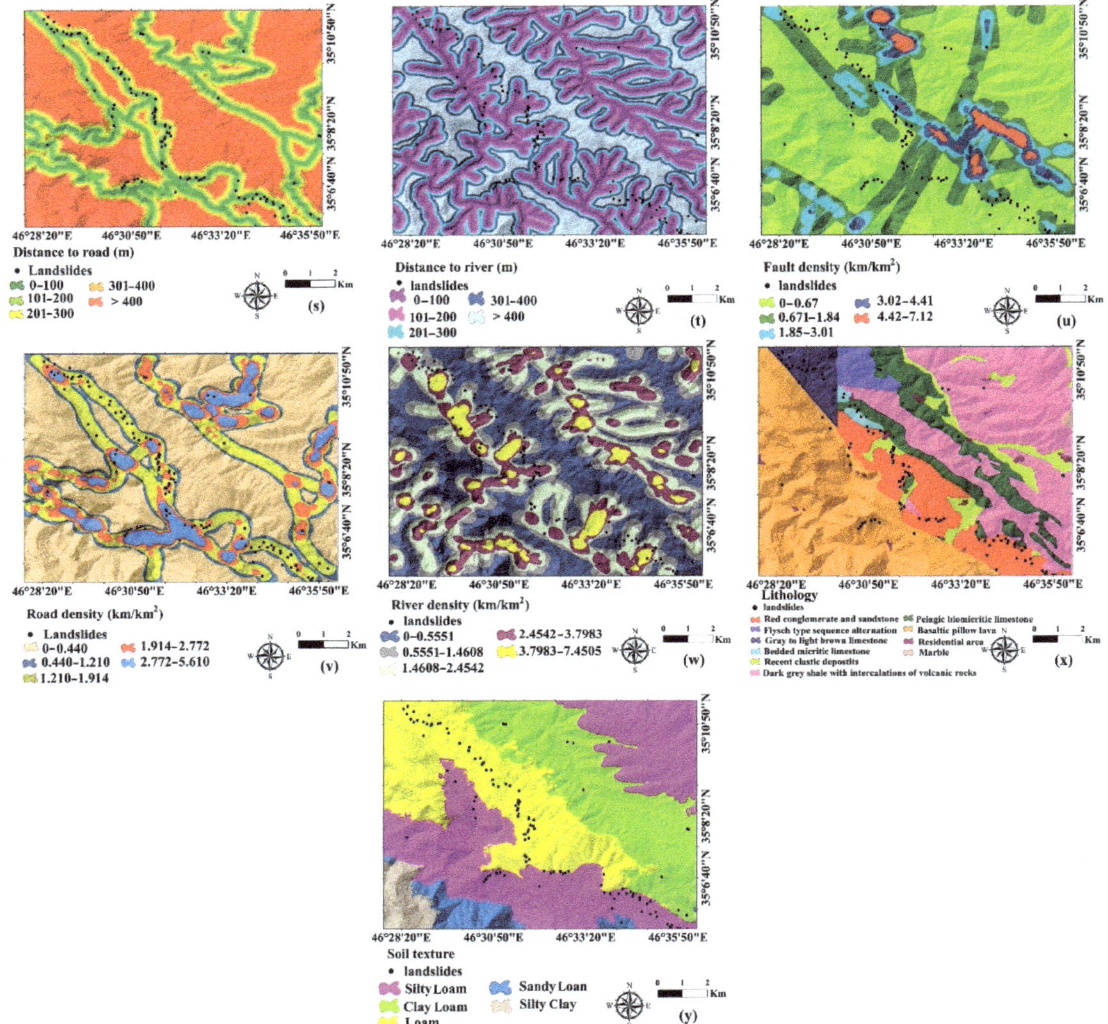

Figure 2. Landslide conditioning factors used in this study: (**a**) slope angle, (**b**) aspect, (**c**) elevation, (**d**) curvature, (**e**) plan curvature, (**f**) profile curvature, (**g**) solar radiation, (**h**) VRM, (**i**) VD, (**j**) SPI, (**k**) TWI, (**l**) TRI, (**m**) TPI, (**n**) LS, (**o**) land use, (**p**) NDVI, (**q**) rainfall, (**r**) distance to fault, (**s**) distance to road, (**t**) distance to river (**u**), fault density, (**v**) road density, (**w**), river density, (**x**) lithology, and (**y**) soil texture.

Slope Angle

Landslide hazard is often linked to slope angle, with shear stresses increasing on steeper slopes [47]. The supply of soil available for sliding often thins dramatically on steeper slopes above 25 degrees [48]. In other words, on high slopes, the type of material is more often stone and outcrops, such that medium slopes are more prone to landslides. This layer in the present study was extracted from the digital elevation model (DEM) and classified into eight intervals: 0–13, 14–22, 23–30, 31–42, and >43 (Figure 2a).

Aspect

Slope direction affects the occurrence of landslides by controlling the parameters related to soil moisture concentration, sunlight, dry winds, rainfall (saturation degree), and discontinuities [49]. This layer was extracted from DEM and categorized into nine classes: flat (−1–39.08), north (39.08–79.16), northeast (79.16–119.24), east (119.246–159.32), southeast (159.32–199.41), south (199.41–239.49), southwest (239.49–279.57), west (279.57–319.65), and northwest (319.65–359.74) (Figure 2b).

Elevation

The influences of elevation on landslides are often displayed as indirect relationships or by means of other factors [50]. The altitude factor of each region is one of the effective layers in creating slope instabilities. This factor indirectly determines many causes of landslides such as annual rainfall, heavy rainfall, temperature, frost changes, ice melting, etc. [51]. Maximum elevation of the region is 2328 m, and the minimum elevation is 850 m, hence the general elevation variance is 1478 m. The elevation map was extracted from DEM and then classified into eight classes: (1) 850–1000, (2) 1000–1200, (3) 1200–1400, (4) 1400–1600, (5) 1600–1800, (6) 1800–2000, (7) 2000–2200, and (8) 2200–2400 (Figure 2c).

Curvature

Curvature maps show the extent to which the surface deviates from the flatness, or in other words, the convexity and concaveness of the slope [52]. The curvature of the slope represents the shape of the topography so that the positive concavity represents the surface where the pixels are convex (Convex, Coves, Hollows), Negative concavity indicates a surface where the pixels are concave (Concave, Noses) and zero indicates a surface that has no slope and is straight (Flat, Straight). These three types of slope shapes have a great effect on slope instability by controlling the concentration and diffusion of surface and subsurface water in the slopes [53]. Convexity and concavity of the slope curvature map using distances between consecutive topographic lines in the GIS were extracted from the DEM of the region and classified into five classes (1) highly concave (−51.20)–(−3.79), concave (−3.79)–(−1.12), (3) flat (−1.12)–(0.54), (4) convex (0.54)–(3.21) and (5) very convex (3.21)–(33.9) (Figure 2d).

Plan Curvature

Plan curvature indicates changes in direction along a curve. This factor affects the divergence and convergence of water and materials containing a landslide in the path of motion. Plan curvature was extracted from DEM and divided into five classes: (1) [(−28.51)–(−1.43)], (2) [(−1.43)–(−0.44)], (3) [(−0.44)–(0.34)], (4) [(0.34)–(1.53)], and (5) [(1.53)–(21.09)] (Figure 2e).

Profile Curvature

Profile curvature is an important factor that affects the stress resistance due to landslides in the path and indicates the intensity of water flow and transportation and deposition processes [54]. The positive values in the transverse curvature of the slope indicate concavity (decrease in flow rate) and the negative values indicate convexity (increase in flow rate) [55]. Profile curvature was extracted from DEM and constructed in five categories: (1) [(−23.05)–(−2.29)], (2) [(−2.29)–(−0.519)], (3) [(−0.519)–(0.272)], (4) [(0.272)–(2.05)], and (5) [(2.05)–(27.4)] (Figure 2f).

Solar Radiation

The average convergence of solar radiation per pixel over a year is called the intensity of solar radiation, which is expressed in kilowatt hours per square meter [56]. The importance of this index is that its larger value indicates more vapor than the soil surface in an area. This index also controls the amount of vegetation on the slope. The less solar radiation that reaches a slope, the more vegetation appears on the slope, and as a result, the slope becomes more stable [57,58]. In the present study, the solar radiation layer was extracted from DEM in ArcGIS and categorized into five classes: (1) 80,000–43,000, (2) 440,000–540,000, (3) 550,000–630,000, (4) 640,000–700,000, and (5) 710,000–810,000 (Figure 2g).

Vector Ruggedness Measure

Vector ruggedness measure (VRM) factor was suggested by Hobson et al. [59]. It provides a way to measure terrain ruggedness as the variation in the three-dimensional orientation of grid cells within a neighborhood: slope and aspect are captured into a single measure and used to decouple terrain ruggedness from just slope or elevation [60]. The VRM map was created from DEM in the SAGA GIS software environment and then it was divided into five classes: (1) 0–0.0302, (2) 0.0303–0.0795, (3) 0.796–0.151, (4) 0.152–0.274, and 0.275–0.699 (Figure 2h).

Valley Depth

The valley depth (VD) factor can also be considered one of the fundamental layers in assessing landslide susceptibility. This index was prepared based on DEM map in the SAGA GIS software, and after exporting to ArcGIS it was classified into five classes: (1) 0–37.9, (2) 38–87.7, (3) 87.8–149, and (4) 150–233 and (5) 234–508 m (Figure 2i).

Stream Power Index

Stream Power Index (SPI) is a criterion derived from the DEM that might affect landslide occurrence, and it reflects the erosive power of slope surface run-off [61,62]. It can be formulated as follows:

$$\mathrm{SPI} = \mathrm{A_s} \times \tan\beta \tag{1}$$

where $\mathrm{A_s}$ is the specific basin area and $\tan\beta$ represents the slope angle. In this study, it was prepared based on DEM in the SAGA GIS software and then exported to ArcGIS software to map. The SPI layer was then extracted in five intervals: (1) 0–1510, (2) 1520–1600, (3) 1610–3110, (4) 3120–26,500, (5) 26,600–390,000 (Figure 2j).

Topographic Wetness Index

Topographic wetness index (TWI) represents a theoretical component of flow accumulation at any point in a watershed or region that is used to describe the spatial pattern of soil moisture [63]. This index is generally used for topographic control over hydrological processes and its high values are generally used in landslide bodies. The TWI can be formulated as follows:

$$\mathrm{TWI} = \mathrm{Ln}\left(\frac{\mathrm{A_s}}{\tan\beta}\right) \tag{2}$$

where $\mathrm{A_s}$ is cumulative drainage upstream area at one point and tan the angle of slope at the point. The TWI was prepared in five classes: (1) 0.0895–2.62, (2) 2.63–3.32, (3) 3.33–4.15, (4) 4.16–6.26, and (5) 6.26–10.70 (Figure 2k).

Terrain Ruggedness Index

Terrain Ruggedness Index (TRI) was introduced by Riley [64], and it is actually the difference in the height of one pixel with the eight pixels around it. Equation (3) is provided to calculate this index:

$$\mathrm{TRI} = \sqrt{\sum_{p=1}^{8} ZMD} \tag{3}$$

where p is the number of pixels in the region and ZMD is the average difference of eight pixels around each pixel. The TRI map was prepared in five classes: (1) 0–2.64, (2) 2.65–4.75, (3) 4.76–7.74, (4) 7.75–13.4, and 13.5–44.9 (Figure 2l).

Topographic Position Index

Topographic position index (TPI) compares the height of each pixel in the digital elevation model with the specified pixel around that pixel [65]. To calculate TPI (Equation (4)), the height of each cell in a digital elevation model compared with the average height of neighboring cells is examined. Finally, the average height decreases from the height value in the center. Areas higher than the surrounding points (hills) are indicated by positive TPI values; negative TPI values denote areas lower than their surroundings (valleys). Zero and near-zero values also illustrate flat areas (where the slope is close to zero) or areas with a fixed slope [66].

$$\mathrm{TPI} = \mathrm{Z}_0 - \textstyle\sum_{n-1} Z_n / n \tag{4}$$

where Z_0 is the point height of the model under evaluation, Z_n is the height of the grid and n is the total number of surrounding points considered in the evaluation. We prepared TPI in five classes: (1) (−75.7)–(−9.77), (2) (−9.77)–(−2.83), (3) (−2.83)–(2.94), (4) (2.94)–(11.03), and (5) (11.03)–(71.7) (Figure 2m).

Slope Length

The slope length (LS) factor, which is a combination of the slope angle and length of the slope, is a fundamental factor in the study of landslides because this factor refers to the sediment transport capacity created by the landslide through the daily (direct) flow. Carrara [67] stated that there is a relationship between landslide density and slope length. Therefore, this factor is examined in this study [67]. Mathematically, this equation is expressed as:

$$\mathrm{LS} = \left(\frac{\mathrm{A_s}}{22.13}\right)^{0.4}\left(\frac{\sin\beta}{0.0896}\right)^{1.3} \tag{5}$$

where $\mathrm{A_s}$ is the specific catchment area and β is the degree of local slope gradient. This index was prepared based on DEM in the SAGA GIS software, and after exporting in the GIS environment it was classified into five classes: (1) 0–6.88, (2) 6.89–13.1, (3) 13.2–19.6, (4) 19.7–28.2, and (5) 28.3–87.8 (Figure 2n).

Land Use/Land Cover

Land use is one of the important indicators in the instability of slopes, and it affects the characteristics of the land and changes its behavior [53]. In this study, the land use layer was prepared and extracted from an Iranian land use map. Land use/cover classes in the current research are dry-farming, semi-dense forest, low-dense forest, semi-dense pasture, dense pasture, and woodland (Figure 2o).

Normalized Difference Vegetation Index

The normalized difference vegetation index (NDVI) factor shows the ability to detect growth and vegetation levels in an area [68,69]. It is obtained by subtracting the reflection values of red band (Red) or visible spectrum (0.6–6.7 μm) and near-infrared band (NIR) (0.7–1/1 μm). Equation (6) is used to calculate this index:

$$\mathrm{NDVI} = (\mathrm{NIR} - \mathrm{RED})/(\mathrm{NIR} + \mathrm{RED}) \tag{6}$$

The minimum and maximum values of this index, respectively, are (−1) and (+1). The NDVI map was produced in five classes: (1) (−0.351)–(−0.064), (2) (−0.064)–(0.008), (3) (0.008)–(0.099), (4) (0.099)–(0.260) and (5) (0.260)–(0.759) (Figure 2p).

Rainfall

Rainfall intensity and duration play a major role in landslide initiation [70]. Here, we obtained the rainfall data of eight meteorological stations from the Iranian Meteorological Organization. A rainfall map of the area was built with the inverse distance weighting

(IDW) method with five classes: (1) 438–440, (2) 440–480, (3) 480–520, (4) and 520–560 (Figure 2q).

Distance to Fault

Large-scale structures such as faults and thrusts can influence the distribution of landslides [71]. In this study, distance to fault was calculated by the "Euclidean Distance" tool in ArcGIS software, in terms of distance from each pixel from the study area to the nearest fault. Based on these results, buffers were constructed around the fault with distances of 100 m, and this map was extracted into five classes: (1) 0–100, (2) 101–200, (3) 201–300, (4) 301–400, and (5) >400 (Figure 2r).

Distance to Roads

Both cut and fill slopes and improper road drainage structures associated with road construction can contribute to slope instability [72]. In this study, distance to road was calculated by the "Euclidean Distance" tool in ArcGIS software, in terms of distance from each pixel from the study area to the nearest road. Distance to roads was mapped with five categories: (1) 0–100, (2) 101–200, (3) 201–300, (4) 301–400, and (5) >400 m (Figure 2s).

Distance to Rivers

Another conditioning factor that directly impacts landslide susceptibility is distance to river. Flowing water is one of the factors increasing the potential for instability in the slopes, playing an effective role in mass movements. Distance to river was calculated by the "Euclidean Distance" tool in the ArcGIS software in meters of each pixel from the study area to the nearest stream line. The map was created with five classes: (1) 0–100, (2) 101–200, (3) 201–300, (4) 301–400, and (5) >400 m (Figure 2t).

Fault Density

Slope instabilities are more likely to occur in areas where the number of faults is high and particularly when the faults are active [73]. Fault density is the ratio of the total length of faults in a given watershed or a given area to the total area of the watershed or the area surrounding those faults [74]. The higher the density of faults in an area, the greater the split in rocks and the reduction in shear strength of rocks and slope constituents due to weathering. As a result, the risk of slope instability and landslides increases on the slopes [75]. Fault density was extracted with five classes: (1) 0–0.67, (2) 0.671–1.84, (3) 1.85–3.01, (4) 3.02–4.41, and (5) 4.42–7.12 km/km^2 (Figure 2u).

Road Density

Road density is the ratio of the total length of roads in a given watershed or a given area to the total area of the watershed or the area surrounding those roads [76]. Although the quality of roads and drainage control are important, road density can also influence landslide occurrence [77]. Road density was calculated using the "Line density" tool in the ArcGIS software for modeling, and the factor was classified into five classes: (1) 0–0.440, (2) 0.440–1.210, (3) 1.210–1.914, (4) 1.914–2.772, and (5) 2.772–5.610 km/km^2 (Figure 2v).

River Density

Another influence controlling landslides is river density [78]. River density is the ratio of the total length of rivers in a given watershed within a given area to the total area of a watershed or area containing those rivers [79]. We used the "Line density" tool in the ArcGIS software to extract five classes of river density: (1) 0–0.5551, (2) 0.5551–1.4608, (3) 1.4608–2.4542, (4) 2.4542–3.7983, (5) 3.7983–7.4505 (Figure 2w).

Lithology

Lithology often strongly influences slope stability [80], in part due to variable strength characteristics of certain bedrock types [81]. Therefore, to determine the susceptibility of various lithological formations to produce landslides, we extracted lithological units of the case study of Kamyaran geology sheet with a scale of 1:100,000. The number of lithological units in the study area was divided into 10 classes (Figure 2x).

Soil Texture

Landslides that involve soils are influenced by the type of soil they occur in [82]. Soil texture influences properties such as permeability and cohesion, which can influence the style of movement [83]. Primarily, landslides change soil features by exposing parent material (the C horizon) by removing organic mats and the horizon A [84]. Changes in soil texture occur when a landslide moves or removes various materials to a specific location [85]. From the study area, 20 soil samples in different lithological units were collected to determine soil texture using the hydrometric method. We used the soil texture triangle to classify textural groups. The soil map was created into five classes: (1) Silty Loam (2) Clay Loam (3) Loam (4) Sandy Loam (5) Silty Clay (Figure 2y).

3. Modeling Process

Figure 3 shows the workflow of our study. In step 1, we collect and interpret landslide-conditioning factors. In step 2, we divide landslide locations into the training and the validating datasets. In step 3, we conduct landslide modeling using the DL (deep learning) model and the three benchmark models (SVM, NBTree, and REPTree). In the DL model, we computed landslide susceptibility index (LSIs) for each pixel of the study in five steps: (i) constructing DBN using RBMs as pretraining on the dataset; (ii) parameter tuning in ELM to obtain the weights matrix from the last restricted Boltzmann machines (RBMs), (iii) fine tuning the training of the whole network by BP, (iv) optimizing the obtained weights from the network by the genetic algorithm (GA), and (v) assigning the optimum weights to the pixel of the study to map the landslide susceptibility. In step 4, we generate the landslide susceptibility maps using the outcomes of step 3. Finally, we compare and validate the performance of the models using a suite of statistical measures.

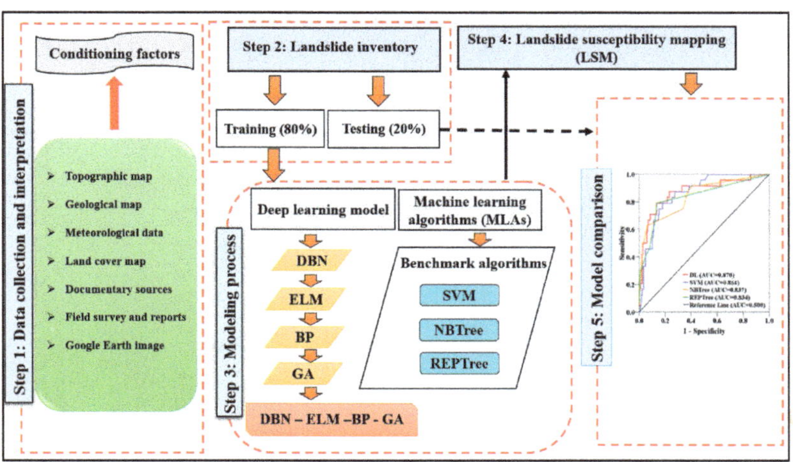

Figure 3. Flowchart of the study.

4. Mathematical Background of the Methods

4.1. Deep Belief Network

One of the most common deep neural networks (DBN) training techniques is the use of unsupervised pretraining, which initializes the network using only unlabeled data. Network initialization has been shown to be a good starting point for fine-tuning with the next observer, and greatly reduces the risk of being trapped at the local minimum according to Kustikova and Druzhkov [86]. One of the methods used to teach deep networking is the deep belief network. The deep belief network [87,88] has become a popular approach in machine learning due to its advantages such as fast inference and the ability to encode richer and higher-order network structures. DBN operates a hierarchical structure with several finite Boltzmann machines, and operates through a layered learning process [89]. A deep belief network with two Boltzmann machines bounded for one problem to n inputs and one output is shown in (Figure 4).

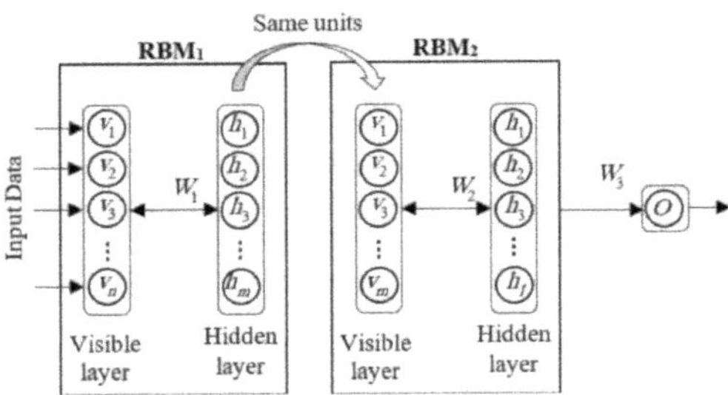

Figure 4. Deep belief network model used in the study.

Usually, with pretraining, the deep belief network training process includes the following steps [90]:

Step 1: Pretraining step: a sequential training of learning modules, greedily, one layer at a time, using unsupervised data;

Step 2: First fine-tuning step: use random weights for the last layer (matrix W3 in Figure 4);

Step 3: Second fine-tuning step: use back propagation to fine-tune the entire network using supervised data.

4.2. Extreme Learning Machine

The extreme learning machine (ELM) [91] was first proposed by Huang in 2004 for the single hidden-layer feedforward neural networks (SLFNs) with the aim of reducing the costs imposed by the post-error propagation procedure during the training process, and then extending to SLFNs where latent layer neurons do not need to be the same. Over the past decade, the extreme learning machine has been extensively studied due to its high productivity, effectiveness, and easy implementation [92]. The ELM has the advantage of a fast learning rate and high generalizability [93]. In ELM, the hidden layer does not need to be adjusted; that is, the connection weights from the input layer to the hidden layer as well as the hidden biases, and neurons are generated randomly without additional adjustment. The efficient least squares method is used to computationally calculate the connection weights from the hidden layer to the output layer [94].

4.3. Structure of the Deep-Learning Model

In the proposed model, for network training, the deep belief network training process mentioned in the DBN section is used; the difference is that in the first fine-tuning step, the ELM is applied to teach the weights between the last hidden layer and the output layer (W3 in Figure 4). The optimal network structure is also derived from GA. The steps of the genetic algorithm are as follows:

Step 1: Chromosome coding and population initialization. The chromosome is directly counted by taking positive integers (to a predetermined population N). The number of genes on each chromosome indicates the number of hidden deep layer layers and the amount of each gene indicates the number of neurons. Chromosome genes are also randomly initialized.

Step 2: Assessment. Each chromosome is trained by the proposed hybrid model using training data. Then, the classification accuracy is calculated and considered as the fit value of that chromosome.

Step 3: Selection. The known mechanism of selecting the roulette wheel has been used to choose the parents for the combination and jump.

Step 4: Combination. To search the problem space, the one-point compound operator, which is one of the most common compound operators in the literature, has been used.

Step 5: Mutation. The mutation operator produces a new chromosome by randomly selecting a gene/layer and decreasing or increasing its amount. The purpose of this operator is to prevent the algorithm from being trapped in the local optimization by discovering new solution spaces.

Step 6: Selection of survivors. After arranging the chromosomes of the current population and the chromosomes resulting from the combination and mutation based on their proportional values, the superior N chromosomes are selected as the survivors.

Step 7: Stop criteria. When the number of generations reaches the predetermined value, the algorithm stops and the best chromosome returns as the answer; otherwise, it returns to step 3. The flowchart of the deep-learning model used in this study is shown in Figure 5.

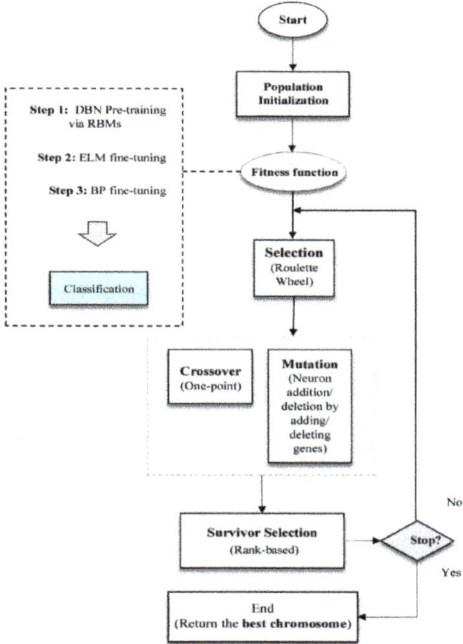

Figure 5. The flowchart of the deep-learning model [39,40].

4.4. Benchmark Methods

4.4.1. Support Vector Machine

The support vector machine (SVM) algorithm is based on the theory of statistical learning that uses the inductive minimization principle of structural error leading to an overall optimal solution [95,96]. In recent years, this algorithm has attracted a lot of attention due to its good classification performance and good generalizability. The SVM includes the two operations, (i) nonlinear mapping of an input vector into a high-dimensional feature space that is hidden from both the input and the output and (ii) construction of an optimal hyperplane to separate the features. The structure of this model is explained as follows:

$$X_i = (i = 1, 2, \ldots, n) \tag{7}$$

The training vectors included two classes of $Y_i = \pm 1$; the purpose of this model is to find a differentiated hyperplane of $-N$ dimensional by the maximum gap. The description is as follows:

$$1/2 = \|W\|^2 \tag{8}$$

Subject to the following constraints.

$$Y_i = ((W\ .\ X_i) + b) \geq 1 \tag{9}$$

where $\|W\|$ is the norm of the normal of the hyperplane, (.) is a specific numerical production and b is a scalar base.

4.4.2. REPTree

The reduced error-pruning tree (REPTree) as a fast decision-tree learning process that combines two kinds of algorithms as a hybrid method involving reduced-error pruning (REP) and decision tree (DT) [97]. The main structure of this method is based on classification and regression problems. The REP minimizes the complexity of tree structure if the DT's performance is high [98]. The REPTree method uses the pruning mechanism to overcome the backward overfitting problem. Additionally, this technique uses the post-pruning method to obtain the minimal version of the most-accurate tree [99].

4.4.3. NBTree

Naïve Bayes tree (NBTree) was used due to its simplicity and linear runtime method, combining the J48 algorithm and the naïve Bayes algorithm [100]. This method is used for classification problems, especially to evaluate and pick the class that maximizes the subsequent class's likelihood. Hence, NBTree can solve problems of big data that relate to the Naïve Bayes algorithm and the data fragmentation of the J48 algorithm. The important distinguishment of this model from other machine-learning methods is that it is based on a minimal training data structure that uses a classification system to evaluate important parameters [101]. To build a Naïve Bayes classifier for detection of landslide occurrence points in the area, NBTree uses information obtained from the root node to a given leaf node down the tree, and then utilizes the training cases that fall into that leaf node [102].

4.5. *Information Gain Ratio*

In the present study, the information gain ratio (IGR) was applied as the basis of judgment for factor selection and to determine important comparative factors for modeling. For landslide susceptibility assessment, selecting the most effective factors as input dataset is fundamental. IGR was proposed by Quinlan [103] to define the quantitative predictive strength of the effective parameters and to select important conditioning factors for modeling. The higher the IGR value, the higher the prediction utility of a factor for modeling [19]. This method enhances the power of prediction of landslides, discarding noise factors with lower IGR. Assuming that the training data T contain n samples, Ci (landslide, nonlandslide) is a classification set of sample data, and the information entropy of the factors is calculated as follows:

$$Info\ (T) = -\sum_{I=1}^{2} \frac{n(C_i, T)}{|T|} log_2 \frac{n(C_i, T)}{|T|} \tag{10}$$

Estimating the amount of information (T_1, T_2 and T_m) from T considering causal factor F takes the form of the following Equation (11):

$$Info\ (T, F) = -\sum_{I=1}^{m} \frac{T_i}{|T|} log_2\ Info\ (T) \tag{11}$$

Eventually, the *IGR* of the landslide causal factor F can be calculated by:

$$IGR\ (Y,F) = \frac{Info\ (T) - Info\ (T,F)}{Split\ Info\ (T,F)} \tag{12}$$

where *Split Info* denotes the potential information produced by dividing the training data *T* into *m* subsets. The formula of *Split Info* is shown as:

$$SplitInfo\ (T,\ F) = -\sum_{I=1}^{m} \frac{|T_i|}{|T|} log_2 \frac{|T_i|}{|T|} \tag{13}$$

If $IGR > 1$, the probability of landslide incidence is higher than average; if $IGR = 0$, the probability of landslide is equal to average; and if $IGR < 0$, the probability of landslide incidence is less than average [104].

4.6. Performance Metrics

To evaluate the performance of all the models, we used a number of statistical index-based metrics: sensitivity (SST), specificity (SPF), accuracy (ACC), F1-measure, and receiver operative characteristic curve (AUC). All statistical metrics were computed based on true positive (TP), true negative (TN), false positive (FP), and false negative (FAN). Table 1 shows the mentioned statistical index-based metrics and their descriptions.

Table 1. Performance metrics and their descriptions to assess the performance of the models.

Metric	Formula	Description
TP	True positive	Number of landslides (positive) that are truly classified as landslide [105].
TN	True negative	Number of nonlandslides (negative) that are truly classified as nonlandslide [106].
FP	False positive	Number of nonlandslides that are incorrectly classified as landslides [107].
FN	False negative	The number of landslides that are incorrectly classified as non-landslides [9].
SST	$\text{SST} = \frac{\text{TP}}{\text{TP+FN}}$	The ratio of landslides that are correctly classified as landslide. This indicates the good predictability of the landslide model for classifying landslides [108].
SPF	$\text{SPF} = \frac{\text{TN}}{\text{FP+TN}}$	The ratio of nonlandslides that are correctly classified as non-landslide. This depicts good predictability of the landslide model for classifying nonlandslides [108].
ACC	$\text{ACC} = \frac{\text{TP+ TN}}{\text{TP+TN+FN+FN}}$	The ratio of landslides and nonlandslides that are correctly classified [109]. This shows how well the landslide model works.
F1-measure	$\text{F1} - \text{measure} = \frac{\text{2TP}}{\text{2TP+FP+FN}}$	F-measure is a way to combine and balance both precision and recall into a single measure [110].
AUC	$\text{AUROC} = \sum \text{TP} + \sum \frac{\text{TN}}{\text{P}} + \text{N}$	The ROC curve is plotted by sensitivity and 1-specificity, respectively, on the *y*-axis and *x*-axis [111]. The area under the ROC curve (AUC) illustrates the power prediction of a model [112].
MSE RMSE	$\text{MSE} = \frac{1}{N}\sum_{i=1}^{N}(x_m - x_p)^2$ $\text{RMSE} = \sqrt{\frac{1}{N}\sum_{i=1}^{N}(x_m - x_p)^2}$	MSE and RMSE measure the difference between measurements (x_m) and predictions (x_p) and indicate modeling error [113]

5. Results

5.1. The Most Important Conditioning Factors

We obtained the relative importance of the factors influencing landslide occurrence based on average merit (AM) as IGR score through the *k*-fold cross-validation technique (Table 2). Results indicated that in the lower folds (1 and 2 folds), the number of removing factors with less predictive power was higher (13 factors) than the higher folds (10-fold; 9 factors). According to Table 2, the results pointed out that from 1-fold to 10-folds cross-validation, distance to road (AM = 0.177), road density (AM = 0.118), lithology (AM = 0.079) and land use (AM = 0.055) were the first four most important factors for landsliding in the study area. These four influencing factors are followed by NDVI (AM = 0.04), elevation (AM = 0.04), soil (AM = 0.031), aspect (AM = 0.025), solar radiation (AM = 0.021), VRM (AM = 0.015), slope angle (AM = 0.014), distance to fault (AM = 0.014), TWI (AM = 0.013), LS (AM = 0.011), TRI (AM = 0.008), and rainfall (AM = 0.006). Further, the results showed that profile curvature, curvature, plan curvature, distance to river, VD, fault density, river density, TPI, and SPI, because of having AM = 0, were removed from the modeling process.

Table 2. Relative importance of landslide conditioning factors measured by information gain ratio technique.

Factor	1-Fold	Factors	2-Folds	Factors	3-Folds	Factors	4-Folds	Factors	5-Folds	Factors	6-Folds	Factors	7-Folds	Factors	8-Folds	Factors	9-Folds	Factors	10-Folds
DRo	0.180	DRo	0.18	DRo	0.177	DRo	0.177	DRo	0.177	DRo	0.177	DRo	0.177	DRo	0.177	DRo	0.177	DRo	0.177
RoD	0.112	RoD	0.112	RoD	0.114	RoD	0.118	RoD	0.118	RoD	0.118	RoD	0.117	RoD	0.118	RoD	0.118	RoD	0.118
Lithology	0.082	Lithology	0.084	Lithology	0.061	Lithology	0.069	Lithology	0.08	Lithology	0.079	Land use	0.055	Lithology	0.076	Lithology	0.073	Lithology	0.079
Land use	0.054	Land use	0.06	Land use	0.055	Land use	0.056	Land use	0.055	Land use	0.055	Lithology	0.07	Land use	0.055	Land use	0.055	Land use	0.055
Elevation	0.0402	Aspect	0.027	NDVI	0.041	NDVI	0.04	Elevation	0.041	NDVI	0.04	Elevation	0.04	Elevation	0.04	Elevation	0.04	NDVI	0.04
NDVI	0.0401	NDVI	0.036	Elevation	0.038	Elevation	0.04	NDVI	0.04	Elevation	0.04	NDVI	0.041	NDVI	0.04	NDVI	0.04	Elevation	0.04
Soil	0.031	Elevation	0.028	Aspect	0.024	Aspect	0.031	Soil	0.031	Soil	0.032	Soil	0.03	Soil	0.031	Soil	0.03	Soil	0.031
Aspect	0.027	SR	0.022	SR	0.021	SR	0.022	Aspect	0.024	Aspect	0.023	Aspect	0.025	Aspect	0.023	Aspect	0.026	Aspect	0.025
SR	0.020	DF	0.017	Soil	0.024	Soil	0.023	SR	0.024	SR	0.022	SR	0.022	SR	0.022	SR	0.021	SR	0.021
VRM	0.015	VRM	0.015	Slope	0.014	Slope	0.015	VRM	0.015	VRM	0.015	VRM	0.015	VRM	0.015	VRM	0.015	VRM	0.015
Slope	0. 014	TWI	0.021	VRM	0.014	VRM	0.014	Slope	0.014	Slope	0.014	DF	0.014	DF	0.014	Slope	0.014	Slope	0.014
DF	0.013	Soil	0.015	TWI	0.013	TWI	0.014	DF	0.015	DF	0.014	Slope	0.014	Slope	0.014	DF	0.014	DF	0.014
TWI	0.012	Curvature	0	DF	0.015	DF	0.012	TWI	0.013	TWI	0.013	TWI	0.013	TWI	0.013	TWI	0.013	TWI	0.013
LS	0.010	PRC	0	Curvature	0.004	Curvature	0.01	LS	0.007	LS	0.009	LS	0.01	LS	0.01	LS	0.01	LS	0.011
TRI	0.009	TRI	0	LS	0.009	LS	0	PRC	0	PRC	0	TRI	0.008	PRC	0	TRI	0.007	TRI	0.008
Rainfall	0.008	PLC	0	PLC	0.004	PLC	0	Curvature	0	Curvature	0.002	PLC	0	PLC	0	PLC	0	Rainfall	0.006
PLC	0	VD	0	TRI	0.008	TRI	0	PLC	0	PLC	0	PRC	0	Curvature	0	Curvature	0.001	PLC	0
Curvature	0	DRi	0	PRC	0	PRC	0	Rainfall	0.004	TRI	0.006	Curvature	0	TRI	0.007	Rainfall	0.006	Curvature	0
PRC	0	Slope	0	DRi	0	DRi	0	TRI	0.006	Rainfall	0.004	Rainfall	0.006	Rainfall	0.005	PRC	0	PRC	0
DRi	0	FD	0	VD	0	VD	0	DRi	0	DRi	0	DRi	0	DRi	0	DRi	0	DRi	0
VD	0	SPI	0	FD	0	FD	0	VD	0	VD	0	VD	0	VD	0	VD	0	VD	0
FD	0	TPI	0	Rainfall	0	Rainfall	0	FD	0.002	FD	0	FD	0	FD	0	FD	0	FD	0
RiD	0	Rainfall	0	TPI	0	TPI	0	TPI	0	TPI	0	TPI	0	RiD	0	RiD	0	RiD	0
TPI	0	LS	0	RiD	0	RiD	0	RiD	0	RiD	0	RiD	0	TPI	0	TPI	0	TPI	0
SPI	0	RiD	0	SPI	0	SPI	0	SPI	0	SPI	0	SPI	0	SPI	0	SPI	0	SPI	0

DRo: Distance to road; RoD: Road density; SR: Solar radiation; DF: Distance to fault; PRC: Profile curvature; PLC: Plan curvature; DRi: Distance to river; FD: Fault density; RiD: River density.

5.2. *Performance of the Deep-Learning Model*

Figure 5 shows the results of training the DL model. Figure 6a,b illustrates how well the landslide (target) and nonlandslide (output) values fitted based on the training and testing datasets, respectively. A well-trained model with a high goodness of fit also has a high agreement between the target and output by the training dataset. However, high prediction accuracy of the model is inferred by the agreement between the target and output of the testing dataset. The two statistical quantitative metrics of MSE (mean squared error) and RMSE (root-mean-square error) show the modeling error of the DL model (Figure 6c,e). The values of MSE and RMSE in the training dataset were 0.0435 and 0.0208, respectively (Figure 6c); however, these values for the testing dataset were 0.079 and 0.281 (Figure 6e). The StD (standard deviation) and mean for the training dataset were, respectively, 0.04 and 0.280, and for the testing dataset these values were 0.01 and 0.208, respectively (Figure 6d,f).

5.2.1. Parameter Tuning

The success rate of a model depends on selecting the optimal value of the parameters of that model. The parameter can be tuned by offline and online approaches. In the offline technique, the values of different parameters are fixed, whereas in the online approach the parameters are dynamically or adaptively controlled and updated [114]. In this study, we used the online parameter tuning approach and the results are shown in Tables 3–5.

Table 3. The optimal value of the genetic algorithm parameters.

Parameter	Optimal Parameter Value
Number of generations	50
Population size	200
Crossover rate	0.8
Mutation rate	0.15
Number of genes	Random in (1, 5)
Value of genes	Random in (1, 200)

Table 4. Optimal parameters of the DBN and BP models.

Parameters	DBN	BP
	Value	Value
Learning rate	1	0.1
# of learning epochs	10	60

#: Number of...

Table 5. The optimal value of parameter of the benchmark methods.

Method	Parameter Value
SVM	Debug: False; BuildLogisticModels: False; c: 1.0; ChecksTurnedOff: False; Debug: False; Epsilon: 1.0×10^{-12}; FilterType; Nonormalization/standardization; Kernel: Poly Kernel; NumFolds: −1; RandomSeed: 1; ToleranceParameter: 0.001
REPTree	Debug: False; MaxDepth: −1; MinNum: 2; MinVarianceProp: 0.001; NoPruning: False: NumFolds: 3; Seed:1
NBTree	Debug: False

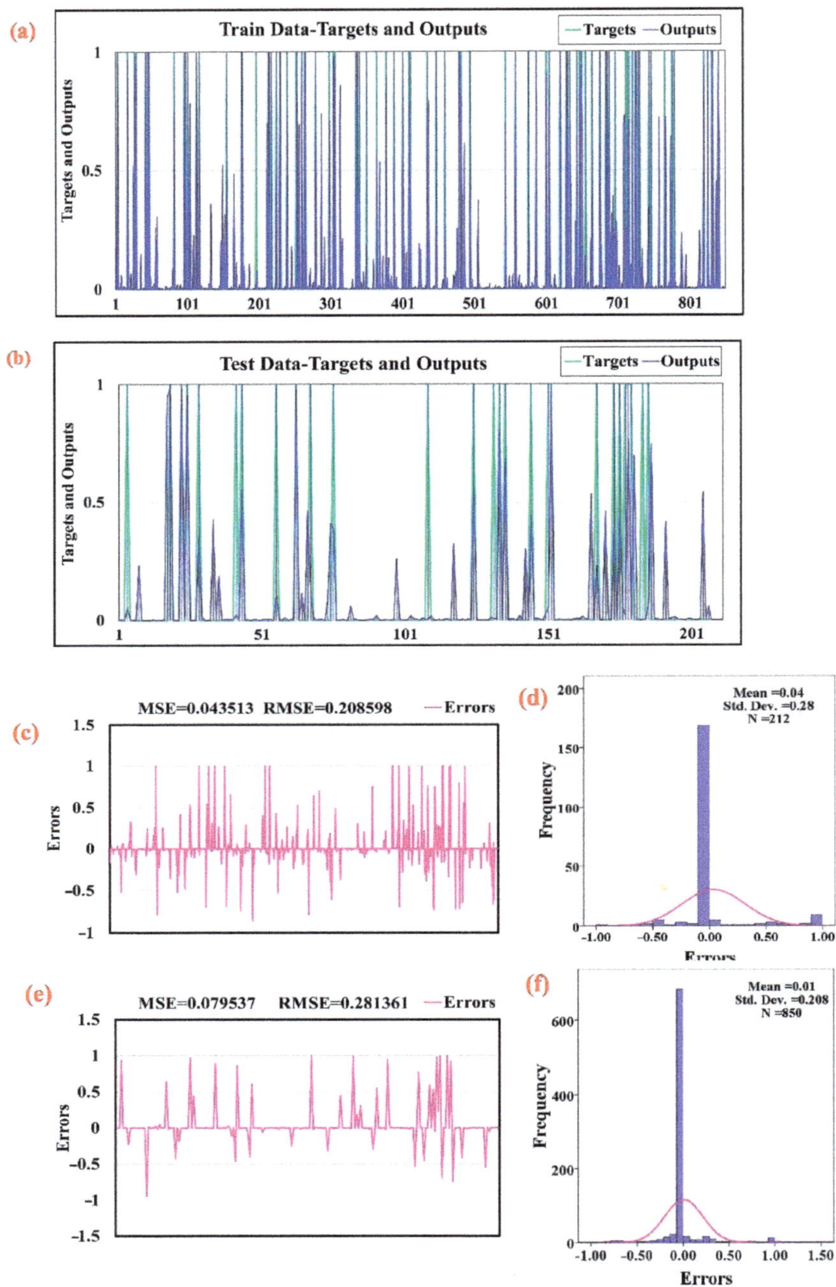

Figure 6. Performance of the DL model: (**a**) target and output for the training dataset, (**b**) target and output for the testing dataset, (**c**) magnitude of the errors for the training dataset, (**d**) distribution of the errors for the training dataset (**e**) magnitude of the errors for the testing dataset, (**f**) distribution of the errors for the testing dataset.

5.2.2. Classification Performance

After selecting the optimal parameter values for each model, we ran the models to obtain the highest evaluation measures but the least error. Our findings for the validation phase using the testing dataset, which are briefly reported below and in Table 6, show that the DL model outclassed and outperformed the three benchmark models. The sensitivity, specificity, accuracy, F1-measure, and AUC metrics were obtained based on the four possibilities from the confusion matrix of TP, TN, FP, and FN (Table 6).

(1) The highest sensitivity (0.667) was obtained by the DL model in which the DL model correctly classified 66.7% of landslides as landslide. It is followed by NBTree, REPTree, and SVM algorithms (Table 6).
(2) The DL model acquired the highest specificity values (0.958), indicating that the DL model correctly classified 95.83% of nonlandslides as nonlandslide. SVM, REPTree, and NBTree yielded the same specificity values based on the validation dataset of 0.953 (Table 6).
(3) The DL model had the highest accuracy on the testing dataset (0.926). It indicated that 92.6% of landslides and nonlandslides were correctly classified, respectively, as landslide and nonlandslide. It was followed by the NBTree (0.917), REPTree (0.903), and SVM (0.894) models (Table 6).
(4) The DL model had the highest value of the F1-measure (0.667), followed by NBTree (0.625), REPTree (0.533) and SVM (0.465) (Table 6).
(5) The DL model yielded the highest value of AUC (0.893) using the testing dataset. It indicated that the DL model had the highest prediction accuracy of all the models including NBTree (0.866), SVM (0.853), and REPTree (0.817) (Table 6).

Table 6. The predictive performance of the deep-learning model and the three benchmark models.

Metric	DL	SVM	REPTree	NBTree
TP	16	10	12	15
TN	184	183	183	183
FP	8	9	9	9
FN	8	14	12	9
Sensitivity	0.667	0.417	0.500	0.625
Specificity	0.958	0.953	0.953	0.953
Accuracy	0.926	0.894	0.903	0.917
F1-mesaure	0.667	0.465	0.533	0.625
AUC	0.893	0.853	0.817	0.866

5.3. *Preparing Landslide Susceptibility Maps*

We ran the DL model and also the three benchmark models (SVM, NBTree, and REPTree) and computed landslide susceptibility index (LSIs) for each pixel of the study area. We then assigned the LSIs from the DL and benchmark machine-learning models to each pixel of the study area to produce landslide susceptibility maps (Figure 7a–d). We classified the maps into the five susceptibility classes: very low susceptibility (VLS), low susceptibility (LS), moderate susceptibility (MS), high susceptibility (HS), and very high susceptibility (VHS). In DL, the range of the classes were, respectively, 0.0000875–0.0092, 0.0312–0.0446, 0.0447–0.125, 0.126–0.333, and 0.333–0.868 (Figure 7a). These classes in SVM (Figure 7b) for the VLS, LS, MS, HS, and VHS were, respectively, 0.001–0.092, 0.0921–0.0293, 0.0294–0.0789, 0.079–0.201, and 0.202–0.5. For NBTree (Figure 7c) these classes were VLS (0.001–0.01299), LS (0.013–0.03783), MS (0.03784–0.09404), HS (0.09405–0.2128), and VHS (0.02129–0.468). Consequently, in REPTree (Figure 7d) the classes were VLS (0–0.03), LS (0.03001–0.0993), MS (0.09931–0.1067), HS (0.1068–0.2556), and VHS (0.2557–0.406).

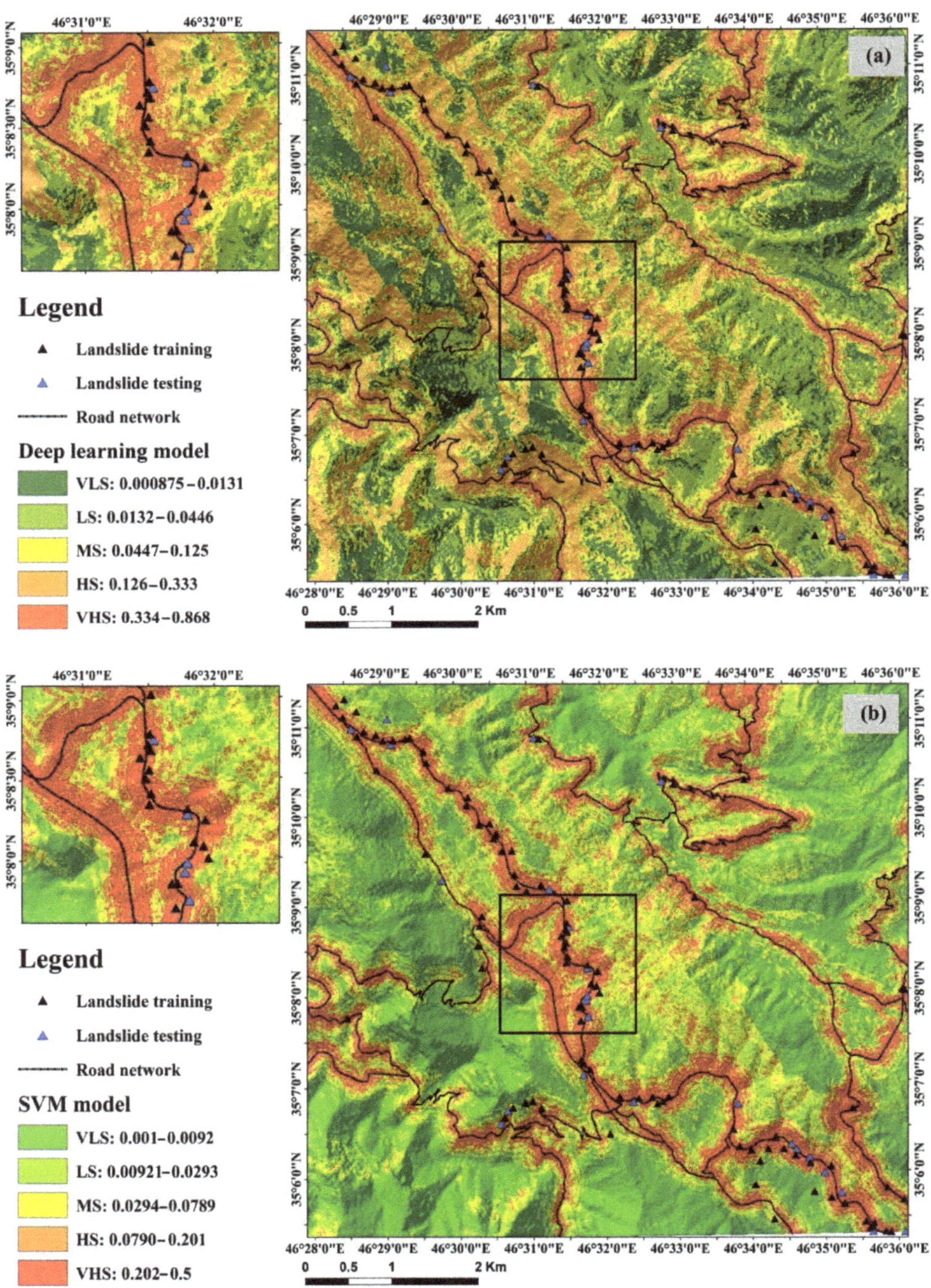

Figure 7. *Cont.*

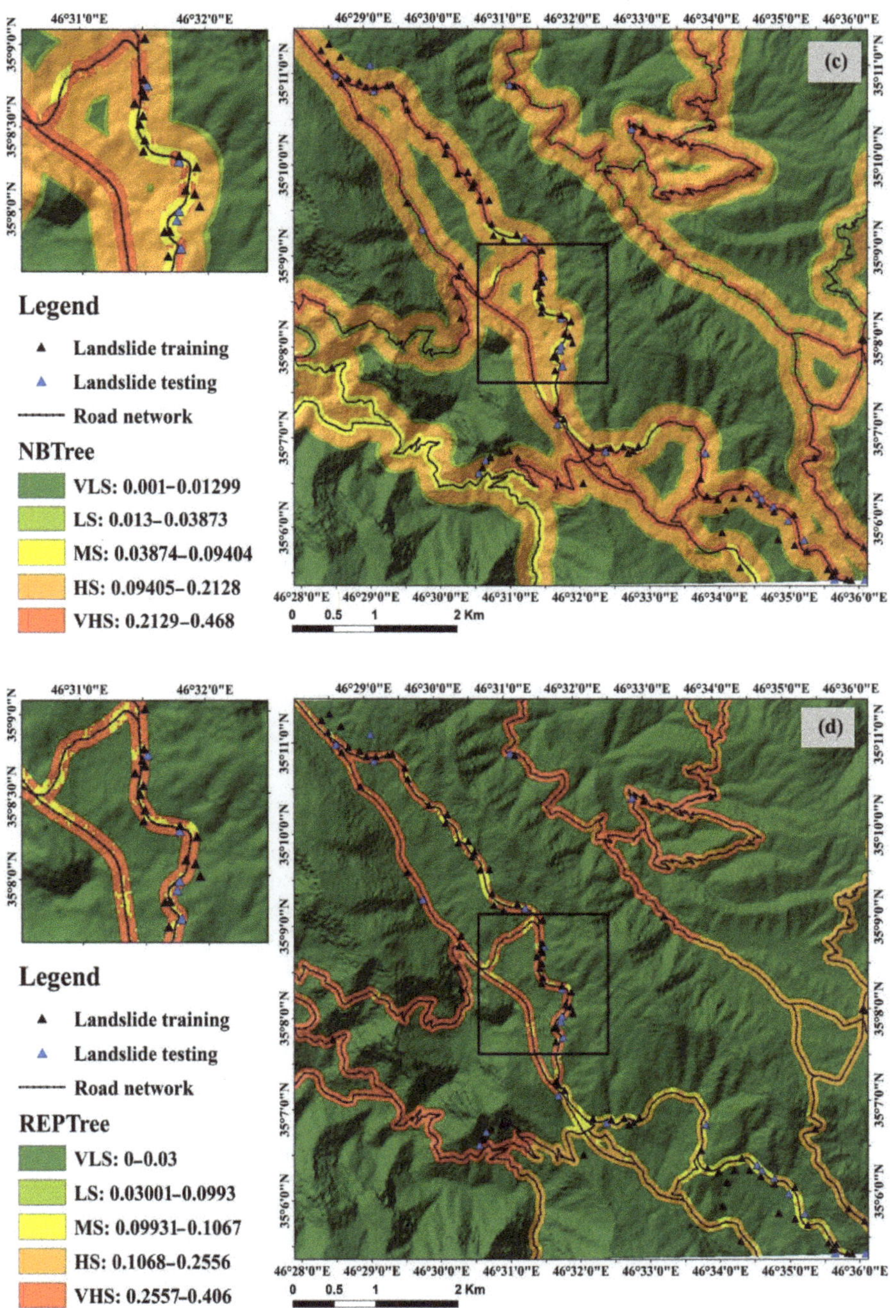

Figure 7. Landslide susceptibility maps produced by the (**a**) DL, (**b**) SVM, (**c**) NBTree, and (**d**) REPTree models.

5.4. Validation and Comparison of the Models

The prediction accuracy of the DL model and the benchmark machine-learning algorithms were assessed by AUC using the testing dataset (Figure 8). As shown in Figure 8, results indicated that the DL model had a high prediction accuracy (AUC = 0.870). In contrast, AUC for the SVM, NBTree, and REPTree models were somewhat lower, at, 0.861, 0.837, and 0.834, respectively (Figure 8). Overall, the DL model was superior compared to the other three benchmark machine-learning models in terms of prediction accuracy.

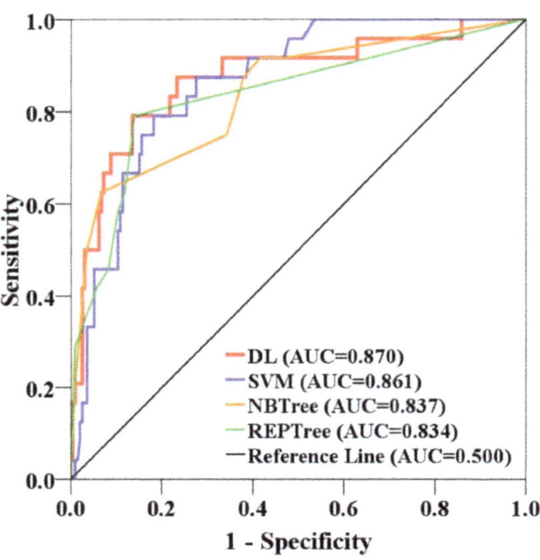

Figure 8. AUC of the models based on the testing dataset.

6. Discussion

In recent years, the demand for accurate prediction of landslides and the production of landslide susceptibility maps has increased, in part due to the improvement of data processing techniques, but also due to the importance of landslide prediction and susceptibility mapping in more effective land-use planning and management. There are numerous approaches and methods for producing landslide susceptibility maps, but machine-learning methods based on GIS-automated techniques offer advantages such as low cost, wide scope, fast analysis, and the option for periodically updating outputs. Each machine-learning method has its specific advantages and disadvantages, and depending on the software capabilities and input data, its outputs may differ from that of other methods. The challenge many researchers face is selecting the most appropriate method. Thus, comparative analysis of the predictive performance of different machine-learning methods is a major topic in the landslide literature [115,116]. With a desire to produce a landslide susceptibility map with high prediction accuracy, we compared the predictive performance of four machine-learning methods. We first investigated the usefulness of the conditioning factors for the modeling using the information gain ratio technique with 10-fold cross-validation. The results revealed the landslides that have occurred in our study area were significantly associated with road networks, such that the distance to roads and road density factors were identified as the most influential landslide-conditioning factors. Jaafari et al. [117] and Schlögl and Matulla [118], and Sultana and Tan [119] have also reported on the strong associations between road networks and the frequency of landslide occurrences. Therefore, these regions should be the priority targets for landslide mitigation measures [119,120].

We measured the predictive performance of the models using several widely used metrics [115,121,122] and found that the DL model has obtained the first rank in all metrics used, and therefore successfully outperformed the benchmark models (i.e., NBTree, REPTree, and SVM) that have been previously used for landslide susceptibility mapping in many regions around the world [121–125].

DL algorithms are powerful types of machine-learning algorithms, which utilize numerous hidden layers to model complex relationships among data for pattern recognition and classification tasks, such as landslide prediction. In contrast to traditional shallow learning algorithms (e.g., backpropagation neural networks, logistic regression, and decision trees) that generate decision boundaries directly based on the original datasets [126], DL algorithms hierarchically analyze the original datasets to extract the most relevant features for the data classification [127].

Despite the infrequent applications of the DL model for the prediction of natural hazards, the superiority of this model to other models derived from machine learning has been confirmed. For example, Wang et al. [128] reported the first application of DL for landslide prediction and achieved better prediction accuracy than that of SVM. Sameen et al. [123] reported that the DL model outperformed ANN and SVM for landslide prediction. Huang et al. [129] reported that the DL model was superior to ANN and SVM for landslide prediction. Dao et al. [130] showed that the DL model could provide a more accurate prediction of landslide susceptibility compared to quadratic discriminant analysis, Fisher's linear discriminant analysis, and ANN. Dou et al. [131] reported that DL provided greater AUCs than the ANN and LR methods for landslide prediction. In a recent study, Mandal et al. [132] demonstrated improved accuracy for landslide prediction using the DL model compared to RF, ANN, and Bagging. Overall, the DL model has proven efficiency for landslide modeling and has been identified as an attractive alternative to traditional machine-learning methods.

7. Conclusions

We illustrated the robustness of a deep-learning model against three benchmark models (SVM, NBTree, and REPTree) for the prediction of landslide susceptibility in Kamyaran city, Kurdistan Province, Iran. First, the landslide inventory map with 118 past landslides was produced using different sources and randomly divided into two groups: 80% for the model training and 20% for the model validation. Next, using the models, the past landslides were linked to 25 conditioning factors. The performance of the models was evaluated using sensitivity, accuracy, specificity, F1-mesaure and AUROC. The results showed that although all models had acceptable performance, the deep-learning model outperformed the other models. This indicates that the DL model can be considered as a promising technique for preparing landslide susceptibility in mountainous areas prone to landsliding. Based on the results obtained from the deep-learning model, an accurate landslide susceptibility map is developed to complement previous research. The findings of this study can be used for future planning, land management, land use allocation, and government policy making, to prevent or reduce landslides in Kamyaran city. In future studies, we suggest integrating the current framework with other individual and metaclassifiers from machine learning to achieve a higher prediction accuracy for landslides, and perhaps other natural hazards.

Author Contributions: Contributed equally to the work, B.G., H.S., A.S., N.A.-A., A.J., V.R.K., M.G., S.R. and A.A.; collected field data and conducted the landslide susceptibility analysis, B.G., H.S. and A.S.; wrote the manuscript, B.G., H.S., A.S., N.A.-A., A.J. and S.R.; provided critical comments in planning this paper and edited the manuscript, V.R.K., M.G. and A.A. All the authors discussed the results and edited the manuscript. All authors have read and agreed to the published version of the manuscript.

Funding: This research was supported by the University of Kurdistan, Iran, based on two grants number 11-99-4469 and 99-11-32657-6.

Acknowledgments: The authors would like to thank the Vice Chancellorship of Research and Technology, University of Kurdistan, Sanandaj, Iran for supplying required data, reports, useful maps, and their nationwide geodatabase to the first author (Bahareh Ghasemian) as a postdoctoral fellowship scheme. The authors also greatly appreciate the assistance of anonymous reviewers for their constructive comments that helped us to improve the paper.

Conflicts of Interest: The authors declare no conflict of interest.

References

1. Benda, L.; Dunne, T. Stochastic forcing of sediment supply to channel networks from landsliding and debris flow. *Water Resour. Res.* **1997**, *33*, 2849–2863. [CrossRef]
2. Niethammer, U.; James, M.R.; Rothmund, S.; Travelletti, J.; Joswig, M. UAV-based remote sensing of the Super-Sauze landslide: Evaluation and results. *Eng. Geol.* **2012**, *128*, 2–11. [CrossRef]
3. Turner, D.; Lucieer, A.; De Jong, S.M. Time Series Analysis of Landslide Dynamics Using an Unmanned Aerial Vehicle (UAV). *Remote Sens.* **2015**, *7*, 1736–1757. [CrossRef]
4. Zare, M.; Pourghasemi, H.R.; Vafakhah, M.; Pradhan, B. Landslide susceptibility mapping at Vaz Watershed (Iran) using an artificial neural network model: A comparison between multilayer perceptron (MLP) and radial basic function (RBF) algorithms. *Arab. J. Geosci.* **2013**, *6*, 2873–2888. [CrossRef]
5. Schwab, J.W.; Geertsema, M. Terrain stability mapping on British Columbia forest lands: An historical perspective. *Nat. Hazards* **2010**, *53*, 63–75. [CrossRef]
6. Shahabi, H.; Hashim, M.; Ahmad, B.B. Remote sensing and GIS-based landslide susceptibility mapping using frequency ratio, logistic regression, and fuzzy logic methods at the central Zab basin, Iran. *Environ. Earth Sci.* **2015**, *73*, 8647–8668. [CrossRef]
7. Goetz, J.; Brenning, A.; Petschko, H.; Leopold, P. Evaluating machine learning and statistical prediction techniques for landslide susceptibility modeling. *Comput. Geosci.* **2015**, *81*, 1–11. [CrossRef]
8. Ghorbanzadeh, O.; Blaschke, T.; Aryal, J.; Gholaminia, K. A new GIS-based technique using an adaptive neuro-fuzzy inference system for land subsidence susceptibility mapping. *J. Spat. Sci.* **2020**, *65*, 401–418. [CrossRef]
9. Tavakkoli Piralilou, S.; Shahabi, H.; Jarihani, B.; Ghorbanzadeh, O.; Blaschke, T.; Gholamnia, K.; Meena, S.R.; Aryal, J. Landslide Detection Using Multi-Scale Image Segmentation and Different Machine Learning Models in the Higher Himalayas. *Remote Sens.* **2019**, *11*, 2575. [CrossRef]
10. Jordan, M.I.; Mitchell, T.M. Machine learning: Trends, perspectives, and prospects. *Science* **2015**, *349*, 255–260. [CrossRef]
11. Vorpahl, P.; Elsenbeer, H.; Märker, M.; Schröder, B. How can statistical models help to determine driving factors of landslides? *Ecol. Model.* **2012**, *239*, 27–39. [CrossRef]
12. Trigila, A.; Frattini, P.; Casagli, N.; Catani, F.; Crosta, G.; Esposito, C.; Iadanza, C.; Lagomarsino, D.; Mugnozza, G.S.; Segoni, S.; et al. Landslide Susceptibility Mapping at National Scale: The Italian Case Study. In *Landslide Science and Practice: Volume 1: Landslide Inventory and Susceptibility and Hazard Zoning*; Margottini, C., Canuti, P., Sassa, K., Eds.; Springer: Berlin/Heidelberg, Germany, 2013; pp. 287–295. [CrossRef]
13. Chen, W.; Yan, X.; Zhao, Z.; Hong, H.; Bui, D.T.; Pradhan, B. Spatial prediction of landslide susceptibility using data mining-based kernel logistic regression, naive Bayes and RBFNetwork models for the Long County area (China). *Bull. Eng. Geol. Environ.* **2019**, *78*, 247–266. [CrossRef]
14. Tsangaratos, P.; Ilia, I.; Hong, H.; Chen, W.; Xu, C. Applying Information Theory and GIS-based quantitative methods to produce landslide susceptibility maps in Nancheng County, China. *Landslides* **2017**, *14*, 1091–1111. [CrossRef]
15. Lei, X.; Chen, W.; Pham, B.T. Performance Evaluation of GIS-Based Artificial Intelligence Approaches for Landslide Susceptibility Modeling and Spatial Patterns Analysis. *ISPRS Int. J. Geo-Inf.* **2020**, *9*, 443. [CrossRef]
16. Wang, L.-J.; Guo, M.; Sawada, K.; Lin, J.; Zhang, J. A comparative study of landslide susceptibility maps using logistic regression, frequency ratio, decision tree, weights of evidence and artificial neural network. *Geosci. J.* **2016**, *20*, 117–136. [CrossRef]
17. Abedini, M.; Ghasemian, B.; Shirzadi, A.; Bui, D.T. A comparative study of support vector machine and logistic model tree classifiers for shallow landslide susceptibility modeling. *Environ. Earth Sci.* **2019**, *78*, 560. [CrossRef]
18. Peethambaran, B.; Anbalagan, R.; Kanungo, D.P.; Goswami, A.; Shihabudheen, K.V. A comparative evaluation of supervised machine learning algorithms for township level landslide susceptibility zonation in parts of Indian Himalayas. *CATENA* **2020**, *195*, 104751. [CrossRef]
19. Dou, J.; Yunus, A.P.; Bui, D.T.; Merghadi, A.; Sahana, M.; Zhu, Z.; Chen, C.-W.; Han, Z.; Pham, B.T. Improved landslide assessment using support vector machine with bagging, boosting, and stacking ensemble machine learning framework in a mountainous watershed, Japan. *Landslides* **2020**, *17*, 641–658. [CrossRef]
20. Hong, H.; Pradhan, B.; Bui, D.T.; Xu, C.; Youssef, A.M.; Chen, W. Comparison of four kernel functions used in support vector machines for landslide susceptibility mapping: A case study at Suichuan area (China). *Geomat. Nat. Hazards Risk* **2017**, *8*, 544–569. [CrossRef]
21. Chen, X.; Chen, W. GIS-based landslide susceptibility assessment using optimized hybrid machine learning methods. *CATENA* **2021**, *196*, 104833. [CrossRef]
22. Avali, V.R.; Cooper, G.F.; Gopalakrishnan, V. Application of Bayesian logistic regression to mining biomedical data. *AMIA Annu. Symp. Proc.* **2014**, *2014*, 266–273.
23. Aditian, A.; Kubota, T.; Shinohara, Y. Comparison of GIS-based landslide susceptibility models using frequency ratio, logistic regression, and artificial neural network in a tertiary region of Ambon, Indonesia. *Geomorphology* **2018**, *318*, 101–111. [CrossRef]
24. Gong, Q.-h.; Zhang, J.-x.; Wang, J. Application of GIS-Based back propagation artificial neural networks and logistic regression for shallow landslide susceptibility mapping in south china-take meijiang river basin as an example. *Open Civ. Eng. J.* **2018**, *12*, 21–34. [CrossRef]

25. Kim, J.-C.; Lee, S.; Jung, H.-S.; Lee, S. Landslide susceptibility mapping using random forest and boosted tree models in Pyeong-Chang, Korea. *Geocarto Int.* **2018**, *33*, 1000–1015. [CrossRef]
26. Dou, J.; Yunus, A.P.; Tien Bui, D.; Merghadi, A.; Sahana, M.; Zhu, Z.; Chen, C.-W.; Khosravi, K.; Yang, Y.; Pham, B.T. Assessment of advanced random forest and decision tree algorithms for modeling rainfall-induced landslide susceptibility in the Izu-Oshima Volcanic Island, Japan. *Sci. Total Environ.* **2019**, *662*, 332–346. [CrossRef]
27. Sevgen, E.; Kocaman, S.; Nefeslioglu, H.A.; Gokceoglu, C. A Novel Performance Assessment Approach Using Photogrammetric Techniques for Landslide Susceptibility Mapping with Logistic Regression, ANN and Random Forest. *Sensors* **2019**, *19*, 3940. [CrossRef]
28. Merghadi, A.; Yunus, A.P.; Dou, J.; Whiteley, J.; ThaiPham, B.; Bui, D.T.; Avtar, R.; Abderrahmane, B. Machine learning methods for landslide susceptibility studies: A comparative overview of algorithm performance. *Earth Sci. Rev.* **2020**, *207*, 103225. [CrossRef]
29. Nguyen, Q.-K.; Tien Bui, D.; Hoang, N.-D.; Trinh, P.T.; Nguyen, V.-H.; Yilmaz, I. A Novel Hybrid Approach Based on Instance Based Learning Classifier and Rotation Forest Ensemble for Spatial Prediction of Rainfall-Induced Shallow Landslides using GIS. *Sustainability* **2017**, *9*, 813. [CrossRef]
30. Hong, H.; Liu, J.; Bui, D.T.; Pradhan, B.; Acharya, T.D.; Pham, B.T.; Zhu, A.X.; Chen, W.; Ahmad, B.B. Landslide susceptibility mapping using J48 Decision Tree with AdaBoost, Bagging and Rotation Forest ensembles in the Guangchang area (China). *CATENA* **2018**, *163*, 399–413. [CrossRef]
31. Luo, X.; Lin, F.; Chen, Y.; Zhu, S.; Xu, Z.; Huo, Z.; Yu, M.; Peng, J. Coupling logistic model tree and random subspace to predict the landslide susceptibility areas with considering the uncertainty of environmental features. *Sci. Rep.* **2019**, *9*, 15369. [CrossRef]
32. Polykretis, C.; Chalkias, C.; Ferentinou, M. Adaptive neuro-fuzzy inference system (ANFIS) modeling for landslide susceptibility assessment in a Mediterranean hilly area. *Bull. Eng. Geol. Environ.* **2019**, *78*, 1173–1187. [CrossRef]
33. Chen, W.; Chen, X.; Peng, J.; Panahi, M.; Lee, S. Landslide susceptibility modeling based on ANFIS with teaching-learning-based optimization and Satin bowerbird optimizer. *Geosci. Front.* **2021**, *12*, 93–107. [CrossRef]
34. Naghibi, S.A.; Pourghasemi, H.R.; Dixon, B. GIS-based groundwater potential mapping using boosted regression tree, classification and regression tree, and random forest machine learning models in Iran. *Environ. Monit. Assess.* **2015**, *188*, 44. [CrossRef] [PubMed]
35. Chen, W.; Shirzadi, A.; Shahabi, H.; Ahmad, B.B.; Zhang, S.; Hong, H.; Zhang, N. A novel hybrid artificial intelligence approach based on the rotation forest ensemble and naïve Bayes tree classifiers for a landslide susceptibility assessment in Langao County, China. *Geomat. Nat. Hazards Risk* **2017**, *8*, 1955–1977. [CrossRef]
36. Mandal, S.; Mondal, S. Artificial neural network (ann) model and landslide susceptibility. In *Statistical Approaches for Landslide Susceptibility Assessment and Prediction*; Springer: Berlin/Heidelberg, Germany, 2019; pp. 123–133.
37. Zhang, K.; Wang, S.; Bao, H.; Zhao, X. Characteristics and influencing factors of rainfall-induced landslide and debris flow hazards in Shaanxi Province, China. *Nat. Hazards Earth Syst. Sci.* **2019**, *19*, 93–105. [CrossRef]
38. Ma, Z.; Mei, G. Deep learning for geological hazards analysis: Data, models, applications, and opportunities. *Earth Sci. Rev.* **2021**, *223*, 103858. [CrossRef]
39. Shirzadi, A.; Asadi, S.; Shahabi, H.; Ronoud, S.; Clague, J.J.; Khosravi, K.; Pham, B.T.; Ahmad, B.B.; Bui, D.T. A novel ensemble learning based on Bayesian Belief Network coupled with an extreme learning machine for flash flood susceptibility mapping. *Eng. Appl. Artif. Intell.* **2020**, *96*, 103971. [CrossRef]
40. Ronoud, S.; Asadi, S. An evolutionary deep belief network extreme learning-based for breast cancer diagnosis. *Soft Comput.* **2019**, *23*, 13139–13159. [CrossRef]
41. Asadi, M.; Goli Mokhtari, L.; Shirzadi, A.; Shahabi, H.; Bahrami, S. A comparison study on the quantitative statistical methods for spatial prediction of shallow landslides (case study: Yozidar-Degaga Route in Kurdistan Province, Iran). *Environ. Earth Sci.* **2022**, *81*, 51. [CrossRef]
42. Guzzetti, F.; Carrara, A.; Cardinali, M.; Reichenbach, P. Landslide hazard evaluation: A review of current techniques and their application in a multi-scale study, Central Italy. *Geomorphology* **1999**, *31*, 181–216. [CrossRef]
43. Dai, F.C.; Lee, C.F. Landslide characteristics and slope instability modeling using GIS, Lantau Island, Hong Kong. *Geomorphology* **2002**, *42*, 213–228. [CrossRef]
44. Pradhan, A.M.S.; Kim, Y.-T. Spatial data analysis and application of evidential belief functions to shallow landslide susceptibility mapping at Mt. Umyeon, Seoul, Korea. *Bull. Eng. Geol. Environ.* **2017**, *76*, 1263–1279. [CrossRef]
45. Galli, M.; Ardizzone, F.; Cardinali, M.; Guzzetti, F.; Reichenbach, P. Comparing landslide inventory maps. *Geomorphology* **2008**, *94*, 268–289. [CrossRef]
46. Jiménez-Perálvarez, J.; Irigaray, C.; El Hamdouni, R.; Chacón, J. Landslide-susceptibility mapping in a semi-arid mountain environment: An example from the southern slopes of Sierra Nevada (Granada, Spain). *Bull. Eng. Geol. Environ.* **2011**, *70*, 265–277. [CrossRef]
47. Dehnavi, A.; Aghdam, I.N.; Pradhan, B.; Varzandeh, M.H.M. A new hybrid model using step-wise weight assessment ratio analysis (SWARA) technique and adaptive neuro-fuzzy inference system (ANFIS) for regional landslide hazard assessment in Iran. *Catena* **2015**, *135*, 122–148. [CrossRef]
48. Hunter, G.; Fell, R. Travel distance angle for "rapid" landslides in constructed and natural soil slopes. *Can. Geotech. J.* **2003**, *40*, 1123–1141. [CrossRef]

49. Ayalew, L.; Yamagishi, H.; Marui, H.; Kanno, T. Landslides in Sado Island of Japan: Part II. GIS-based susceptibility mapping with comparisons of results from two methods and verifications. *Eng. Geol.* **2005**, *81*, 432–445. [CrossRef]
50. Bhandary, N.P.; Dahal, R.K.; Timilsina, M.; Yatabe, R. Rainfall event-based landslide susceptibility zonation mapping. *Nat. Hazards* **2013**, *69*, 365–388. [CrossRef]
51. Mandal, S.; Maiti, R. *Semi-Quantitative Approaches for Landslide Assessment and Prediction*; Springer: Berlin/Heidelberg, Germany, 2015.
52. Minár, J.; Evans, I.S.; Jenčo, M. A comprehensive system of definitions of land surface (topographic) curvatures, with implications for their application in geoscience modelling and prediction. *Earth Sci. Rev.* **2020**, *211*, 103414. [CrossRef]
53. Sidle, R.C.; Ochiai, H. *Landslides Processes, Prediction, and Land Use. Water Resources Monograph 18*; American Geophysical Union: Washington, DC, USA, 2006; pp. 322–326.
54. Ercanoglu, M.; Gokceoglu, C. Assessment of landslide susceptibility for a landslide-prone area (north of Yenice, NW Turkey) by fuzzy approach. *Environ. Geol.* **2002**, *41*, 720–730.
55. Kornejady, A.; Ownegh, M.; Bahremand, A. Landslide susceptibility assessment using maximum entropy model with two different data sampling methods. *Catena* **2017**, *152*, 144–162. [CrossRef]
56. Brown, M.E. The Compositions of Kuiper Belt Objects. *Annu. Rev. Earth Planet. Sci.* **2012**, *40*, 467–494. [CrossRef]
57. Beguería, S. Changes in land cover and shallow landslide activity: A case study in the Spanish Pyrenees. *Geomorphology* **2006**, *74*, 196–206. [CrossRef]
58. Chao, L.; Zhang, K.; Wang, J.; Feng, J.; Zhang, M. A Comprehensive Evaluation of Five Evapotranspiration Datasets Based on Ground and GRACE Satellite Observations: Implications for Improvement of Evapotranspiration Retrieval Algorithm. *Remote Sens.* **2021**, *13*, 2414. [CrossRef]
59. Hobson, D.; Curry, R.; Beare, A.; Ward-Gardner, A. The role of serum haemagglutination-inhibiting antibody in protection against challenge infection with influenza A2 and B viruses. *Epidemiol. Infect.* **1972**, *70*, 767–777. [CrossRef]
60. Walbridge, S.; Slocum, N.; Pobuda, M.; Wright, D.J. Unified geomorphological analysis workflows with Benthic Terrain Modeler. *Geosciences* **2018**, *8*, 94. [CrossRef]
61. Poudyal, C.P.; Chang, C.; Oh, H.-J.; Lee, S. Landslide susceptibility maps comparing frequency ratio and artificial neural networks: A case study from the Nepal Himalaya. *Environ. Earth Sci.* **2010**, *61*, 1049–1064. [CrossRef]
62. Li, Z.-J.; Zhang, K. Comparison of three GIS-based hydrological models. *J. Hydrol. Eng.* **2008**, *13*, 364–370. [CrossRef]
63. Zhang, K.; Chao, L.; Wang, Q.; Huang, Y.; Liu, R.; Hong, Y.; Tu, Y.; Qu, W.; Ye, J. Using multi-satellite microwave remote sensing observations for retrieval of daily surface soil moisture across China. *Water Sci. Eng.* **2019**, *12*, 85–97. [CrossRef]
64. Riley, S.J. *Integration of Environmental, Biological, and Human Dimensions for Management of Mountain Lions (*Puma concolor*) in Montana*; Cornell University Ithaca: New York, NY, USA, 1998.
65. Kramm, T.; Hoffmeister, D.; Curdt, C.; Maleki, S.; Khormali, F.; Kehl, M. Accuracy assessment of landform classification approaches on different spatial scales for the Iranian loess plateau. *ISPRS Int. J. Geo. Inf.* **2017**, *6*, 366. [CrossRef]
66. Weiss, A. Topographic position and landforms analysis. In Proceedings of the Poster Presentation, ESRI User Conference, San Diego, CA, USA, 9–13 July 2001.
67. Carrara, A. Digital terrain analysis for land evaluation. *Geol. Appl. E Idrogeol.* **1978**, *13*, 69–127.
68. Huang, S.; Tang, L.; Hupy, J.P.; Wang, Y.; Shao, G. A commentary review on the use of normalized difference vegetation index (NDVI) in the era of popular remote sensing. *J. For. Res.* **2021**, *32*, 1–6. [CrossRef]
69. Kim, Y.; Kimball, J.S.; Zhang, K.; Didan, K.; Velicogna, I.; McDonald, K.C. Attribution of divergent northern vegetation growth responses to lengthening non-frozen seasons using satellite optical-NIR and microwave remote sensing. *Int. J. Remote Sens.* **2014**, *35*, 3700–3721. [CrossRef]
70. Espizua, L.E.; Bengochea, J.D. Landslide hazard and risk zonation mapping in the Rio Grande Basin, Central Andes of Mendoza, Argentina. *Mt. Res. Dev.* **2002**, *22*, 177–185. [CrossRef]
71. Roback, K.; Clark, M.K.; West, A.J.; Zekkos, D.; Li, G.; Gallen, S.F.; Chamlagain, D.; Godt, J.W. The size, distribution, and mobility of landslides caused by the 2015 Mw7. 8 Gorkha earthquake, Nepal. *Geomorphology* **2018**, *301*, 121–138. [CrossRef]
72. Fell, R.; Hartford, D. Landslide risk management. In *Landslide Risk Assessment*; Routledge: London, UK, 2018; pp. 51–109.
73. Radbruch-Hall, D.H. *Maps Showing Areal Slope Stability in Part of the Northern Coast Ranges, California*; Department of the Interior, United States Geological Survey: Washington, DC, USA, 1976.
74. Yu, L.; Liu, H. Feature selection for high-dimensional data: A fast correlation-based filter solution. In Proceedings of the 20th International Conference on Machine Learning (ICML-03), Washington, DC, USA, 21–24 August 2003; pp. 856–863.
75. Ghosh, S.; Günther, A.; Carranza, E.J.M.; van Westen, C.J.; Jetten, V.G. Rock slope instability assessment using spatially distributed structural orientation data in Darjeeling Himalaya (India). *Earth Surf. Processes Landf.* **2010**, *35*, 1773–1792. [CrossRef]
76. Guthrie, R. The effects of logging on frequency and distribution of landslides in three watersheds on Vancouver Island, British Columbia. *Geomorphology* **2002**, *43*, 273–292. [CrossRef]
77. Lin, Y.; Hu, X.; Zheng, X.; Hou, X.; Zhang, Z.; Zhou, X.; Qiu, R.; Lin, J. Spatial variations in the relationships between road network and landscape ecological risks in the highest forest coverage region of China. *Ecol. Indic.* **2019**, *96*, 392–403. [CrossRef]
78. Gokceoglu, C. Discussion on "Landslide hazard zonation of the Khorshrostam area, Iran" by A. Uromeihy and MR Mahdavifar, Bull Eng Geol Environ 58: 207–213. *Bull. Eng. Geol. Environ.* **2001**, *60*, 79–80. [CrossRef]

79. Benda, L.; Andras, K.; Miller, D.; Bigelow, P. Confluence effects in rivers: Interactions of basin scale, network geometry, and disturbance regimes. *Water Resour. Res.* **2004**, *40*. [CrossRef]
80. Sarkar, S.; Kanungo, D.; Mehrotra, G. Landslide hazard zonation: A case study in Garhwal Himalaya, India. *Mt. Res. Dev.* **1995**, *15*, 301–309. [CrossRef]
81. Jiang, S.; Zuo, Y.; Yang, M.; Feng, R. Reconstruction of the Cenozoic tectono-thermal history of the Dongpu Depression, Bohai Bay Basin, China: Constraints from apatite fission track and vitrinite reflectance data. *J. Pet. Sci. Eng.* **2021**, *205*, 108809. [CrossRef]
82. Geertsema, M.; Highland, L.; Vaugeouis, L. Environmental impact of landslides. In *Landslides–Disaster Risk Reduction*; Springer: Berlin/Heidelberg, Germany, 2009; pp. 589–607.
83. Zuo, Y.; Jiang, S.; Wu, S.; Xu, W.; Zhang, J.; Feng, R.; Yang, M.; Zhou, Y.; Santosh, M. Terrestrial heat flow and lithospheric thermal structure in the Chagan Depression of the Yingen-Ejinaqi Basin, north central China. *Basin Res.* **2020**, *32*, 1328–1346. [CrossRef]
84. Geertsema, M.; Pojar, J.J. Influence of landslides on biophysical diversity—a perspective from British Columbia. *Geomorphology* **2007**, *89*, 55–69. [CrossRef]
85. Schwab, J.W.; Geertsema, M.; Blais-Stevens, A. The Khyex River landslide of November 28, 2003, Prince Rupert British Columbia Canada. *Landslides* **2004**, *1*, 243–246. [CrossRef]
86. Kustikova, V.; Druzhkov, P. A survey of deep learning methods and software for image classification and object detection. In Proceedings of the OGRW2014—9th Open German-Russian Workshop on Pattern Recognition and Image Understanding, Koblenz, Germany, 1–5 December 2014.
87. Lv, Z.; Li, Y.; Feng, H.; Lv, H. Deep Learning for Security in Digital Twins of Cooperative Intelligent Transportation Systems. *IEEE Trans. Intell. Transp. Syst.* **2021**, 1–10. [CrossRef]
88. Liu, K.; Ke, F.; Huang, X.; Yu, R.; Lin, F.; Wu, Y.; Ng, D.W.K. DeepBAN: A Temporal Convolution-Based Communication Framework for Dynamic WBANs. *IEEE Trans. Commun.* **2021**, *69*, 6675–6690. [CrossRef]
89. Zhou, W.; Guo, Q.; Lei, J.; Yu, L.; Hwang, J.-N. IRFR-Net: Interactive recursive feature-reshaping network for detecting salient objects in RGB-D images. *IEEE Trans. Neural Netw. Learn. Syst.* **2021**, 1–13. [CrossRef] [PubMed]
90. Le, Q.V. A tutorial on deep learning part 2: Autoencoders, convolutional neural networks and recurrent neural networks. *Google Brain* **2015**, *20*, 1–20.
91. Huang, G.-B.; Zhu, Q.-Y.; Siew, C.-K. Extreme learning machine: A new learning scheme of feedforward neural networks. In Proceedings of the 2004 IEEE International Joint Conference on Neural Networks (IEEE Cat. No. 04CH37541), Budapest, Hungary, 25–29 July 2004; pp. 985–990.
92. Wang, Y.; Dou, Y.; Liu, X.; Lei, Y. PR-ELM: Parallel regularized extreme learning machine based on cluster. *Neurocomputing* **2016**, *173*, 1073–1081. [CrossRef]
93. Zhou, G.; Bao, X.; Ye, S.; Wang, H.; Yan, H. Selection of Optimal Building Facade Texture Images From UAV-Based Multiple Oblique Image Flows. *IEEE Trans. Geosci. Remote Sens.* **2020**, *59*, 1534–1552. [CrossRef]
94. Qu, B.-Y.; Lang, B.; Liang, J.J.; Qin, A.K.; Crisalle, O.D. Two-hidden-layer extreme learning machine for regression and classification. *Neurocomputing* **2016**, *175*, 826–834. [CrossRef]
95. Yao, X.; Tham, L.; Dai, F. Landslide susceptibility mapping based on support vector machine: A case study on natural slopes of Hong Kong, China. *Geomorphology* **2008**, *101*, 572–582. [CrossRef]
96. Xie, W.; Zhang, R.; Zeng, D.; Shi, K.; Zhong, S. Strictly dissipative stabilization of multiple-memory Markov jump systems with general transition rates: A novel event-triggered control strategy. *Int. J. Robust Nonlinear Control.* **2020**, *30*, 1956–1978. [CrossRef]
97. Quinlan, J.R. Generating production rules from decision trees. *IJCAI* **1987**, *87*, 304–307.
98. Mohamed, W.N.H.W.; Salleh, M.N.M.; Omar, A.H. A comparative study of reduced error pruning method in decision tree algorithms. In Proceedings of the 2012 IEEE International Conference on Control System, Computing and Engineering, Penang, Malaysia, 23–25 November 2012; pp. 392–397.
99. Quinlan, J.R. Simplifying decision trees. *Int. J. Man-Mach. Stud.* **1987**, *27*, 221–234. [CrossRef]
100. Farid, D.M.; Zhang, L.; Rahman, C.M.; Hossain, M.A.; Strachan, R. Hybrid decision tree and naïve Bayes classifiers for multi-class classification tasks. *Expert Syst. Appl.* **2014**, *41*, 1937–1946. [CrossRef]
101. Pham, B.T.; Bui, D.T.; Pourghasemi, H.R.; Indra, P.; Dholakia, M. Landslide susceptibility assesssment in the Uttarakhand area (India) using GIS: A comparison study of prediction capability of naïve bayes, multilayer perceptron neural networks, and functional trees methods. *Theor. Appl. Climatol.* **2017**, *128*, 255–273. [CrossRef]
102. Wang, S.; Jiang, L.; Li, C. Adapting naive Bayes tree for text classification. *Knowl. Inf. Syst.* **2015**, *44*, 77–89. [CrossRef]
103. Quinlan, J.R. Induction of decision trees. *Mach. Learn.* **1986**, *1*, 81–106. [CrossRef]
104. Luo, X.; Lin, F.; Zhu, S.; Yu, M.; Zhang, Z.; Meng, L.; Peng, J. Mine landslide susceptibility assessment using IVM, ANN and SVM models considering the contribution of affecting factors. *PLoS ONE* **2019**, *14*, e0215134. [CrossRef] [PubMed]
105. Chen, W.; Xie, X.; Peng, J.; Shahabi, H.; Hong, H.; Bui, D.T.; Duan, Z.; Li, S.; Zhu, A.-X. GIS-based landslide susceptibility evaluation using a novel hybrid integration approach of bivariate statistical based random forest method. *Catena* **2018**, *164*, 135–149. [CrossRef]
106. Chen, W.; Xie, X.; Peng, J.; Wang, J.; Duan, Z.; Hong, H. GIS-based landslide susceptibility modelling: A comparative assessment of kernel logistic regression, Naïve-Bayes tree, and alternating decision tree models. *Geomat. Nat. Hazards Risk* **2017**, *8*, 950–973. [CrossRef]

107. Zhang, T.; Han, L.; Zhang, H.; Zhao, Y.; Li, X.; Zhao, L. GIS-based landslide susceptibility mapping using hybrid integration approaches of fractal dimension with index of entropy and support vector machine. *J. Mt. Sci.* **2019**, *16*, 1275–1288. [CrossRef]
108. Pham, B.T.; Bui, D.T.; Dholakia, M.; Prakash, I.; Pham, H.V. A comparative study of least square support vector machines and multiclass alternating decision trees for spatial prediction of rainfall-induced landslides in a tropical cyclones area. *Geotech. Geol. Eng.* **2016**, *34*, 1807–1824. [CrossRef]
109. Bennett, N.D.; Croke, B.F.; Guariso, G.; Guillaume, J.H.; Hamilton, S.H.; Jakeman, A.J.; Marsili-Libelli, S.; Newham, L.T.; Norton, J.P.; Perrin, C. Characterising performance of environmental models. *Environ. Model. Softw.* **2013**, *40*, 1–20. [CrossRef]
110. Konishi, T.; Suga, Y. Landslide detection using COSMO-SkyMed images: A case study of a landslide event on Kii Peninsula, Japan. *Eur. J. Remote Sens.* **2018**, *51*, 205–221. [CrossRef]
111. Hosmer, D.W.; Lemeshow, S. *Applied Logistic Regression*; John Wiley & Sons: New York, NY, USA, 2000.
112. Tien Bui, D.; Tuan, T.A.; Klempe, H.; Pradhan, B.; Revhaug, I. Spatial prediction models for shallow landslide hazards: A comparative assessment of the efficacy of support vector machines, artificial neural networks, kernel logistic regression, and logistic model tree. *Landslides* **2016**, *13*, 361–378. [CrossRef]
113. Ma, Z.; Zheng, W.; Chen, X.; Yin, L. Joint embedding VQA model based on dynamic word vector. *PeerJ Comput. Sci.* **2021**, *7*, e353. [CrossRef]
114. Kohavi, R. A study of cross-validation and bootstrap for accuracy estimation and model selection. *IJCAI* **1995**, *14*, 1137–1145.
115. Tanyu, B.F.; Abbaspour, A.; Alimohammadlou, Y.; Tecuci, G. Landslide susceptibility analyses using Random Forest, C4.5, and C5.0 with balanced and unbalanced datasets. *CATENA* **2021**, *203*, 105355. [CrossRef]
116. Morales, B.; Lizama, E.; Somos-Valenzuela, M.A.; Lillo-Saavedra, M.; Chen, N.; Fustos, I. A comparative machine learning approach to identify landslide triggering factors in northern Chilean Patagonia. *Landslides* **2021**, *18*, 2767–2784. [CrossRef]
117. Jaafari, A.; Rezaeian, J.; Omrani, M.S. Spatial prediction of slope failures in support of forestry operations safety. *Croat. J. For. Eng.* **2017**, *38*, 107–118.
118. Schlögl, M.; Matulla, C. Potential future exposure of European land transport infrastructure to rainfall-induced landslides throughout the 21st century. *Nat. Hazards Earth Syst. Sci.* **2018**, *18*, 1121–1132. [CrossRef]
119. Sultana, N.; Tan, S. Landslide mitigation strategies in southeast Bangladesh: Lessons learned from the institutional responses. *Int. J. Disaster Risk Reduct.* **2021**, *62*, 102402. [CrossRef]
120. Nefeslioglu, H.A.; Gorum, T. The use of landslide hazard maps to determine mitigation priorities in a dam reservoir and its protection area. *Land Use Policy* **2020**, *91*, 104363. [CrossRef]
121. Pham, B.T.; Jaafari, A.; Nguyen-Thoi, T.; Van Phong, T.; Nguyen, H.D.; Satyam, N.; Masroor, M.; Rehman, S.; Sajjad, H.; Sahana, M. Ensemble machine learning models based on Reduced Error Pruning Tree for prediction of rainfall-induced landslides. *Int. J. Digit. Earth* **2021**, *14*, 575–596. [CrossRef]
122. Shirzadi, A.; Soliamani, K.; Habibnejhad, M.; Kavian, A.; Chapi, K.; Shahabi, H.; Chen, W.; Khosravi, K.; Pham, B.T.; Pradhan, B.; et al. Novel GIS based machine learning algorithms for shallow landslide susceptibility mapping. *Sensors* **2018**, *18*, 3777. [CrossRef]
123. Sameen, M.I.; Pradhan, B.; Lee, S. Application of convolutional neural networks featuring Bayesian optimization for landslide susceptibility assessment. *CATENA* **2019**, *186*, 104249. [CrossRef]
124. Chang, K.-T.; Merghadi, A.; Yunus, A.P.; Pham, B.T.; Dou, J. Evaluating scale effects of topographic variables in landslide susceptibility models using GIS-based machine learning techniques. *Sci. Rep.* **2019**, *9*, 12296. [CrossRef]
125. Tien Bui, D.; Shirzadi, A.; Shahabi, H.; Geertsema, M.; Omidvar, E.; Clague, J.J.; Thai Pham, B.; Dou, J.; Talebpour Asl, D.; Bin Ahmad, B. New ensemble models for shallow landslide susceptibility modeling in a semi-arid watershed. *Forests* **2019**, *10*, 743. [CrossRef]
126. Nhu, V.-H.; Hoang, N.-D.; Nguyen, H.; Ngo, P.T.T.; Bui, T.T.; Hoa, P.V.; Samui, P.; Bui, D.T. Effectiveness assessment of Keras based deep learning with different robust optimization algorithms for shallow landslide susceptibility mapping at tropical area. *CATENA* **2020**, *188*, 104458. [CrossRef]
127. Goodfellow, I.; Bengio, Y.; Courville, A. *Deep Learning*; MIT Press: Cambridge, MA, USA, 2016.
128. Wang, Y.; Fang, Z.; Hong, H. Comparison of convolutional neural networks for landslide susceptibility mapping in Yanshan County, China. *Sci. Total Environ.* **2019**, *666*, 975–993. [CrossRef] [PubMed]
129. Huang, F.; Zhang, J.; Zhou, C.; Wang, Y.; Huang, J.; Zhu, L. A deep learning algorithm using a fully connected sparse autoencoder neural network for landslide susceptibility prediction. *Landslides* **2020**, *17*, 217–229. [CrossRef]
130. Dao, D.V.; Jaafari, A.; Bayat, M.; Mafi-Gholami, D.; Qi, C.; Moayedi, H.; Phong, T.V.; Ly, H.-B.; Le, T.-T.; Trinh, P.T.; et al. A spatially explicit deep learning neural network model for the prediction of landslide susceptibility. *CATENA* **2020**, *188*, 104451. [CrossRef]
131. Dou, J.; Yunus, A.P.; Merghadi, A.; Shirzadi, A.; Nguyen, H.; Hussain, Y.; Avtar, R.; Chen, Y.; Pham, B.T.; Yamagishi, H. Different sampling strategies for predicting landslide susceptibilities are deemed less consequential with deep learning. *Sci. Total Environ.* **2020**, *720*, 137320. [CrossRef]
132. Mandal, K.; Saha, S.; Mandal, S. Applying deep learning and benchmark machine learning algorithms for landslide susceptibility modelling in Rorachu river basin of Sikkim Himalaya, India. *Geosci. Front.* **2021**, *12*, 101203. [CrossRef]

Article

Landslide Susceptibility Mapping Using Machine Learning Algorithm Validated by Persistent Scatterer In-SAR Technique

Muhammad Afaq Hussain [1], Zhanlong Chen [1,*], Ying Zheng [1], Muhammad Shoaib [2], Safeer Ullah Shah [3], Nafees Ali [4] and Zeeshan Afzal [5]

1 School of Geography and Information Engineering, China University of Geosciences (Wuhan), Wuhan 430074, China; khanafaq121@cug.edu.cn (M.A.H.); 20161002229@cug.edu.cn (Y.Z.)
2 State Key Laboratory of Hydraulic Engineering, Simulation and Safety, School of Civil Engineering, Tianjin University, Tianjin 300072, China; xs4shoaib@tju.edu.cn
3 Ministry of Climate Change, Islamabad 44000, Pakistan; safeershah@uop.edu.pk
4 Chinese Academy of Sciences, Beijing 100045, China; nafeesali@mails.ucas.ac.cn
5 Key State Laboratory of Information Engineering in Surveying, Mapping and Remote Sensing (LIESMARS), Wuhan University, Wuhan 430079, China; zeeshanafzal@whu.edu.cn
* Correspondence: chenzl@cug.edu.cn

Abstract: Landslides are the most catastrophic geological hazard in hilly areas. The present work intends to identify landslide susceptibility along Karakorum Highway (KKH) in Northern Pakistan, using landslide susceptibility mapping (LSM). To compare and predict the connection between causative factors and landslides, the random forest (RF), extreme gradient boosting (XGBoost), k nearest neighbor (KNN) and naive Bayes (NB) models were used in this research. Interferometric synthetic aperture radar persistent scatterer interferometry (PS-InSAR) technology was used to explore the displacement movement of retrieved models. Initially, 332 landslide areas alongside the Karakorum Highway were found to generate the landslide inventory map using various data. The landslides were categorized into two sections for validation and training, of 30% and 70%. For susceptibility mapping, thirteen landslide-condition factors were created. The area under curve (AUC) of the receiver operating characteristic (ROC) curve technique was utilized for accuracy comparison, yielding 83.08, 82.15, 80.31, and 72.92% accuracy for RF, XGBoost, KNN, and NB, respectively. The PS-InSAR technique demonstrated a high deformation velocity along the line of sight (LOS) in model-sensitive areas. The PS-InSAR technique was used to evaluate the slope deformation velocity, which can be used to improve the LSM for the research region. The RF technique yielded superior findings, integrating with the PS-InSAR outcomes to provide the region with a new landslide susceptibility map. The enhanced model will help mitigate landslide catastrophes, and the outcomes may help ensure the roadway's safe functioning in the study region.

Keywords: CPEC; random forest; landslides; susceptibility; PS-InSAR; ArcGIS

Citation: Hussain, M.A.; Chen, Z.; Zheng, Y.; Shoaib, M.; Shah, S.U.; Ali, N.; Afzal, Z. Landslide Susceptibility Mapping Using Machine Learning Algorithm Validated by Persistent Scatterer In-SAR Technique. *Sensors* **2022**, *22*, 3119. https://doi.org/10.3390/s22093119

Academic Editor: Francesca Cigna

Received: 11 March 2022
Accepted: 17 April 2022
Published: 19 April 2022

1. Introduction

The China–Pakistan Economic Corridor (CPEC) demonstrates the flagship project of the "One Belt, One Road" policy. It is also thought to hold Pakistan's financial prospects, which are receiving a lot of interest. The Karakoram Highway was built in 1974–1978 and inaugurated in 1979 and runs alongside the CPEC in Northern Pakistan. It is regularly closed for a few months each year because of landslides.

Local topography, tectonic features, geomorphology, landcover, geology and human interference all have an influence on the spatial likelihood of landslides, which is then examined to determine landslide susceptibility (LS) [1]. Landslide vulnerability assessment models frequently assume that historical and current landslide conditions would be constant in the future [2]. LS methodologies can be quantitative or qualitative; quantitative methods evaluate the likelihood of landslide incidence in a susceptible zone, whereas qualitative methods introduce subjectivity into illustrative susceptibility zonation [1,3]. The analytical hierarchy model (AHP) [4–6], weight of evidence model [7,8], frequency ratio [4,5,9] and certainty factor [4] are all commonly used landslide susceptibility models. A growing trend is to compare the outcomes of implementing two or more models and the result is a landslide susceptibility model (LSM). However, most studies still use only one model for LSM [1,10]. Reichenbach et al. [1] suggest using numerous models to assess landslides and developing an "optimal" zonation map to reduce risk prediction errors and its integrity to be used for land-use planning. Our literature review of the investigated area demonstrates several statistical approaches for LS such as frequency ratio and weight of evidence [11,12], AHP, and Scoops3D [13], the weighted overlay technique, and the AHP [14] were used in the research region. Several investigations [15,16] provide bivariate analyses that measure the geographical links between particular variables and landslides that influence their occurrence. However, the key disadvantages of these models are that they change the ambiguity of risk processes, are typically static, incorporate geometrical assumptions, and are costly and difficult for the gathering of hydrological and geotechnical data, especially when examining vast and different locations.

In recent years, advances in ML algorithms, computing power, and geospatial innovations have made it easier to create landslide susceptibility (LS) maps [17]. The precision of LS maps can be improved using machine learning algorithms. Knowledge-based methods [18], multivariate logistic regression methods [19–21] and multivariate binary logistic regression [22] have all been presented in recent papers. General linear model [23,24], quadratic discriminant analysis [10,24], boosted regression tree [23,25], random forest [26–29], multivariate adaptive regression splines [30,31], classification and regression tree [23,32], support vector machine [33–35], naïve Bayes [36,37], generalized additive model [24,32], neuro-fuzzy and adaptive neuro-fuzzy inference [38–40], fuzzy logic [41], artificial neural networks [42–47], maximum entropy [48,49] and decision tree [19,50,51] are some of the ML models used in LSM. Qing et al. [52] used various ML techniques for LSM alongside the China–Pakistan Karakoram Highway. In two South Korean catchments, Pradhan and Kim [53] compared the precision consequences of deep neural network (83.71%), and XGBoost (76.73%) approaches for LS mapping. Merghadi et al. [54] assessed the performance and competency of various ML techniques in the literature and discovered that tree-based ensemble optimization algorithms outcompete other ML algorithms. In a comparison analysis, Sahin [55] found that CatBoost had the best precision (85%), followed by XGBoost (83.36%) since the proportion of samples of the model was determined by Catboost was more precisely anticipated than other models. The primary advantages of ML and probabilistic processes are their objective statistical foundation, repeatability, capacity to quantitatively analyze the effect of variables on landslide evolution, and capacity to update them regularly. Machine learning models can be built using a variety of landslide-conditioning factors (slope, aspect, and elevation). Several studies on landslide susceptibility evaluation have been undertaken using remote sensing and GIS techniques [56–58].

Furthermore, remote sensing (RS) is an effective method for determining the motion of landslides [59–61]. It provides a solution in surveys or enhanced detection in places where catastrophic landslides occur frequently and quickly [62–65]. Furthermore, interferometric algorithms to radar images effectively map large-scale landslide mapping and detection. It may aid in the development of landslide inventory maps. In particular, decrypt ADInSAR and PS-InSAR [66,67], coherence pixel technique [68], SqeeSAR [69], small baseline subset [70,71], Stanford method for persistent scatterers [72], stable point network [73,74], and interferometric point target analysis [75] have created various useful case research. As noted in past studies [75–77], these approaches are involved with mapping and identifying landslide occurrences.

A diversity of researchers in northern Pakistan has analyzed landslides using historical records, field observations, tectonic characteristics, and geological data [78–81]. Previous studies [14,82–85] concentrated on probabilistic and statistical relationships and regression interpretation of landslides with parameters. For the first time, the PS-InSAR approach evaluated the surface displacement in the study area using RF, XGBoost KNN, and NB models, making it a distinctive method of identifying landslide movements. Persistent scatterer interferometry (PSI), interferometric synthetic aperture radar (InSAR), and area under curve (AUC) of ROC techniques were used along the KKH to assess displacements and the precision of the models used. Single landslides in hazardous areas can be identified and defined using PS-InSAR. Landslides can also be detected using a spatial statistical method based on a multitemporal assessment of SAR images that calculate slow landslide movements [71].

The current work seeks to develop a susceptibility model and a complete visually interpreted landslide inventory utilizing recently developed ML models, including RF, XGBoost, KNN, and NB. The second goal is to quantify the deformation velocities of slow-moving landslides using PS-InSAR to identify high-susceptibility zones for future landslide disaster management. The third goal is to select the most susceptible model based on accuracy and AUC value and then combine it with PS-InSAR outcomes to produce a new landslide susceptibility map for the research area. These prediction approaches will help lead future development and land management efforts in the area. These susceptibility maps will aid in avoiding and limiting human and economic losses along this critical corridor.

2. Methods

2.1. Study Area

The research region is 178 km long and has a 5 km radius buffer zone along KKH (Figure 1). The KKH in northern Pakistan is a crucial component of the CPEC; nevertheless, it is frequently disrupted due to several hydro-climatological and geological risks along the route. Landslides are the most common and devastating to highways, human lives, and economic activity.

The research region experiences harsh winters and mild summers. The region's annual rainfall ranges from 120 mm to 130 mm: the maximum and minimum temperatures vary from 16 °C to −21 °C (Meteorological Department of Pakistan). The lithology of various sources with thicknesses of up to 100 m is irregularly scattered [86,87]. The majority of these deposits are weakly consolidated, making them conducive to landslides in the form of rockfalls and debris flows [80]. The combination of complex topography, high erosion rates, human causes, and active tectonics makes this area one of the most susceptible to landslides.

2.2. Geological Setting of the Area

The rocks in the region are mostly Paleozoic, Proterozoic, and Mesozoic in age (Figure 2). According to the geological map prepared by Searle et al. [88], the study area is comprised of the following lithology.

The Chilas complex in the study area comprises mafic and ultramafic plutonic rocks and Kohistan batholiths composed of granodiorite, granite, and diorite. The Gilgit complex metasedimentary rocks are slates, minor phyllite, quartzite, and dolomite limestone. Komila amphibolite comprises of plutonic and meta plutonic rocks with intrusion of diorite granodiorite and granite. In Paleozoic metasedimentary rocks are marble, dolomite, and quartzite.

2.3. Landslide Susceptibility Mapping

Geological maps, remote sensing data, and meteorological data were gathered from various sources for the study (Table 1). The Alaska Satellite Facility dataset contained an Advanced Land Observing Satellite, Phased Array type L-band Synthetic Aperture Radar DEM (Digital Elevation Model) with a resolution of 12.5 m (https://search.asf.alaska.edu/ (accessed on 20 January 2022)). Sentinel-2 images with a resolution of 10 m were extracted from the USGS (https://earthexplorer.usgs.gov (accessed on 20 January 2022)) dataset to create a landcover map for the research area. The geological map for the area was digitized in the ArcGIS environment to comprehend surficial geological characteristics. PS-InSAR processing was used to compute the deformation velocity using Sentinel-1 (https://search.asf.alaska.edu/ (accessed on 8 February)) (31 images in descending path and 33 images in ascending path). Figure 3 depicts the approach used in the investigation.

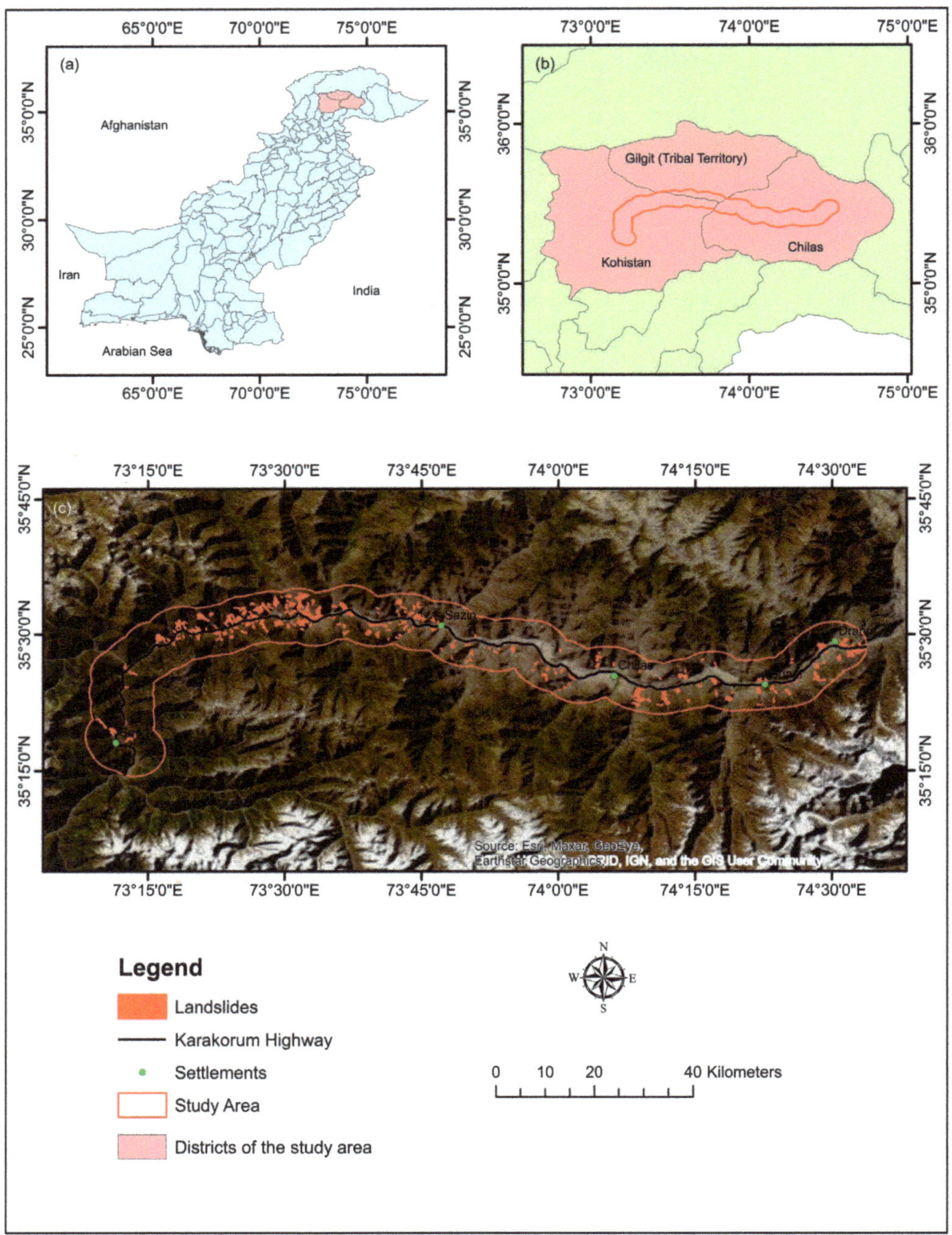

Figure 1. The 541 band combination Landsat image showing the area under investigation. (**a**) Pakistan, (**b**) District boundaries, (**c**) Study area in red outline.

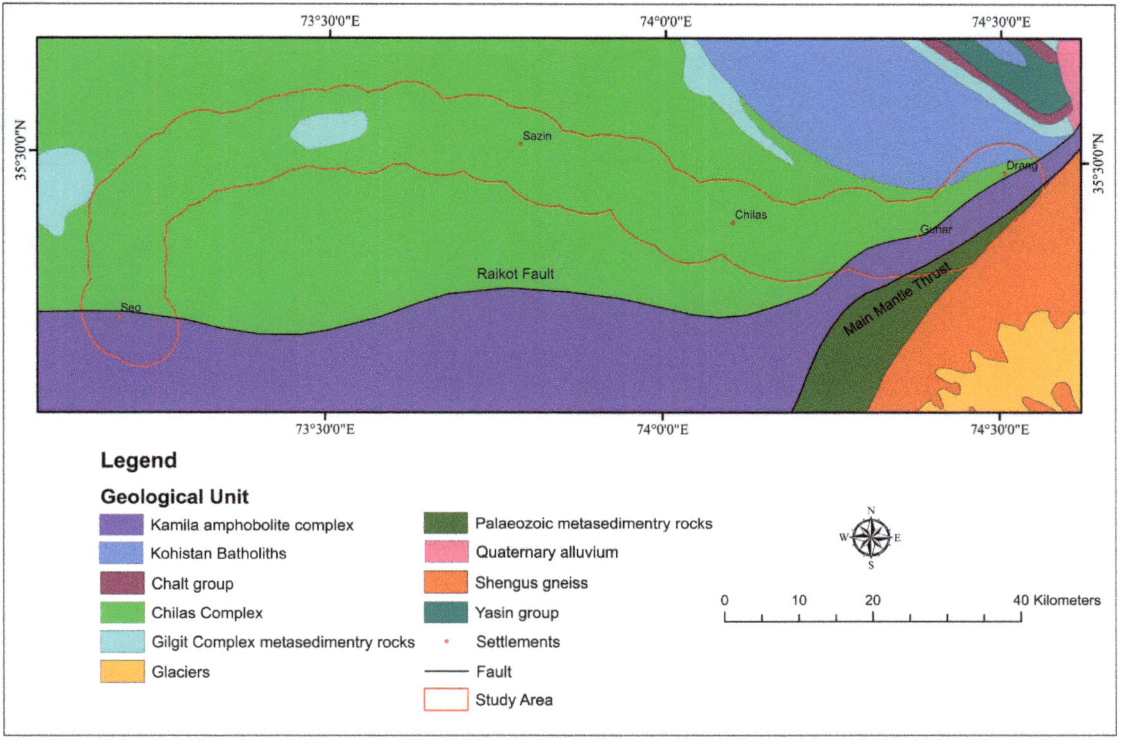

Figure 2. Regional geological map of the study area.

Table 1. Information on landslide conditioning factors.

S.NO	Factors	Description/Extraction	Category
1	Elevation, aspect, curvature, slope, profile curvature, TWI, plan curvature, roughness	ALOS-PALSAR DEM (https://search.asf.alaska.edu/ (accessed on 20 January 2022))	Topography
2	Geology, distance to fault	Geological Survey of Pakistan	Geology
3	Landcover	Land cover classes (https://earthexplorer.usgs.gov) (accessed on 20 January 2022) (Sentinel-2 images)	Conditioning factor
4	NDVI	Normalized Different Vegetation Index (Landsat-8, 2021)	Landcover
5	Precipitation	Annual rainfall (Pakistan Metrological Department)	Triggered factor

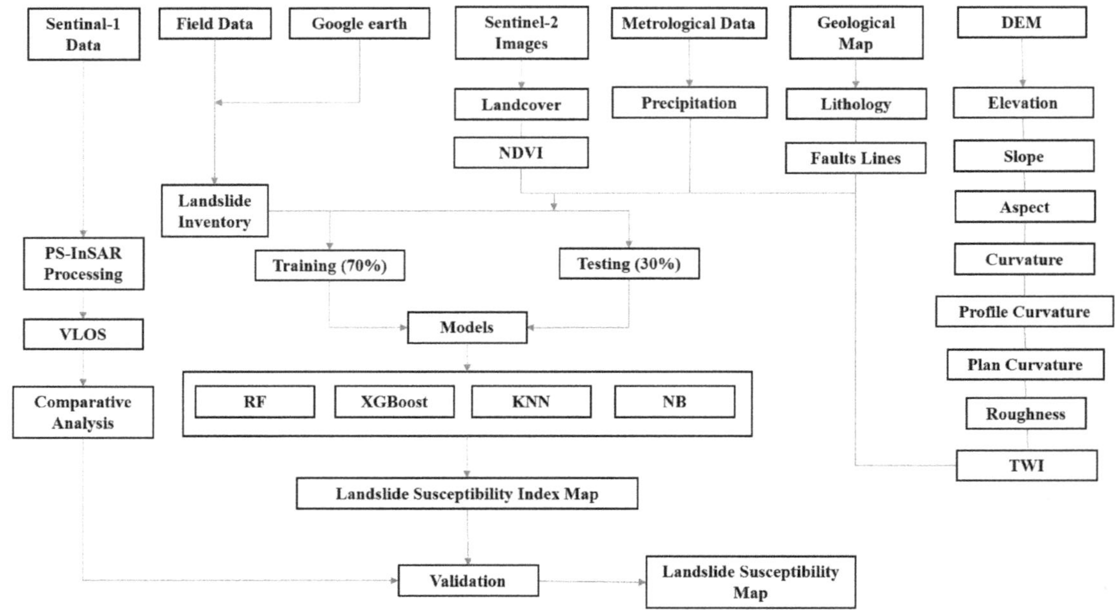

Figure 3. Flow chart of research.

2.4. Landslide Inventory

The landslide inventory is the stage in estimating susceptibility since it provides details on all sorts of past landslides in the research area. This is the most important stage since the required precision of the landslide inventory to fine tune the models influences the LSM accuracy [15,89,90]. As a result, the more accurate and high-quality the landslide inventory, the more improved the prediction execution of the SM [89]. The evaluation of landslide hazard begins with creating realistic and detailed landslide inventory maps that show the type of landslide, geographic extension, the date of the event, and location [89,91]. The produced landslide inventory maps are then associated with contributing geo-environmental parameters such as land cover, topography, geology, geomorphology, and other factors to assess the likelihood of terrain causing a landslide allocated to a susceptibility level [1,9,92–94].

Inventory maps contain information on all active and historical landslide distributions based on field surveying, aerial image interpretation, and previous report data [80]. In this research, we were using actual data of landslide occurrences obtained from the Geological Survey of Pakistan (GSP) publications [95–97], Frontier Works Organization road clearance logs, a research article [98], and Google Earth imagery to produce a multitemporal landslide inventory along the highway. On the other hand, the landslide inventory was created by the visual interpretation of Sentinel-2 photos with 10 m resolution (2020) and from Google Earth and was validated using earlier reports and a field assessment of the research area. Polygon shapes were constructed on satellite images for clearly visible landslides (based on GSP and FWO data). Debris flow (188 locations) and scree slopes (51 locations) are mostly found in the research area as a result of unconsolidated sediments on barren mountains and rainfall, although rock falls (93 locations) are also common as a result of seismic activity and toe cutting of steep slopes by anthropogenic activities for various causes (Figure 4). There were 332 landslides mapped, shown in Figure 5. Of these landslides, 30% (100 landslides) and 70% (232 landslides) were chosen for training for model validation [99].

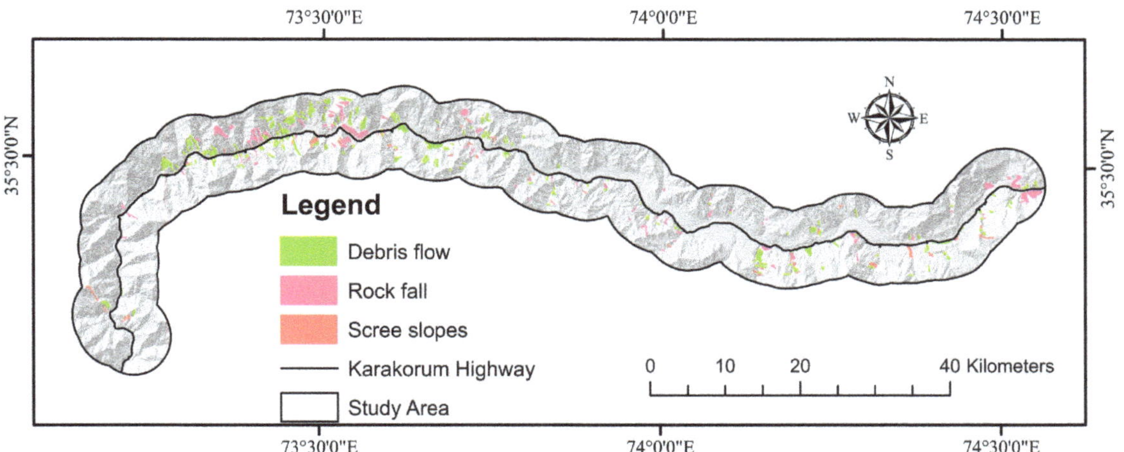

Figure 4. Showing the debris flow, rockfall, and scree slopes in the research area.

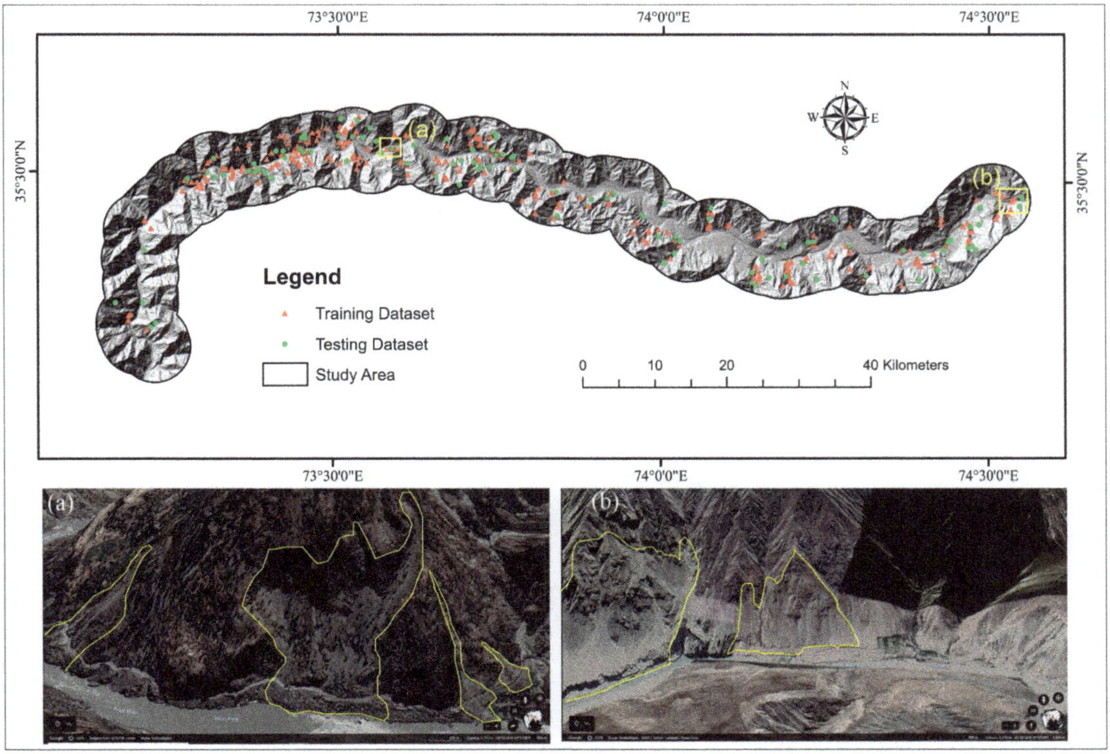

Figure 5. Landslide inventory of study area. (**a**) Shows the rockfall and scree slopes near the Sazin area, (**b**) rockfall and debris flow near the Drang area.

2.5. Landslide Causative Factors

Landslides' spatial distribution is influenced by triggering, and conditioning variables were chosen based on the region's morphology, geology, hydrology, and anthropogenic activities. There are no general criteria for choosing independent factors for LSM [71]. The concepts are that the factors must be non-redundant, non-uniform, operational, and measurable [100]. ArcGIS is commonly used to extract important susceptibility conditioning factors from digital elevation models, including elevation, slope, profile curvature, aspect, curvature, and topographic wetness index (TWI) [101]. Land cover, geology, precipitation, roughness, normalized difference vegetation index (NDVI), distance to faults, TWI, slope, plan curvature, curvature, elevation, profile curvature, and aspect were all utilized to estimate the landslides' disaster susceptibility in the research area (Table 1). All of these maps were converted to a 12.5 × 12.5 m pixel raster format for the models, which was up to digital elevation model resolution. In the resampling method, the cell size of each factor was kept at 12.5 m so that the overlay assessment would obtain the pixels at the same scale, and the output was also the same scale. The maps were digitized at various scales, and the pixel resolution was kept at 12.5 m while converting them to raster format. The DEM with a pixel of 12.5 m was used to extract the majority of the factors, and all other factors were brought to a similar resolution. The thirteen landslide factors are depicted in Figures 6 and 7.

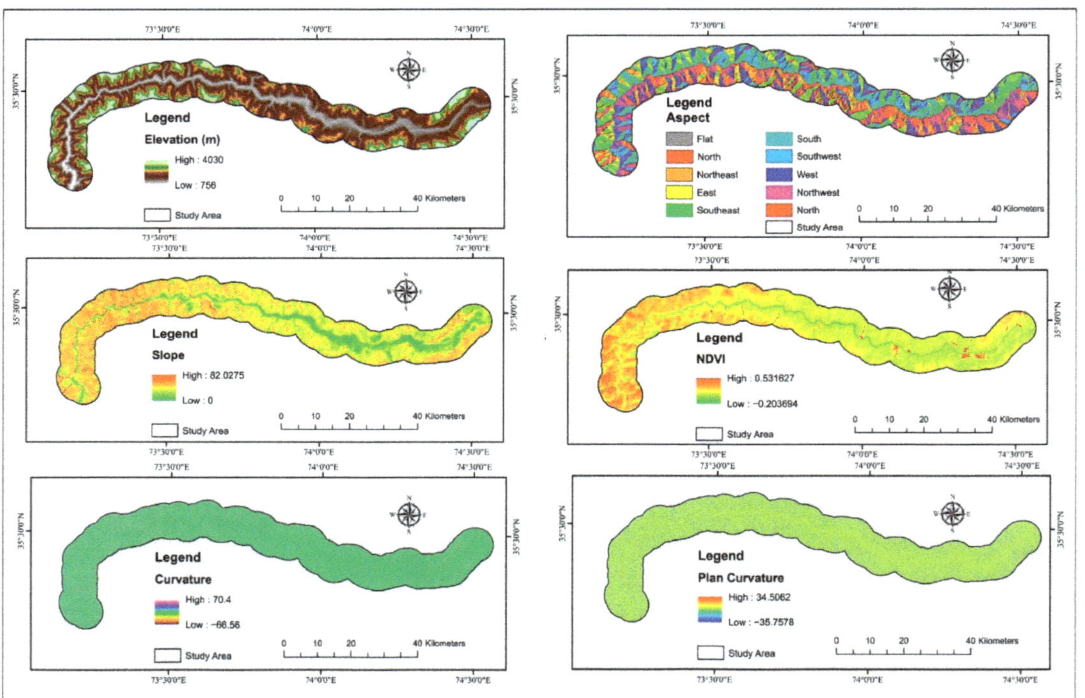

Figure 6. Landslides factors used in the research study.

The modeling procedure included machine learning model fitting, identification, and development.

I. The model unit in this investigation was the grid unit (12.5 m). The spatial resolution of DEM and RS data corresponds to 12.5 m, and all assessment variables have been recalculated at this level.

II. A condition property reached thirteen causative variables and a landslide decision attribute (1 indicates landslides, 0 indicates non-landslides), with each row creating an object.

III. Each column represents an object's attribute and has been converted into training (70%) and testing the two-dimensional matrix (30%). Training data were used to assemble the models, and test data was used to make forecasts.

IV. The landslide susceptibility index maps were created using the forecast values of every model unit per group. The findings of the four algorithms were exported into GIS.

V. The Jenks natural breaks [102] classifications were used to categorize LS: very low, low, moderate, high, and very high. The ROC curve and the area under the ROC curve were used to test the four models.

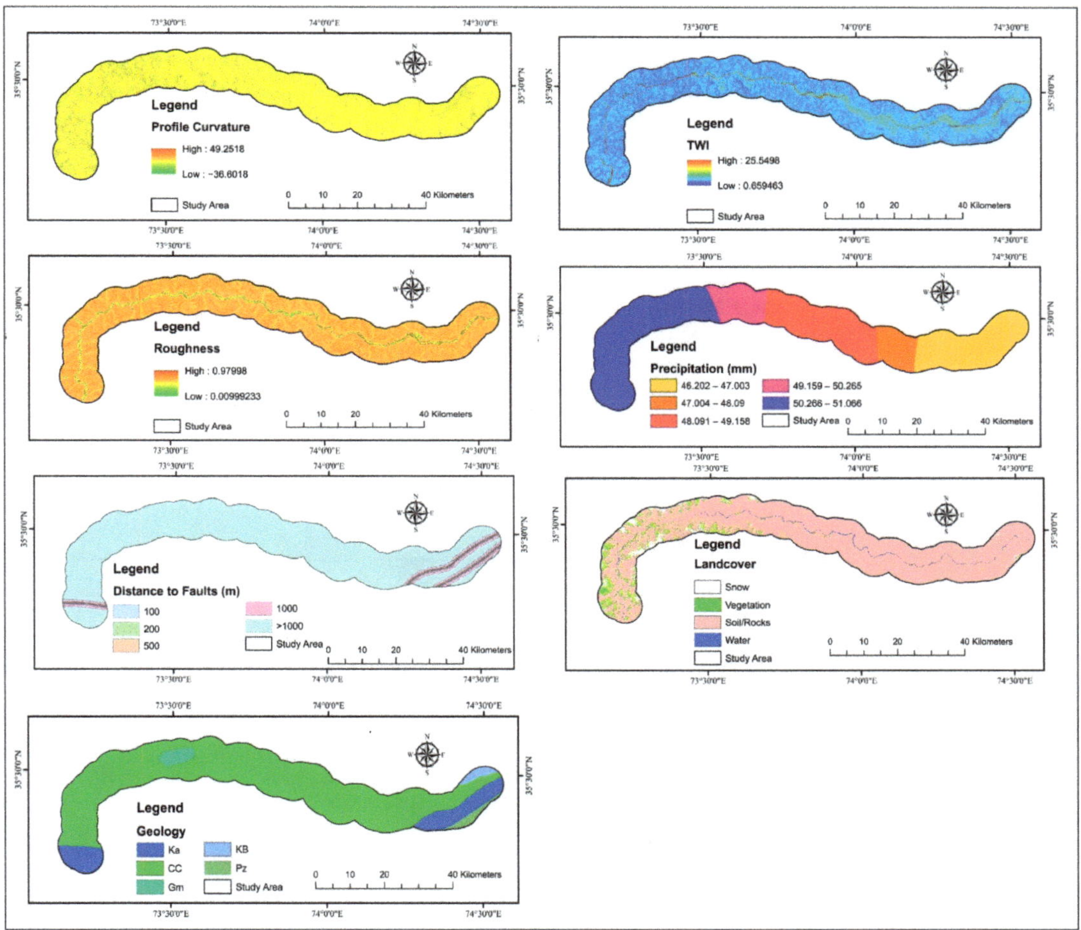

Figure 7. Landslides factors used in the research area.

2.6. RF

One of the most widely used methods for regression and classification is the random forest, which was designed by [103]. RF has a lot of important features for classification tasks. Because RF is a non-parametric, non-linear approach, it can handle big datasets with numerical and category data and complicated nonlinearity and interactions between factors. Secondly, it can deal with the situation with more predictors than data and integrate the connection between different predictors. Third, random forest can manage missing values while maintaining precision for missing data.

Furthermore, unlike other ML approaches such as support vector machines and artificial neural networks, RF does not need extensive hyper-parameter tuning. In many instances, utilizing the default parameter values yields good results. When compared to other tree-ensemble approaches (boosting), random forest is relatively fast. Decision trees are built during the training phase, and the output class is based on the classification or regression mode of each individual tree's decision trees. In order to train random forests, the general method of bootstrap aggregation (also known as "bagging") is used for tree learners. This bootstrapping approach improves model performance by reducing the model's variance without raising the model's bias [104]. Random forest has been extensively utilized for classification applications and large-scale mapping in LSM [26,28] ecology [105], flood mapping [106] and soil science [107].

R statistical software was used to develop the RF model [108]. Because the analysis in the RF model was grid-based, gridded cells (12.5 by 12.5 m) were derived from the randomly shaped sample spatial polygons of landslides and non-landslides, respectively.

The RF model operates by developing numerous decorrelated decision trees as a base learner, with replacement, utilizing a percentage of randomly chosen landslide-predicting variables and landslide observation. Every tree was trained using two-thirds of the randomly chosen training samples, while the remaining one-third of the training samples, called out-of-bag (OOB), was utilized to verify the prediction result. Finally, a pixel was assigned to a class using the majority vote or mode rule [109]. In this research, this model has employed the "randomForest" package in R-studio.

Table 2 lists the three significant parameters: the number of features that are appropriate for dividing (mtry), the minimum number of a sample can also be taken arbitrarily in each bootstrap sample to balance any tree with recursive portioning (ntree) [110].

Table 2. Parameters used in RF model.

Parameters	Values
Node size	14
mtry	10
ntree	500

2.7. XGBoost

According to Stanford statistics professor Fridman, the gradient boosting algorithm was developed in 2001 to estimate gradient descent approaches [111]. As the supervised classification model in this work, the XGBoost approach [112] was applied. The approach was invented by the gradient tree boosting algorithm [113,114], which is a powerful machine learning approach. It employs the regularized boosting strategy to prevent overfitting and improve model precision. XGBoost provides scalability for various scenarios, sparse data handling, thorough documentation, minimal computing resource requirements, good performance (i.e., speed), and easy implementation [112]. The approach was chosen since it has won several data science contests [112]. Further adjustments to the approach are needed for extremely unbalanced datasets (e.g., [115]).

Algorithms that boost or lift data are known as "lifting tree models" or "XGBoost". Their key innovations are summarized below [113].

I. They optimize their loss function.
II. The candidate split value may be quickly and accurately generated using the parallel approximation histogram method.
III. In addition to a novel sparsity-aware linear tree learning algorithm, they offer an efficient cache-aware block structure for out-of-core tree learning.

In this research, this model has employed the "XGBoost" package in R-studio. Several model preview parameters must be selected for the XGBoost model. User-friendly settings are needed for three of the most important ones: colsample_bytree (column ratio subsamples when each tree is constructed), nrounds (maximum number of iterations boosting), and subsamples (the training instance subsample ratio); (Table 3).

Table 3. Parameters used in XGBoost model.

Parameters	Values
nround	210
colsample_bytree	1
subsample	1
max_depth	6
eta	0.05
gamma	0

2.8. KNN

The KNN algorithm is one of the most fundamental machine learning techniques. It has recently been used in several other disciplines, including LSM [116,117]. KNN uses the k nearest training examples in the components space as input. When it comes to classification difficulties, class membership possibilities describe the degree of uncertainty with which a particular given item may be assigned to any given class [118]. The attributes of the nearby data points are used to classify a data point using a KNN algorithm [119]. It is a more effective version of the ball tree idea [120] that may be used in bigger dimensions. The approach is commonly employed in SM applications [116], and the categorization of a data point's nearest neighbors determines the chance of it being assigned to any class [118]. The data point chooses the categorization that classifies the greatest number of neighbors. The number of K will be determined through a tuning procedure to obtain better outcomes.

According to Chen et al. [121], they propose that in KNN, objects are evaluated based on the opinions of a majority of their immediate neighbors. The highest consistent closeness of its adjacent neighbors is used to assign the item. If k = 1, the object is solely transferred to the single contiguous neighbor's class.

2.9. NB

NB is a statistical classifier predicated on the Bayesian principle [122]. The Bayes theorem enables this methodology, which is a classification method. The NB maintains that each attribute impacts classification outcomes individually to make estimating the posterior likelihood of observed instances in training data easier [123]. The conditional self-reliance assumption holds that all variables are completely self-sufficient of one another given the output class [124].

The NB technique's most notable benefits include its robustness to noise and irrelevant variables, ease of use, and lack of reliance on time-consuming iterative procedures [125]. Numerous studies have used the NB approach for LSM [36,37,126]. The following equation can be used to estimate the spatial prediction of landslides using NB:

$$yNB = P(yi)\prod_{i=1}^{n} P\left(\frac{xi}{yi}\right) \tag{1}$$

where $P(xi/yi)$ is the conditional probability of each attribute and $P(yi)$ is the prior probability of target class yi (landslide).

2.10. PS-InSAR

PS-InSAR is an enhanced InSAR technology designed for gradual deformation monitoring or long-term displacement. InSAR is a time series-based method that is broadly classified into two classes: small baseline (SBAS) approaches that focus on spatial correlation and dispersed scattering and PS-InSAR techniques that work on the locations of persistent scatterers (PS) [127]. PSI is a multi-interferometric SAR technique that can estimate ground movement with millimeter precision [128]. The PS-InSAR process uses multitemporal SAR images wrapping the same region to analyze the consistency of the phase and amplitude, which identifies the pixels that are less influenced by spatiotemporal decorrelation and then determines specific deformation details on the constituents of the phase which must be collectively evaluated and modeled to eliminate inconsistencies [129].

We employed Sentinel-1 C-band SAR pictures recorded along both ascending and descending orbit tracks in this investigation. To complete the analysis in C-band data, the PSI [68] requires at least 20 SAR pictures [130]. The PSI monitors surface displacement over months or years, accounting for signal noise, atmospheric, and topographic impacts. This sensor has a ground resolution of around 20 m in the azimuth direction and 5 m in the range direction [131]. This sensor has several acquisition modes, including interferometric wide (IW), wave (Wave), extra-wide swath (EW), and strip map (SM). This study gathered images from the Sentinel-1A IW sensor and analyzed them in SARPROZ software (12 days of temporal resolution). The line of sight (LOS) displacement velocity (V_{LOS}) was determined using 0.7 as the coherence threshold in PS-InSAR processing, as shown in Table 4. The InSAR approach computes surface deformation values along the LOS; however, the deformation rate in the LOS direction is inadequate for representing the actual slope displacement [71]. The following equation was used to determine slope velocity (Vslope), which is actually deformation velocity [132]:

$$\text{Vslope} = \frac{\text{VLOS}}{\cos \varnothing} \tag{2}$$

where VLOS is deformation and Ø is the incident angle.

Table 4. Details of PS-InSAR processing.

Specification	Ascending	Descending
Temporal range	1 May 2020–20 May 2021	14 May 2020–9 May 2021
No. of images	33	31
No. of PS/DS	526,815	450,990
Minimum VLOS (mm/year)	−98	−34
Maximum VLOS (mm/year)	31	73

Finally, the calculated result was used for comparative analysis with susceptibility models generated by RF, XGBoost, KNN, and NB methods. The Vslope points were converted into 12.5 × 12.5 grid cells to provide a more precise LSM-like ML model and integrated to enhance the susceptibility degree of those cells defined by ground deformation, minimizing missed alerts, while cells stable and consisting of high susceptibility degrees according to SAR interferometry were not altered [128].

3. Results

3.1. The Significance of Landslide Variables

To compute the significance of the landslide variables in this study, we utilized R-Studio Software. In comparison, the RF model performed better in estimating the relevance of each element in causing landslides.

Figure 8, using origin software, depicts the significance of the factors using the RF model. The slope and elevation had the greatest impact, according to Figure 8, and profile curvature, roughness, distance to fault, and NDVI were almost equal on landslides in the research region. The slope is critical for landslides in the region (Figure 8); it encourages landslides and makes an area susceptible to landslides. Weathered rocks and medium height frequently define high elevation zones, and slopes are usually overlaid by thin colluvium, making them more prone to landslides [112]. The barren ground is in close contact with climatological factors such as sunlight and precipitation, causing rock deterioration and increasing the likelihood of landslides [133]. Because shear zones and active faults strongly influence landslide activities in the region, the buffer class nearest to the fault line is more susceptible [14]. The bulk of the debris flow, rockfalls, and other slides in the area are caused by monsoon rains [134]. Annual average precipitation data were utilized in this study, which found that while precipitation was not a significant causal factor, there were more landslides in locations with high precipitation (Figure 8). The aspect and plan curvature had a minor impact on the landslide in the studied region. The majority of landslides in the research area are northward facing and south-facing. Arabameri et al. [135] employed RF models for LSM in Iran and found that aspect has a minor impact on LSM. The Komila amphibolite and Gilgit complex metasedimentary rocks are the most vulnerable formations in the study area [11,12,14,101]. The rocks in research area are highly fractured and deformed.

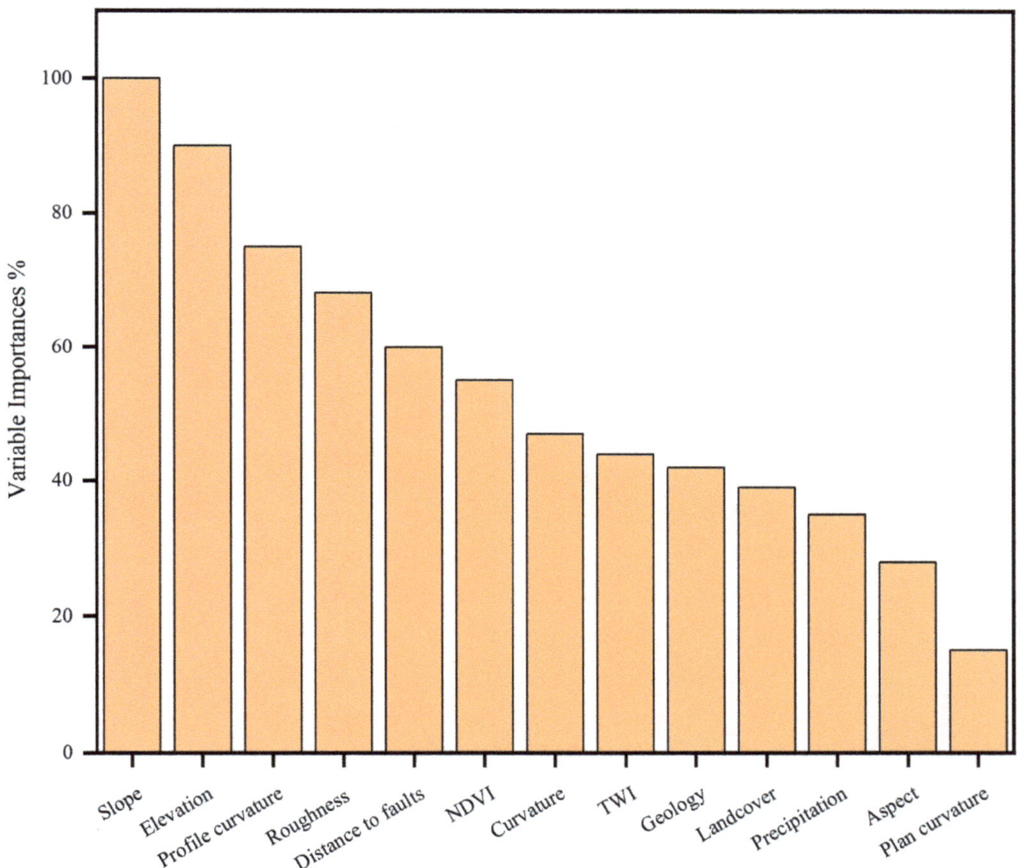

Figure 8. Factors important in the research region using the RF model.

The outcomes of employing the four LSM models obtained using the LPI are depicted in Figure 9. The greater the LPI, the more probable it is that a landslide may happen [136]. The likelihood value of LS was categorized into five classes using the natural break (Jenks) [102] method: very high, high, moderate, low, and very low (Figure 10).

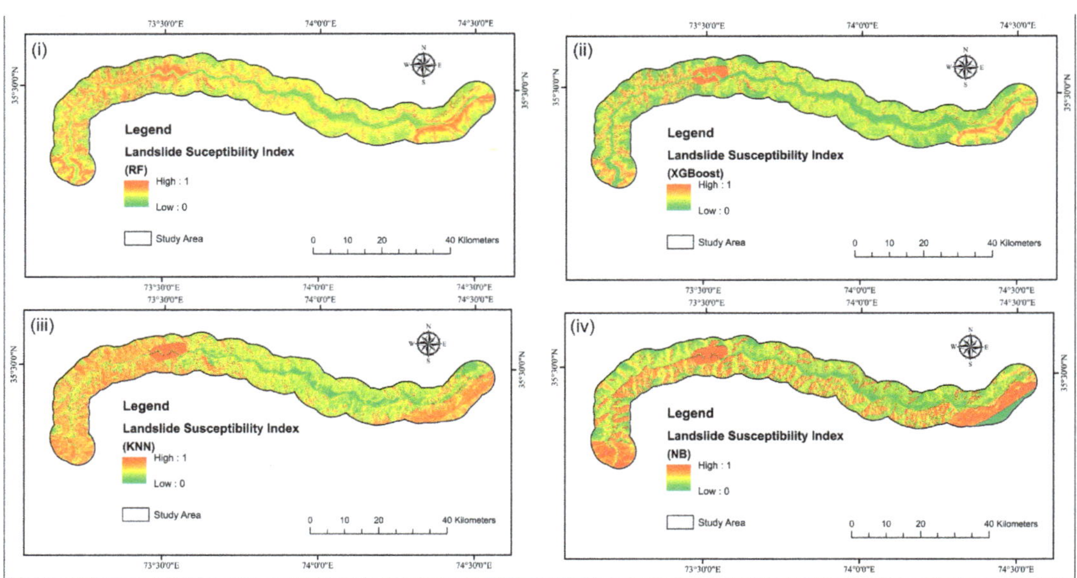

Figure 9. Landslide susceptibility index maps (**i**) RF, (**ii**) XGBoost, (**iii**) KNN, (**iv**) NB.

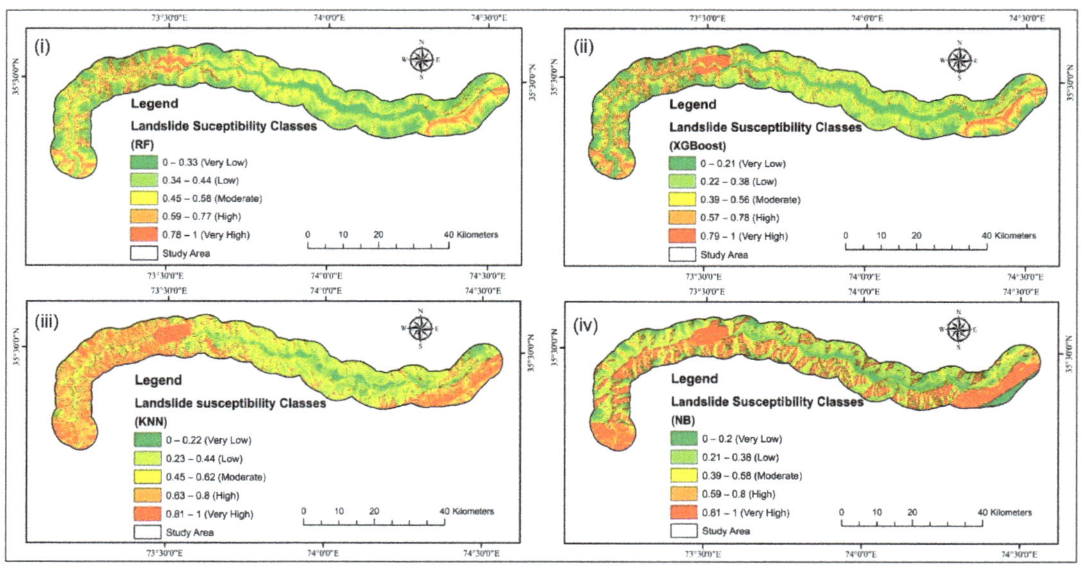

Figure 10. LSM using (**i**) RF, (**ii**) XGBoost, (**iii**) KNN, and (**iv**) NB models.

The precision of the maps was evaluated using a confusion matrix, as suggested by [137]. A confusion matrix illustrates the capabilities of the RF, XGBoost, KNN, and NB models during the training stage (Table 5). The RF model shows a high accuracy (0.830) in the research area. Validation was accomplished using the ROC approach [36]. A ROC curve is created in this approach by plotting "sensitivity" versus "specificity" on cut-off values, but it does not fully explain the model's efficiency; so, the AUC of the ROC curve was utilized to analyze the quantitative functioning of the models [138]. A larger proportion of the area below the curve suggests that the model is more accurate. In contrast, a smaller percentage of the area below the curve shows that the model is less accurate in predicting future occurrences of the phenomena [139]. The AUC of the prediction rate curve was determined to be 88.83, 87.44, 83.38, and 72.80% for the RF, XGBoost, KNN, and NB models, respectively (Figure 11).

Table 5. Confusion matrix of models.

Models	Observation	Predicted		Accuracy
		No	Yes	
RF	No	35	12	0.830
	Yes	43	235	
XGBoost	No	33	13	0.821
	Yes	45	234	
KNN	No	32	18	0.803
	Yes	46	229	
NB	No	39	49	0.729
	Yes	39	198	

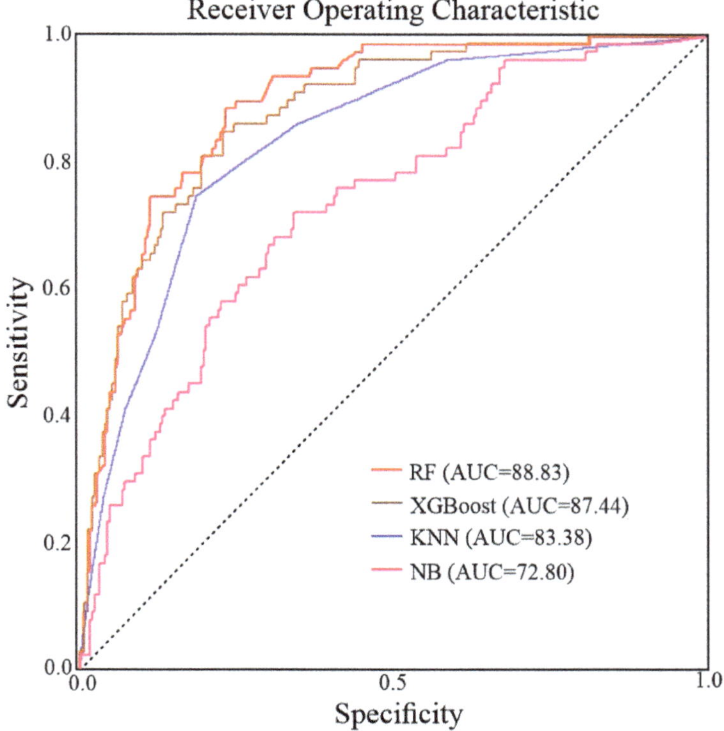

Figure 11. Receiver operating characteristic plots of models.

3.2. PS-InSAR Based Validation

PS-InSAR approaches were utilized to evaluate and verify the models by checking the displacement in the area. The Interferometric Synthetic Aperture Radar (InSAR) approach has been well documented for identifying and tracking mass movements during the previous decade due to its extensive high spatial–temporal resolution, spatial coverage, and operating capacity in all-weather conditions [93]. Many PS-InSAR studies have been conducted to determine the temporal or spatial landslide deformation patterns or the kinematic resolution of slow-moving landslides to quantify the scale of slow-moving landslides [140]. The estimated result was compared to the RF model's susceptibility model.

The line of sight (LOS) displacement velocity (V_{LOS}) was determined using 0.7 as the coherence threshold in PS-InSAR processing (Figure 12). PS-InSAR was also shown to be a useful technique for monitoring slow landslide movement in non-vegetation regions. The InSAR approach computes surface deformation values along the LOS; however, the deformation rate in the LOS direction is inadequate for representing the actual slope displacement [71].

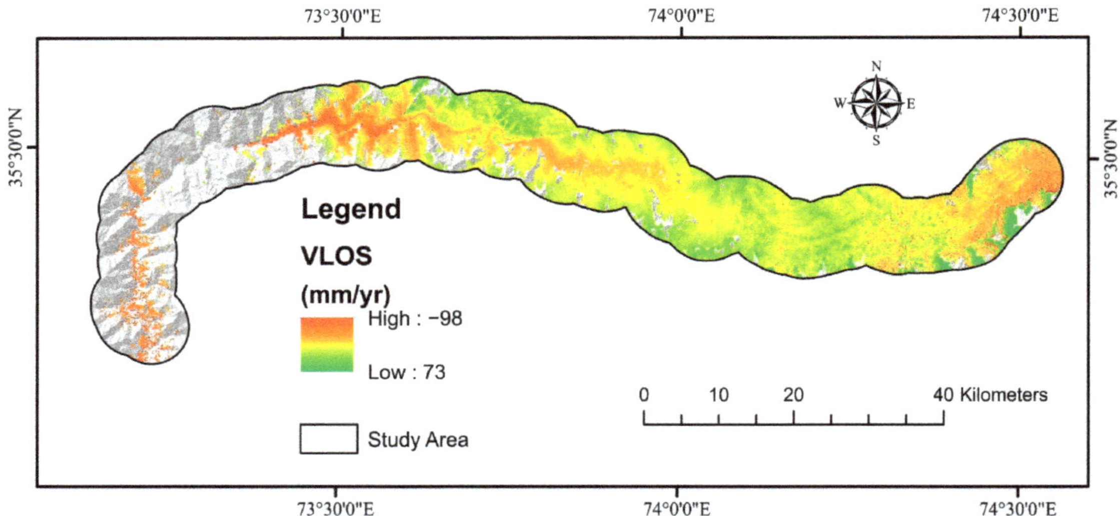

Figure 12. Landslide deformation velocity map along LOS direction for ascending and descending paths using PS-InSAR.

PS-InSAR analysis was performed for both descending and ascending geometries, with VLOS approved deformation in the region. The total number of PS/DS target points acquired with LOS direction deformation results varying from −98 to 73 mm/year was obtained. Using the transformation formula, the VLOS was changed to Vslope. The greatest slope deformation velocity was determined to be −100 mm/year. VLOS indicates just one direction's deformation based on the satellite's LOS, which is determined to evaluate slope orientation velocity (Vslope). Because most landslides or ground surface displacements occur along the direction of steep terrain in the event of landslide assertion, Vslope is the main ingredient employed to define landslide advancement. The calculated Vslope for ascending and descending pathways was added together (Figure 13). The only displacements in RF's highly sensitive zone-produced susceptible model were depicted in an ultimate deformation map (Figure 14). The PSI findings revealed that most of the mapped landslides were in deforming zones, although slow-moving landslides were forecasted more precisely because of the Sentinel-1A sensor's extended revisiting period.

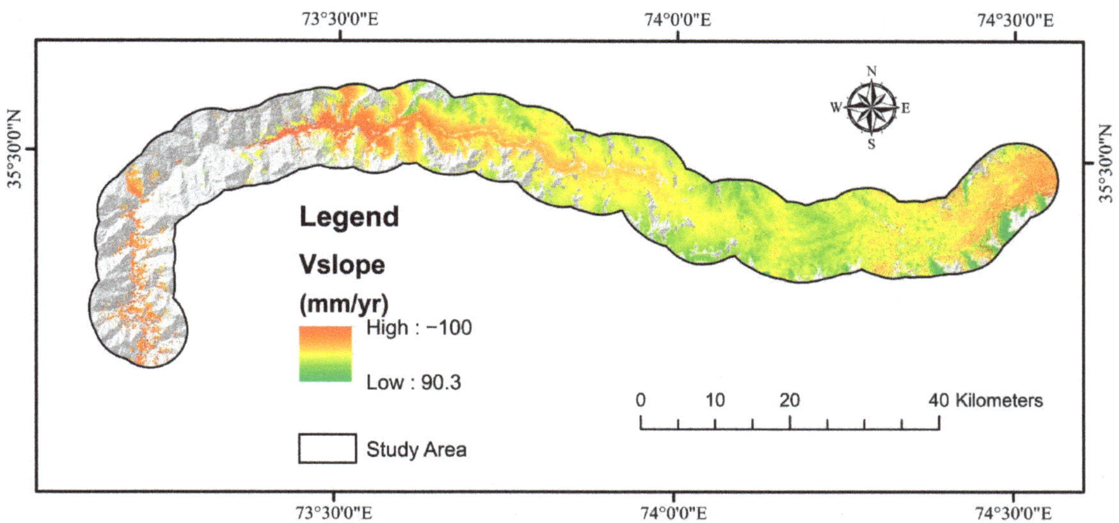

Figure 13. Showing the deformation velocity along slope direction for both ascending and descending paths using PS-InSAR.

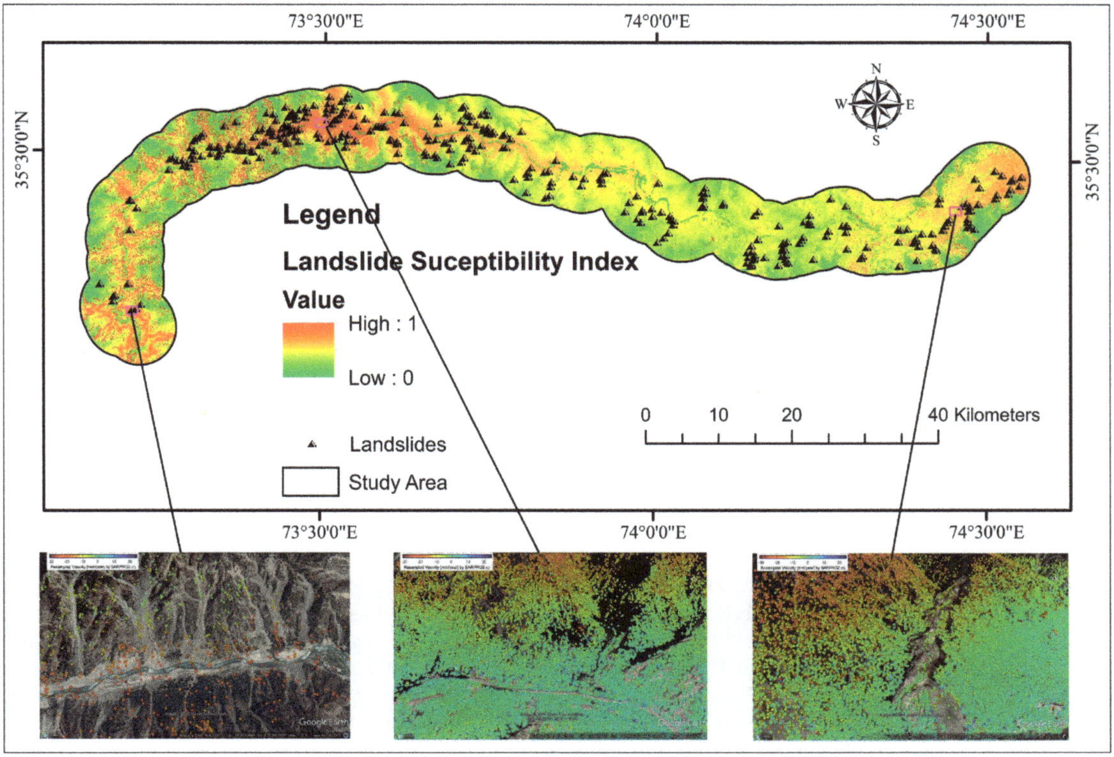

Figure 14. PSI distribution of LOS deformation velocity on Google Earth using RF landslide susceptibility index.

Finally, the RF-based LSM was combined with Vslope to improve the region's precise susceptibility map. The Vslope points were converted into 12.5 × 12.5 grid cells to provide a more precise LSM-like RF model and integrated to enhance the susceptibility degree of those cells defined by ground deformation, minimizing missed alerts, while cells stable and consisting of high susceptibility degrees according to SAR interferometry were not altered [128]. The contingency matrix was used to improve an LSM for the region to a Vslope and RF-based susceptibility model (Figure 15). In other words, the degree of difference for each cell was evaluated using the newly created LSM, which was generated using the RF model.

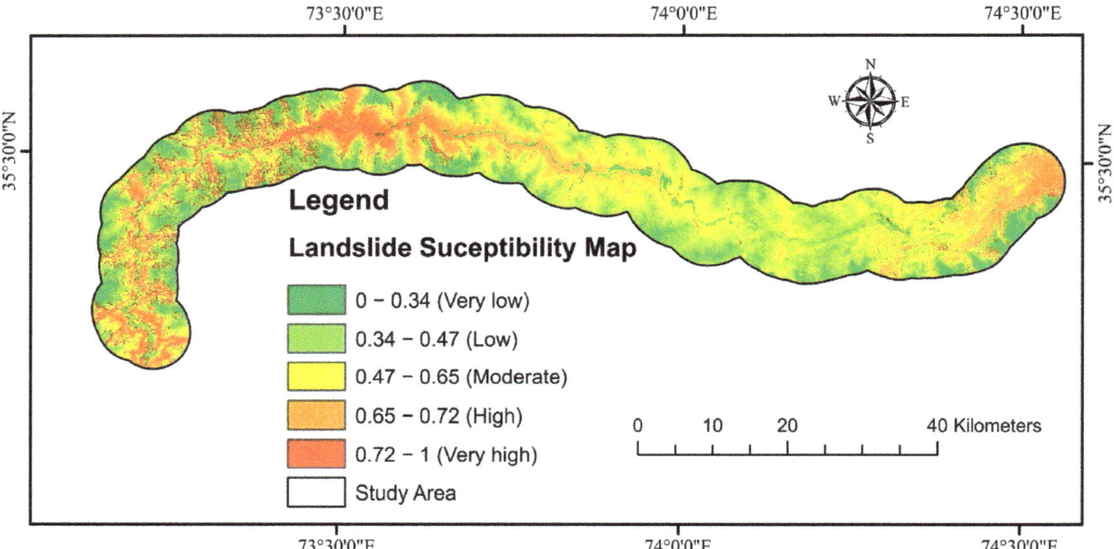

Figure 15. Final LSM via Vslope.

4. Discussion

The findings demonstrate that the RF, XGBoost, KNN, and NB data-mining techniques have comparable precision for LSM along the KKH, with the RF outperforming the others in terms of AUC value and accuracy. Our results conform to the consequences outlined in other work [11,12,14,52,101].

The overall precision values found in this research (RF, 83.0%) were compared to Youssef et al. [23]; it was discovered that the precision values found in this study were higher than the RF (81.2%) in other ML models of the revealed research. Sevgen et al. [29] compared ANN, logistic regression, and RF for LSM and found that the RF model shows the best classification precision with respect to ANN and LR. Taalab et al. [26] evaluated the RF algorithm for landslide in northwest Italy and found that the RF model performed well compared to other tree-based models. Chen et al. [141] reassembled the random forest (RF), logistic model tree (LMT), and classification and regression tree (CART) models to map LS. The LMT (74%) and CART (73%) models showed slightly lower precision values than the RF model (77%); RF performed better in LSM. Zhang et al. [142] demonstrated that the random RF model outperformed the C5.0 decision tree model by comparing it to the C5.0 decision tree model. The RF technique has an advantage over other ML models. It can use multiple input parameters without removing them and provide a limited number of classes with good forecast precision [143]. This model's categorization precision is determined by the training dataset's type, scale, number, and accuracy. The combination of all appropriate parameters boosts the precision of this model. Furthermore, compared to other models, RF has a greater capacity to implement a large number of data [144]. Arabameri et al. [135]

employed RF models for spatial modeling of gully erosion in Iran and found that RF performed best. Zhang et al. [145], when compared to neural networks, decision trees, and the RF, obtained the best results for debris flow susceptibility with the RF method in Shigatse Area, China. In areas with debris flow and rockfalls, we discovered that the RF model is the best predictive technique for LSM.

Recently, InSAR approaches for producing and updating landslide inventories have been created [146]. The findings of the InSAR techniques are thought to be more precise [147], yielding susceptibility maps with high production precision. The LOS velocity statistics only reveal the velocities along the slopes in the highly sensitive zones of both models. PS-InSAR was also shown to be a useful technique for monitoring slow landslide movement in non-vegetation regions. The ROC curve AUC was used to evaluate prediction capabilities, and it predicts 88.83, 87.44, 83.38, and 72.80% for RF, XGBoost, KNN, and NB, respectively, confirming the model's accuracy. The collected susceptibility map was categorized into five groups using the Jenks natural breaks [102]: very low, low, moderate, high, and very high. In comparison, the RF model performed better in estimating the relevance of each element in causing landslides.

LSM was performed using the RF, XGBoost, KNN, and NB in this work. Nonetheless, several limitations caused misclassification in the results, such as (1) the accuracy of the landslide inventory and (2) the accuracy of data connected to each landslide variable. Because of the severe environment along KKH, only 332 landslides have been mapped for this research region. It resulted in considerable misclassification inaccuracies for the LSM, emphasizing the significance of upgrading the LSM utilizing PS-InSAR results. Surprisingly, when paired with the PS-InSAR data, the novel LSM reduced misclassification in which landscape altered by slope deformity was categorized as extremely low and very low.

Another problem is that the landslide susceptibility mapping merely represents anticipated landslide dispersion in regions rather than interactive displacement processes through time. Variations in landslide behavior over time, on the other hand, are a serious challenge for decision-makers [148]. In conjunction with the PS-InSAR outputs, a new landslide susceptibility map can depict the real conditions of landslides. It can be designed for quantitative hazard assessment and preliminary landslide mapping at the province level [149].

The LSM creates a susceptibility map for landslides, identifies the important variables that cause landslides, and evaluates the effect and their contribution [27,150]. Land cover, geology, slope, precipitation, NDVI, distance to faults, elevation, curvature, plan curvature, TWI, profile curvature, roughness, and aspect were all utilized to estimate the probability of landslides disaster in the research area. The main contributors of landslides in the area are slope, elevation. The slope is critical for landslides in the region (Figure 8); it encourages landslides and makes an area susceptible to landslides. Weathered rocks and medium height frequently define high elevation zones, and slopes are usually overlaid by thin colluvium, making them more prone to landslides [112]. Because shear zones and active faults strongly influence landslide activities in the region, the buffer class nearest to the fault line is more susceptible [14].

Previous studies such as [13] in this area relied mainly on statistical models, and a considerable number of landslides were missing in the inventory. As a result of the inadequate landslide inventory, the LSM is ineffective. This work focused on complete mapping of landslides to identify primary landslide triggers and define high susceptible zones using the PS-InSAR approach, which will be used in the future to mitigate landslide risks in the region.

5. Conclusions

Landslides are one of Pakistan's most devastating natural disasters, generating major risks to lives and socioeconomic damage each year. So far, the process of landslide mapping has been highly difficult and volatile to perform correct and quick estimation of landslides in most places. Multiple attempts have been made to improve reliability based on many forecast models for mapping the landslide susceptibility, targeting different locations for this goal. Decision-makers must construct more relevant landslide susceptibility maps to improve the prediction model's performance. This study concentrated on complete landslide mapping to determine the fundamental causes of landslides and designate high-risk zones, which will be useful in the future to mitigate landslide threats in the region. The study's distinctive feature is that it provides more accurate LSM by employing ML models verified by PS-InSAR processing.

This study used RF, XGBoost, KNN, and NB ML algorithms to enhance the LSM of the Karakoram Highway using the PS-InSAR approach. This study assessed the vulnerability using elevation, precipitation, slope, land cover, roughness, NDVI, curvature, distance to faults, plan curvature, aspect, profile curvature, geology, and TWI. Slope, elevation, and profile curvature are the primary causes of landslides in the region. The susceptibility model created will be used to identify zones for construction growth and improved management planning along the KKH. The LSM illustrates the just forecast landslide distribution in regions, not the dynamic displacement process over time. Variations in landslide activity eventually, on the other hand, are a main consideration for decision-makers. The newly developed LSM, when merged with the PS-InSAR results, may show the true situation of landslides and should be utilized for quantitative hazard analysis and preparatory landslide mapping at the regional level. Geotechnical and other slope stabilization procedures are necessary to minimize future landslide catastrophes in an environment. We conclude that our approach can give valuable insights into highway safety measures.

Author Contributions: Conceptualization, M.A.H.; methodology, M.A.H. and Y.Z.; software, M.A.H. and Y.Z.; validation, M.S. and Z.A.; formal analysis, S.U.S.; investigation, N.A.; resources, Z.C.; data curation, M.A.H.; writing—original draft preparation, M.A.H.; writing—review and editing, S.U.S.; visualization, M.S.; supervision, Z.C.; project administration, Z.C.; funding acquisition, Z.C. All authors have read and agreed to the published version of the manuscript.

Funding: This research was funded by the National Natural Science Foundation of China (No. 41871305); National key R & D program of China (No.2017YFC0602204); Fundamental Research Funds for the Central Universities, China University of Geo-sciences (Wuhan) (No. CUGQY1945); Opening Fund of Key Laboratory of Geological Survey and Evaluation of Ministry of Education; and Fundamental Research Funds for the Central Universities (No. GLAB2019ZR02).

Institutional Review Board Statement: Not applicable.

Informed Consent Statement: Not applicable.

Data Availability Statement: The data presented in the study are available on request from the first and corresponding author. The data are not publicly available due to the thesis that is being prepared from these data.

Acknowledgments: The author appreciates Javaid Iqbal for his recommendations and suggestions during the writing of the manuscript. Additionally, we are thankful to Muzamil Hassan for providing Sarproz Software for the analysis.

Conflicts of Interest: The authors declare no conflict of interest.

References

1. Reichenbach, P.; Rossi, M.; Malamud, B.D.; Mihir, M.; Guzzetti, F. A review of statistically-based landslide susceptibility models. *Earth-Sci. Rev.* **2018**, *180*, 60–91. [CrossRef]
2. Zhou, C.; Yin, K.; Cao, Y.; Ahmed, B.; Li, Y.; Catani, F.; Pourghasemi, H.R. Landslide susceptibility modeling applying machine learning methods: A case study from Longju in the Three Gorges Reservoir area, China. *Comput. Geosci.* **2018**, *112*, 23–37. [CrossRef]
3. Guzzetti, F.; Carrara, A.; Cardinali, M.; Reichenbach, P. Landslide hazard evaluation: A review of current techniques and their application in a multi-scale study, Central Italy. *Geomorphology* **1999**, *31*, 181–216. [CrossRef]
4. Wu, Y.; Li, W.; Liu, P.; Bai, H.; Wang, Q.; He, J.; Liu, Y.; Sun, S. Application of analytic hierarchy process model for landslide susceptibility mapping in the Gangu County, Gansu Province, China. *Environ. Earth Sci.* **2016**, *75*, 422. [CrossRef]
5. Wang, Q.; Li, W. A GIS-based comparative evaluation of analytical hierarchy process and frequency ratio models for landslide susceptibility mapping. *Phys. Geogr.* **2017**, *38*, 318–337. [CrossRef]

6. Nguyen, T.T.N.; Liu, C.-C. A new approach using AHP to generate landslide susceptibility maps in the Chen-Yu-Lan Watershed, Taiwan. *Sensors* **2019**, *19*, 505. [CrossRef]
7. Neuhäuser, B.; Damm, B.; Terhorst, B. GIS-based assessment of landslide susceptibility on the base of the weights-of-evidence model. *Landslides* **2012**, *9*, 511–528. [CrossRef]
8. Razavizadeh, S.; Solaimani, K.; Massironi, M.; Kavian, A. Mapping landslide susceptibility with frequency ratio, statistical index, and weights of evidence models: A case study in northern Iran. *Environ. Earth Sci.* **2017**, *76*, 499. [CrossRef]
9. Khan, H.; Shafique, M.; Khan, M.A.; Bacha, M.A.; Shah, S.U.; Calligaris, C. Landslide susceptibility assessment using Frequency Ratio, a case study of northern Pakistan. *Egypt. J. Remote Sens. Space Sci.* **2019**, *22*, 11–24. [CrossRef]
10. Rossi, M.; Guzzetti, F.; Reichenbach, P.; Mondini, A.C.; Peruccacci, S. Optimal landslide susceptibility zonation based on multiple forecasts. *Geomorphology* **2010**, *114*, 129–142. [CrossRef]
11. Hussain, M.L.; Shafique, M.; Bacha, A.S.; Chen, X.-Q.; Chen, H.-Y. Landslide inventory and susceptibility assessment using multiple statistical approaches along the Karakoram highway, northern Pakistan. *J. Mt. Sci.* **2021**, *18*, 583–598. [CrossRef]
12. Iqbal, J.; Cui, P.; Hussain, M.L.; Pourghasemi, H.R.; Pradhan, B. Landslide susceptibility assessment along the dubair-dud ishal section of the karakoram highway, northwestern himalayas, pakistan. *Acta Geodyn. Geomater* **2021**, *18*, 137–155. [CrossRef]
13. Rashid, B.; Iqbal, J.; Su, L.-J. Landslide susceptibility analysis of Karakoram highway using analytical hierarchy process and scoops 3D. *J. Mt. Sci.* **2020**, *17*, 1596–1612. [CrossRef]
14. Ali, S.; Biermanns, P.; Haider, R.; Reicherter, K. Landslide susceptibility mapping by using a geographic information system (GIS) along the China–Pakistan Economic Corridor (Karakoram Highway), Pakistan. *Nat. Hazards Earth Syst. Sci.* **2019**, *19*, 999–1022. [CrossRef]
15. Regmi, A.D.; Devkota, K.C.; Yoshida, K.; Pradhan, B.; Pourghasemi, H.R.; Kumamoto, T.; Akgun, A. Application of frequency ratio, statistical index, and weights-of-evidence models and their comparison in landslide susceptibility mapping in Central Nepal Himalaya. *Arab. J. Geosci.* **2014**, *7*, 725–742. [CrossRef]
16. Shirzadi, A.; Chapi, K.; Shahabi, H.; Solaimani, K.; Kavian, A.; Ahmad, B.B. Rock fall susceptibility assessment along a mountainous road: An evaluation of bivariate statistic, analytical hierarchy process and frequency ratio. *Environ. Earth Sci.* **2017**, *76*, 152. [CrossRef]
17. Achour, Y.; Pourghasemi, H.R. How do machine learning techniques help in increasing accuracy of landslide susceptibility maps? *Geosci. Front.* **2020**, *11*, 871–883. [CrossRef]
18. Kumar, R.; Anbalagan, R. Landslide susceptibility mapping using analytical hierarchy process (AHP) in Tehri reservoir rim region, Uttarakhand. *J. Geol. Soc. India* **2016**, *87*, 271–286. [CrossRef]
19. Tien Bui, D.; Pradhan, B.; Lofman, O.; Revhaug, I. Landslide susceptibility assessment in vietnam using support vector machines, decision tree, and Naive Bayes Models. *Math. Probl. Eng.* **2012**, *2012*, 974638. [CrossRef]
20. Park, S.; Choi, C.; Kim, B.; Kim, J. Landslide susceptibility mapping using frequency ratio, analytic hierarchy process, logistic regression, and artificial neural network methods at the Inje area, Korea. *Environ. Earth Sci.* **2013**, *68*, 1443–1464. [CrossRef]
21. Tengtrairat, N.; Woo, W.L.; Parathai, P.; Aryupong, C.; Jitsangiam, P.; Rinchumphu, D. Automated landslide-risk prediction using web gis and machine learning models. *Sensors* **2021**, *21*, 4620. [CrossRef] [PubMed]
22. Mandal, S.; Mandal, K. Modeling and mapping landslide susceptibility zones using GIS based multivariate binary logistic regression (LR) model in the Rorachu river basin of eastern Sikkim Himalaya, India. *Modeling Earth Syst. Environ.* **2018**, *4*, 69–88. [CrossRef]
23. Youssef, A.M.; Pourghasemi, H.R.; Pourtaghi, Z.S.; Al-Katheeri, M.M. Landslide susceptibility mapping using random forest, boosted regression tree, classification and regression tree, and general linear models and comparison of their performance at Wadi Tayyah Basin, Asir Region, Saudi Arabia. *Landslides* **2016**, *13*, 839–856. [CrossRef]
24. Pourghasemi, H.R.; Rahmati, O. Prediction of the landslide susceptibility: Which algorithm, which precision? *Catena* **2018**, *162*, 177–192. [CrossRef]
25. Park, S.; Kim, J. Landslide susceptibility mapping based on random forest and boosted regression tree models, and a comparison of their performance. *Appl. Sci.* **2019**, *9*, 942. [CrossRef]
26. Taalab, K.; Cheng, T.; Zhang, Y. Mapping landslide susceptibility and types using Random Forest. *Big Earth Data* **2018**, *2*, 159–178. [CrossRef]
27. Hussain, M.A.; Chen, Z.; Wang, R.; Shoaib, M. PS-InSAR-Based Validated Landslide Susceptibility Mapping along Karakorum Highway, Pakistan. *Remote Sens.* **2021**, *13*, 4129. [CrossRef]
28. Hussain, M.A.; Chen, Z.; Wang, R.; Shah, S.U.; Shoaib, M.; Ali, N.; Xu, D.; Ma, C. Landslide Susceptibility Mapping using Machine Learning Algorithm. *Civ. Eng. J.* **2022**, *8*, 209–224. [CrossRef]
29. Sevgen, E.; Kocaman, S.; Nefeslioglu, H.A.; Gokceoglu, C.J.S. A novel performance assessment approach using photogrammetric techniques for landslide susceptibility mapping with logistic regression, ANN and random forest. *Sensors* **2019**, *19*, 3940. [CrossRef]
30. Felicísimo, Á.M.; Cuartero, A.; Remondo, J.; Quirós, E. Mapping landslide susceptibility with logistic regression, multiple adaptive regression splines, classification and regression trees, and maximum entropy methods: A comparative study. *Landslides* **2013**, *10*, 175–189. [CrossRef]

31. Conoscenti, C.; Ciaccio, M.; Caraballo-Arias, N.A.; Gómez-Gutiérrez, Á.; Rotigliano, E.; Agnesi, V. Assessment of susceptibility to earth-flow landslide using logistic regression and multivariate adaptive regression splines: A case of the Belice River basin (western Sicily, Italy). *Geomorphology* **2015**, *242*, 49–64. [CrossRef]
32. Vorpahl, P.; Elsenbeer, H.; Märker, M.; Schröder, B. How can statistical models help to determine driving factors of landslides? *Ecol. Model.* **2012**, *239*, 27–39. [CrossRef]
33. Kalantar, B.; Pradhan, B.; Naghibi, S.A.; Motevalli, A.; Mansor, S. Assessment of the effects of training data selection on the landslide susceptibility mapping: A comparison between support vector machine (SVM), logistic regression (LR) and artificial neural networks (ANN). *Geomat. Nat. Hazards Risk* **2018**, *9*, 49–69. [CrossRef]
34. Ma, J.; Wang, Y.; Niu, X.; Jiang, S.; Liu, Z. A comparative study of mutual information-based input variable selection strategies for the displacement prediction of seepage-driven landslides using optimized support vector regression. *Stoch. Environ. Res. Risk Assess.* **2022**, *1*, 1–21. [CrossRef]
35. Ghasemian, B.; Shahabi, H.; Shirzadi, A.; Al-Ansari, N.; Jaafari, A.; Kress, V.R.; Geertsema, M.; Renoud, S.; Ahmad, A. A robust deep-learning model for landslide susceptibility mapping: A case study of Kurdistan Province, Iran. *Sensors* **2022**, *22*, 1573. [CrossRef]
36. Pham, B.T.; Pradhan, B.; Bui, D.T.; Prakash, I.; Dholakia, M. A comparative study of different machine learning methods for landslide susceptibility assessment: A case study of Uttarakhand area (India). *Environ. Model. Softw.* **2016**, *84*, 240–250. [CrossRef]
37. Pham, B.T.; Tien Bui, D.; Pourghasemi, H.R.; Indra, P.; Dholakia, M. Landslide susceptibility assesssment in the Uttarakhand area (India) using GIS: A comparison study of prediction capability of naïve bayes, multilayer perceptron neural networks, and functional trees methods. *Theor. Appl. Climatol.* **2017**, *128*, 255–273. [CrossRef]
38. Aghdam, I.N.; Varzandeh, M.H.M.; Pradhan, B. Landslide susceptibility mapping using an ensemble statistical index (Wi) and adaptive neuro-fuzzy inference system (ANFIS) model at Alborz Mountains (Iran). *Environ. Earth Sci.* **2016**, *75*, 553. [CrossRef]
39. Dehnavi, A.; Aghdam, I.N.; Pradhan, B.; Varzandeh, M.H.M. A new hybrid model using step-wise weight assessment ratio analysis (SWARA) technique and adaptive neuro-fuzzy inference system (ANFIS) for regional landslide hazard assessment in Iran. *Catena* **2015**, *135*, 122–148. [CrossRef]
40. Mehrabi, M.; Pradhan, B.; Moayedi, H.; Alamri, A. Optimizing an adaptive neuro-fuzzy inference system for spatial prediction of landslide susceptibility using four state-of-the-art metaheuristic techniques. *Sensors* **2020**, *20*, 1723. [CrossRef]
41. Kumar, R.; Anbalagan, R. Landslide susceptibility zonation in part of Tehri reservoir region using frequency ratio, fuzzy logic and GIS. *J. Earth Syst. Sci.* **2015**, *124*, 431–448. [CrossRef]
42. Aditian, A.; Kubota, T.; Shinohara, Y. Comparison of GIS-based landslide susceptibility models using frequency ratio, logistic regression, and artificial neural network in a tertiary region of Ambon, Indonesia. *Geomorphology* **2018**, *318*, 101–111. [CrossRef]
43. Arnone, E.; Francipane, A.; Scarbaci, A.; Puglisi, C.; Noto, L.V. Effect of raster resolution and polygon-conversion algorithm on landslide susceptibility mapping. *Environ. Model. Softw.* **2016**, *84*, 467–481. [CrossRef]
44. Bui, D.T.; Moayedi, H.; Kalantar, B.; Osouli, A.; Pradhan, B.; Nguyen, H.; Rashid, A.S.A. A novel swarm intelligence—Harris hawks optimization for spatial assessment of landslide susceptibility. *Sensors* **2019**, *19*, 3590. [CrossRef] [PubMed]
45. Moayedi, H.; Osouli, A.; Tien Bui, D.; Foong, L.K. Spatial landslide susceptibility assessment based on novel neural-metaheuristic geographic information system based ensembles. *Sensors* **2019**, *19*, 4698. [CrossRef]
46. Roshani, M.; Sattari, M.A.; Ali, P.J.M.; Roshani, G.H.; Nazemi, B.; Corniani, E.; Nazemi, E. Application of GMDH neural network technique to improve measuring precision of a simplified photon attenuation based two-phase flowmeter. *Flow Meas. Instrum.* **2020**, *75*, 101804. [CrossRef]
47. Charandabi, S.E.; Kamyar, K. Prediction of Cryptocurrency Price Index Using Artificial Neural Networks: A Survey of the Literature. *Eur. J. Bus. Manag. Res.* **2021**, *6*, 17–20. [CrossRef]
48. Park, N.-W. Using maximum entropy modeling for landslide susceptibility mapping with multiple geoenvironmental data sets. *Environ. Earth Sci.* **2015**, *73*, 937–949. [CrossRef]
49. Kornejady, A.; Ownegh, M.; Bahremand, A. Landslide susceptibility assessment using maximum entropy model with two different data sampling methods. *Catena* **2017**, *152*, 144–162. [CrossRef]
50. Wu, X.; Ren, F.; Niu, R. Landslide susceptibility assessment using object mapping units, decision tree, and support vector machine models in the Three Gorges of China. *Environ. Earth Sci.* **2014**, *71*, 4725–4738. [CrossRef]
51. Shirzadi, A.; Soliamani, K.; Habibnejhad, M.; Kavian, A.; Chapi, K.; Shahabi, H.; Chen, W.; Khosravi, K.; Thai Pham, B.; Pradhan, B. Novel GIS based machine learning algorithms for shallow landslide susceptibility mapping. *Sensors* **2018**, *18*, 3777. [CrossRef] [PubMed]
52. Qing, F.; Zhao, Y.; Meng, X.; Su, X.; Qi, T.; Yue, D. Application of Machine Learning to Debris Flow Susceptibility Mapping along the China–Pakistan Karakoram Highway. *Remote Sens.* **2020**, *12*, 2933. [CrossRef]
53. Pradhan, A.M.S.; Kim, Y.-T. Rainfall-induced shallow landslide susceptibility mapping at two adjacent catchments using advanced machine learning algorithms. *ISPRS Int. J. Geo-Inf.* **2020**, *9*, 569. [CrossRef]
54. Merghadi, A.; Yunus, A.P.; Dou, J.; Whiteley, J.; ThaiPham, B.; Bui, D.T.; Avtar, R.; Abderrahmane, B. Machine learning methods for landslide susceptibility studies: A comparative overview of algorithm performance. *Earth-Sci. Rev.* **2020**, *207*, 103225. [CrossRef]
55. Sahin, E.K. Comparative analysis of gradient boosting algorithms for landslide susceptibility mapping. *Geocarto Int.* **2020**, *35*, 1–25. [CrossRef]

56. Yan, F.; Zhang, Q.; Ye, S.; Ren, B. A novel hybrid approach for landslide susceptibility mapping integrating analytical hierarchy process and normalized frequency ratio methods with the cloud model. *Geomorphology* **2019**, *327*, 170–187. [CrossRef]
57. Pourghasemi, H.R.; Gayen, A.; Panahi, M.; Rezaie, F.; Blaschke, T. Multi-hazard probability assessment and mapping in Iran. *Sci. Total Environ.* **2019**, *692*, 556–571. [CrossRef]
58. Nohani, E.; Moharrami, M.; Sharafi, S.; Khosravi, K.; Pradhan, B.; Pham, B.T.; Lee, S.; Melesse, A.M. Landslide susceptibility mapping using different GIS-based bivariate models. *Water* **2019**, *11*, 1402. [CrossRef]
59. Scaioni, M.; Longoni, L.; Melillo, V.; Papini, M. Remote sensing for landslide investigations: An overview of recent achievements and perspectives. *Remote Sens.* **2014**, *6*, 9600–9652. [CrossRef]
60. Corsini, A.; Borgatti, L.; Cervi, F.; Dahne, A.; Ronchetti, F.; Sterzai, P. Estimating mass-wasting processes in active earth slides–earth flows with time-series of High-Resolution DEMs from photogrammetry and airborne LiDAR. *Nat. Hazards Earth Syst. Sci.* **2009**, *9*, 433–439. [CrossRef]
61. Lai, J.-S.; Tsai, F. Improving GIS-based landslide susceptibility assessments with multi-temporal remote sensing and machine learning. *Sensors* **2019**, *19*, 3717. [CrossRef] [PubMed]
62. Schlögel, R.; Doubre, C.; Malet, J.-P.; Masson, F. Landslide deformation monitoring with ALOS/PALSAR imagery: A D-InSAR geomorphological interpretation method. *Geomorphology* **2015**, *231*, 314–330. [CrossRef]
63. Intrieri, E.; Gigli, G.; Mugnai, F.; Fanti, R.; Casagli, N. Design and implementation of a landslide early warning system. *Eng. Geol.* **2012**, *147*, 124–136. [CrossRef]
64. Lotfi, F.; Semiari, O. Performance Analysis and Optimization of Uplink Cellular Networks with Flexible Frame Structure. In Proceedings of the 2021 IEEE 93rd Vehicular Technology Conference (VTC2021-Spring), Helsinki, Finland, 25–28 April 2021; pp. 1–5.
65. Lotfi, F.; Semiari, O.; Saad, W. Semantic-Aware Collaborative Deep Reinforcement Learning Over Wireless Cellular Networks. *arXiv* **2011**, arXiv:2111.12064.
66. Ferretti, A.; Prati, C.; Rocca, F. Nonlinear subsidence rate estimation using permanent scatterers in differential SAR interferometry. *IEEE Trans. Geosci. Remote Sens.* **2000**, *38*, 2202–2212. [CrossRef]
67. Colesanti, C.; Ferretti, A.; Prati, C.; Rocca, F. Monitoring landslides and tectonic motions with the Permanent Scatterers Technique. *Eng. Geol.* **2003**, *68*, 3–14. [CrossRef]
68. Mora, O.; Mallorqui, J.J.; Broquetas, A. Linear and nonlinear terrain deformation maps from a reduced set of interferometric SAR images. *IEEE Trans. Geosci. Remote Sens.* **2003**, *41*, 2243–2253. [CrossRef]
69. Ferretti, A.; Fumagalli, A.; Novali, F.; Prati, C.; Rocca, F.; Rucci, A. A new algorithm for processing interferometric data-stacks: SqueeSAR. *IEEE Trans. Geosci. Remote Sens.* **2011**, *49*, 3460–3470. [CrossRef]
70. Berardino, P.; Fornaro, G.; Lanari, R.; Sansosti, E. A new algorithm for surface deformation monitoring based on small baseline differential SAR interferograms. *IEEE Trans. Geosci. Remote Sens.* **2002**, *40*, 2375–2383. [CrossRef]
71. Zhao, F.; Meng, X.; Zhang, Y.; Chen, G.; Su, X.; Yue, D. Landslide susceptibility mapping of karakorum highway combined with the application of SBAS-InSAR technology. *Sensors* **2019**, *19*, 2685. [CrossRef]
72. Hooper, A.; Segall, P.; Zebker, H. Persistent scatterer interferometric synthetic aperture radar for crustal deformation analysis, with application to Volcán Alcedo, Galápagos. *J. Geophys. Res. Solid Earth* **2007**, *112*. [CrossRef]
73. Crosetto, M.; Biescas, E.; Duro, J.; Closa, J.; Arnaud, A. Generation of advanced ERS and Envisat interferometric SAR products using the stable point network technique. *Photogramm. Eng. Remote Sens.* **2008**, *74*, 443–450. [CrossRef]
74. Herrera, G.; Notti, D.; García-Davalillo, J.C.; Mora, O.; Cooksley, G.; Sánchez, M.; Arnaud, A.; Crosetto, M. Analysis with C-and X-band satellite SAR data of the Portalet landslide area. *Landslides* **2011**, *8*, 195–206. [CrossRef]
75. Strozzi, T.; Wegmuller, U.; Keusen, H.R.; Graf, K.; Wiesmann, A. Analysis of the terrain displacement along a funicular by SAR interferometry. *IEEE Geosci. Remote Sens. Lett.* **2006**, *3*, 15–18. [CrossRef]
76. Lu, P.; Stumpf, A.; Kerle, N.; Casagli, N. Object-oriented change detection for landslide rapid mapping. *IEEE Geosci. Remote Sens. Lett.* **2011**, *8*, 701–705. [CrossRef]
77. Agostini, A.; Tofani, V.; Nolesini, T.; Gigli, G.; Tanteri, L.; Rosi, A.; Cardellini, S.; Casagli, N. A new appraisal of the Ancona landslide based on geotechnical investigations and stability modelling. *Q. J. Eng. Geol. Hydrogeol.* **2014**, *47*, 29–43. [CrossRef]
78. Jones, D.; Brunsden, D.; Goudie, A. A preliminary geomorphological assessment of part of the Karakoram Highway. *Q. J. Eng. Geol.* **1983**, *16*, 331–355. [CrossRef]
79. Bishop, M.P.; Shroder Jr, J.F.; Hickman, B.L.; Copland, L. Scale-dependent analysis of satellite imagery for characterization of glacier surfaces in the Karakoram Himalaya. *Geomorphology* **1998**, *21*, 217–232. [CrossRef]
80. Derbyshire, E.; Fort, M.; Owen, L.A. Geomorphological hazards along the Karakoram highway: Khunjerab pass to the Gilgit River, northernmost Pakistan (Geomorphologische hazards entlang des Karakorum highway: Khunjerab Paß bis zum Gilgit River, nördlichstes Pakistan). *Erdkunde* **2001**, *1*, 49–71. [CrossRef]
81. Korup, O.; Clague, J.J.; Hermanns, R.L.; Hewitt, K.; Strom, A.L.; Weidinger, J.T. Giant landslides, topography, and erosion. *Earth Planet. Sci. Lett.* **2007**, *261*, 578–589. [CrossRef]
82. Rahman, M.; Ahmed, B.; Di, L. Landslide initiation and runout susceptibility modeling in the context of hill cutting and rapid urbanization: A combined approach of weights of evidence and spatial multi-criteria. *J. Mt. Sci.* **2017**, *14*, 1919–1937. [CrossRef]
83. Bacha, A.S.; Shafique, M.; van der Werff, H. Landslide inventory and susceptibility modelling using geospatial tools, in Hunza-Nagar valley, northern Pakistan. *J. Mt. Sci.* **2018**, *15*, 1354–1370. [CrossRef]

84. Ahmed, M.F.; Rogers, J.D.; Ismail, E.H. A regional level preliminary landslide susceptibility study of the upper Indus river basin. *Eur. J. Remote Sens.* **2014**, *47*, 343–373. [CrossRef]
85. Basharat, M.; Shah, H.R.; Hameed, N. Landslide susceptibility mapping using GIS and weighted overlay method: A case study from NW Himalayas, Pakistan. *Arab. J. Geosci.* **2016**, *9*, 292. [CrossRef]
86. Owen, L. Wet-sediment deformation of Quaternary and recent sediments in the Skardu Basin, Karakoram Mountains, Pakistan. In Proceedings of the Glaciotectonics: Forms and Processes; Various Meetings of the Glaciotectonics Work Group. Routledge: Norflok, UK, 1988; pp. 123–147.
87. Hewitt, K. Quaternary moraines vs catastrophic rock avalanches in the Karakoram Himalaya, northern Pakistan. *Quat. Res.* **1999**, *51*, 220–237. [CrossRef]
88. Searle, M.; Khan, M.A.; Fraser, J.; Gough, S.; Jan, M.Q. The tectonic evolution of the Kohistan-Karakoram collision belt along the Karakoram Highway transect, north Pakistan. *Tectonics* **1999**, *18*, 929–949. [CrossRef]
89. Guzzetti, F.; Mondini, A.C.; Cardinali, M.; Fiorucci, F.; Santangelo, M.; Chang, K.-T. Landslide inventory maps: New tools for an old problem. *Earth-Sci. Rev.* **2012**, *112*, 42–66. [CrossRef]
90. Paliaga, G.; Luino, F.; Turconi, L.; Faccini, F. Inventory of geo-hydrological phenomena in Genova municipality (NW Italy). *J. Maps* **2019**, *15*, 28–37. [CrossRef]
91. Shafique, M.; van der Meijde, M.; Khan, M.A. A review of the 2005 Kashmir earthquake-induced landslides; from a remote sensing prospective. *J. Asian Earth Sci.* **2016**, *118*, 68–80. [CrossRef]
92. Ilia, I.; Tsangaratos, P. Applying weight of evidence method and sensitivity analysis to produce a landslide susceptibility map. *Landslides* **2016**, *13*, 379–397. [CrossRef]
93. Chen, W.; Pourghasemi, H.R.; Panahi, M.; Kornejady, A.; Wang, J.; Xie, X.; Cao, S. Spatial prediction of landslide susceptibility using an adaptive neuro-fuzzy inference system combined with frequency ratio, generalized additive model, and support vector machine techniques. *Geomorphology* **2017**, *297*, 69–85. [CrossRef]
94. Riaz, M.T.; Basharat, M.; Hameed, N.; Shafique, M.; Luo, J. A data-driven approach to landslide-susceptibility mapping in mountainous terrain: Case study from the Northwest Himalayas, Pakistan. *Nat. Hazards Rev.* **2018**, *19*, 05018007. [CrossRef]
95. Fayaz, A.; Latif, M.; Khan, K. *Landslide Evaluation and Stabilization Between Gilgit ans Thakot along the Karakoram Highway*; Geological Survey of Pakistan: Islamabad, Pakistan, 1985.
96. Khan, K.; Fayaz, A.; Latif, M.; Wazir, A. *Rock and Debris Slides between Khunjrab Pass and Gilgit along the Karakoram Highway*; Geological Survey of Pakistan: Islamabad, Pakistan, 1986.
97. Khan, K.; Fayaz, A.; Hussain, M.; Latif, M. *Landslides Problems and Their Mitigation along the Karakoram Highway*; Geological Survey of Pakistan: Islamabad, Pakistan, 2003.
98. Hewitt, K. Catastrophic landslides and their effects on the Upper Indus streams, Karakoram Himalaya, northern Pakistan. *Geomorphology* **1998**, *26*, 47–80. [CrossRef]
99. Arabameri, A.; Pradhan, B.; Rezaei, K.; Sohrabi, M.; Kalantari, Z. GIS-based landslide susceptibility mapping using numerical risk factor bivariate model and its ensemble with linear multivariate regression and boosted regression tree algorithms. *J. Mt. Sci.* **2019**, *16*, 595–618. [CrossRef]
100. Ayalew, L.; Yamagishi, H. The application of GIS-based logistic regression for landslide susceptibility mapping in the Kakuda-Yahiko Mountains, Central Japan. *Geomorphology* **2005**, *65*, 15–31. [CrossRef]
101. Hussain, M.A.; Chen, Z.; Kalsoom, I.; Asghar, A.; Shoaib, M. Landslide Susceptibility Mapping Using Machine Learning Algorithm: A Case Study Along Karakoram Highway (KKH), Pakistan. *J. Indian Soc. Remote Sens.* **2022**, *1*, 239. [CrossRef]
102. Roy, J.; Saha, S.; Arabameri, A.; Blaschke, T.; Bui, D.T. A novel ensemble approach for landslide susceptibility mapping (LSM) in Darjeeling and Kalimpong districts, West Bengal, India. *Remote Sens.* **2019**, *11*, 2866. [CrossRef]
103. Breiman, L. Random forests. *Mach. Learn.* **2001**, *45*, 5–32. [CrossRef]
104. Hastie, T.; Tibshirani, R.; Friedman, J.H.; Friedman, J.H. *The Elements of Statistical Learning: Data Mining, Inference, and Prediction*; Springer: New York, NY, USA, 2009; Volume 2, pp. 1–758.
105. Akar, Ö.; Güngör, O. Integrating multiple texture methods and NDVI to the Random Forest classification algorithm to detect tea and hazelnut plantation areas in northeast Turkey. *Int. J. Remote Sens.* **2015**, *36*, 442–464. [CrossRef]
106. Feng, Q.; Liu, J.; Gong, J. Urban flood mapping based on unmanned aerial vehicle remote sensing and random forest classifier—A case of Yuyao, China. *Water* **2015**, *7*, 1437–1455. [CrossRef]
107. Hengl, T.; Heuvelink, G.B.; Kempen, B.; Leenaars, J.G.; Walsh, M.G.; Shepherd, K.D.; Sila, A.; MacMillan, R.A.; Mendes de Jesus, J.; Tamene, L. Mapping soil properties of Africa at 250 m resolution: Random forests significantly improve current predictions. *PLoS ONE* **2015**, *10*, e0125814. [CrossRef] [PubMed]
108. Liaw, A. *Package Random Forest*; University of California: Berkeley, CA, USA, 2006.
109. Ghimire, B.; Rogan, J.; Galiano, V.R.; Panday, P.; Neeti, N. An evaluation of bagging, boosting, and random forests for land-cover classification in Cape Cod, Massachusetts, USA. *GIScience Remote Sens.* **2012**, *49*, 623–643. [CrossRef]
110. Kim, J.-C.; Lee, S.; Jung, H.-S.; Lee, S. Landslide susceptibility mapping using random forest and boosted tree models in Pyeong-Chang, Korea. *Geocarto Int.* **2018**, *33*, 1000–1015. [CrossRef]
111. Nelson, T.A.; Nijland, W.; Bourbonnais, M.L.; Wulder, M.A. Regression tree modeling of spatial pattern and process interactions. In *Mapping Forest Landscape Patterns*; Springer: Berlin/Heidelberg, Germany, 2017; pp. 187–212.

112. Chen, T.; Guestrin, C. Xgboost: A scalable tree boosting system. In Proceedings of the 22nd Acm Sigkdd International Conference on Knowledge Discovery and Data Mining, San Francisco, CA, USA, 13–17 August 2016; pp. 785–794.
113. Friedman, J.H. Stochastic gradient boosting. *Comput. Stat. Data Anal.* **2002**, *38*, 367–378. [CrossRef]
114. Friedman, J.H. Greedy function approximation: A gradient boosting machine. *Ann. Stat.* **2001**, *29*, 1189–1232. [CrossRef]
115. Wang, C.; Deng, C.; Wang, S. Imbalance-XGBoost: Leveraging weighted and focal losses for binary label-imbalanced classification with XGBoost. *Pattern Recognit. Lett.* **2020**, *136*, 190–197. [CrossRef]
116. Marjanovic, M.; Bajat, B.; Kovacevic, M. Landslide susceptibility assessment with machine learning algorithms. In Proceedings of the 2009 International Conference on Intelligent Networking and Collaborative Systems, Barcelona, Spain, 4–6 November 2009; pp. 273–278.
117. Miner, A.; Vamplew, P.; Windle, D.; Flentje, P.; Warner, P. A Comparative Study of Various Data Mining Techniques as Applied to the Modeling of Landslide Susceptibility on the Bellarine Peninsula, Victoria, Australia. In Proceedings of the 11th IAEG Congress of the International Association of Engineering Geology and the Environment, Auckland, New Zealand, 5–10 September 2010.
118. Bröcker, J.; Smith, L.A. Increasing the reliability of reliability diagrams. *Weather Forecast.* **2007**, *22*, 651–661. [CrossRef]
119. Pedregosa, F.; Varoquaux, G.; Gramfort, A.; Michel, V.; Thirion, B.; Grisel, O.; Blondel, M.; Prettenhofer, P.; Weiss, R.; Dubourg, V. Scikit-learn: Machine learning in Python. *J. Mach. Learn. Res.* **2011**, *12*, 2825–2830.
120. Omohundro, S.M. *Five Balltree Construction Algorithms*; International Computer Science Institute Berkeley: Berkeley, CA, USA, 1989.
121. Chen, J.S.; Huang, H.Y.; Hsu, C.Y. A kNN based position prediction method for SNS places. In Proceedings of the Asian Conference on Intelligent Information and Database Systems, Cham, Germany, 23 March 2020; pp. 266–273.
122. Soria, D.; Garibaldi, J.M.; Ambrogi, F.; Biganzoli, E.M.; Ellis, I.O. A 'non-parametric'version of the naive Bayes classifier. *Knowl.-Based Syst.* **2011**, *24*, 775–784. [CrossRef]
123. Domingos, P.; Pazzani, M. Beyond independence: Conditions for the optimality of the simple Bayesian classifier. In Proceedings of the Thirteenth International Conference on Machine Learning, Miami, FL, USA, 4–7 December 2013; pp. 105–112.
124. Soria, D.; Garibaldi, J.M.; Biganzoli, E.; Ellis, I.O. A comparison of three different methods for classification of breast cancer data. In Proceedings of the 2008 Seventh International Conference on Machine Learning and Applications, San Diego, CA, USA, 11–13 November 2008; pp. 619–624.
125. Wu, X.; Kumar, V.; Ross Quinlan, J.; Ghosh, J.; Yang, Q.; Motoda, H.; McLachlan, G.J.; Ng, A.; Liu, B.; Yu, P.S. Top 10 algorithms in data mining. *Knowl. Inf. Syst.* **2008**, *14*, 1–37. [CrossRef]
126. Chen, Y.-R.; Chen, J.-W.; Hsieh, S.-C.; Ni, P.-N. The application of remote sensing technology to the interpretation of land use for rainfall-induced landslides based on genetic algorithms and artificial neural networks. *IEEE J. Sel. Top. Appl. Earth Obs. Remote Sens.* **2009**, *2*, 87–95. [CrossRef]
127. Singh, H.; Pandey, A. Land deformation monitoring using optical remote sensing and PS-InSAR technique nearby Gangotri glacier in higher Himalayas. *Modeling Earth Syst. Environ.* **2021**, *7*, 221–233. [CrossRef]
128. Ciampalini, A.; Raspini, F.; Lagomarsino, D.; Catani, F.; Casagli, N. Landslide susceptibility map refinement using PSInSAR data. *Remote Sens. Environ.* **2016**, *184*, 302–315. [CrossRef]
129. Zhou, C.; Cao, Y.; Yin, K.; Wang, Y.; Shi, X.; Catani, F.; Ahmed, B. Landslide characterization applying sentinel-1 images and InSAR technique: The muyubao landslide in the three Gorges Reservoir Area, China. *Remote Sens.* **2020**, *12*, 3385. [CrossRef]
130. Crosetto, M.; Devanthéry, N.; Cuevas-González, M.; Monserrat, O.; Crippa, B. Exploitation of the full potential of PSI data for subsidence monitoring. *Proc. Int. Assoc. Hydrol. Sci.* **2015**, *372*, 311–314. [CrossRef]
131. Yagüe-Martínez, N.; Prats-Iraola, P.; Gonzalez, F.R.; Brcic, R.; Shau, R.; Geudtner, D.; Eineder, M.; Bamler, R. Interferometric processing of Sentinel-1 TOPS data. *IEEE Trans. Geosci. Remote Sens.* **2016**, *54*, 2220–2234. [CrossRef]
132. Yastika, P.; Shimizu, N.; Abidin, H. Monitoring of long-term land subsidence from 2003 to 2017 in coastal area of Semarang, Indonesia by SBAS DInSAR analyses using Envisat-ASAR, ALOS-PALSAR, and Sentinel-1A SAR data. *Adv. Space Res.* **2019**, *63*, 1719–1736. [CrossRef]
133. Xu, C.; Xu, X.; Dai, F.; Xiao, J.; Tan, X.; Yuan, R. Landslide hazard mapping using GIS and weight of evidence model in Qingshui river watershed of 2008 Wenchuan earthquake struck region. *J. Earth Sci.* **2012**, *23*, 97–120. [CrossRef]
134. Malek, Ž.; Zumpano, V.; Schröter, D.; Glade, T.; Balteanu, D.; Micu, M. Scenarios of land cover change and landslide susceptibility: An example from the buzau subcarpathians, romania. In *Engineering Geology for Society and Territory*; Springer: Berlin/Heidelberg, Germany, 2015; Volume 5, pp. 743–746.
135. Arabameri, A.; Pradhan, B.; Pourghasemi, H.R.; Rezaei, K.; Kerle, N. Spatial modelling of gully erosion using GIS and R programing: A comparison among three data mining algorithms. *Appl. Sci.* **2018**, *8*, 1369. [CrossRef]
136. Rahim, I.; Ali, S.M.; Aslam, M. GIS Based landslide susceptibility mapping with application of analytical hierarchy process in District Ghizer, Gilgit Baltistan Pakistan. *J. Geosci. Environ. Prot.* **2018**, *6*, 34–49. [CrossRef]
137. Kavzoglu, T.; Sahin, E.K.; Colkesen, I. Landslide susceptibility mapping using GIS-based multi-criteria decision analysis, support vector machines, and logistic regression. *Landslides* **2014**, *11*, 425–439. [CrossRef]
138. Song, Y.; Niu, R.; Xu, S.; Ye, R.; Peng, L.; Guo, T.; Li, S.; Chen, T. Landslide susceptibility mapping based on weighted gradient boosting decision tree in Wanzhou section of the Three Gorges Reservoir Area (China). *ISPRS Int. J. Geo-Inf.* **2019**, *8*, 4. [CrossRef]

139. Rehman, A.; Song, J.; Haq, F.; Mahmood, S.; Ahamad, M.I.; Basharat, M.; Sajid, M.; Mehmood, M.S. Multi-Hazard Susceptibility Assessment Using the Analytical Hierarchy Process and Frequency Ratio Techniques in the Northwest Himalayas, Pakistan. *Remote Sens.* **2022**, *14*, 554. [CrossRef]
140. Aslan, G.; Foumelis, M.; Raucoules, D.; De Michele, M.; Bernardie, S.; Cakir, Z. Landslide mapping and monitoring using persistent scatterer interferometry (PSI) technique in the French Alps. *Remote Sens.* **2020**, *12*, 1305. [CrossRef]
141. Chen, W.; Xie, X.; Wang, J.; Pradhan, B.; Hong, H.; Bui, D.T.; Duan, Z.; Ma, J. A comparative study of logistic model tree, random forest, and classification and regression tree models for spatial prediction of landslide susceptibility. *Catena* **2017**, *151*, 147–160. [CrossRef]
142. Zhang, K.; Wu, X.; Niu, R.; Yang, K.; Zhao, L. The assessment of landslide susceptibility mapping using random forest and decision tree methods in the Three Gorges Reservoir area, China. *Environ. Earth Sci.* **2017**, *76*, 405. [CrossRef]
143. Immitzer, M.; Atzberger, C.; Koukal, T. Tree species classification with random forest using very high spatial resolution 8-band WorldView-2 satellite data. *Remote Sens.* **2012**, *4*, 2661–2693. [CrossRef]
144. Yu, K.; Yao, X.; Qiu, Q.; Liu, J. Landslide spatial prediction based on random forest model. *Trans. CSAM* **2016**, *47*, 338–345.
145. Zhang, Y.; Ge, T.; Tian, W.; Liou, Y.-A. Debris flow susceptibility mapping using machine-learning techniques in Shigatse area, China. *Remote Sens.* **2019**, *11*, 2801. [CrossRef]
146. Piacentini, D.; Devoto, S.; Mantovani, M.; Pasuto, A.; Prampolini, M.; Soldati, M. Landslide susceptibility modeling assisted by Persistent Scatterers Interferometry (PSI): An example from the northwestern coast of Malta. *Nat. Hazards* **2015**, *78*, 681–697. [CrossRef]
147. Hakim, W.L.; Achmad, A.R.; Lee, C.-W. Land subsidence susceptibility mapping in jakarta using functional and meta-ensemble machine learning algorithm based on time-series InSAR data. *Remote Sens.* **2020**, *12*, 3627. [CrossRef]
148. Xie, Z.; Chen, G.; Meng, X.; Zhang, Y.; Qiao, L.; Tan, L. A comparative study of landslide susceptibility mapping using weight of evidence, logistic regression and support vector machine and evaluated by SBAS-InSAR monitoring: Zhouqu to Wudu segment in Bailong River Basin, China. *Environ. Earth Sci.* **2017**, *76*, 313. [CrossRef]
149. Chalkias, C.; Ferentinou, M.; Polykretis, C. GIS-based landslide susceptibility mapping on the Peloponnese Peninsula, Greece. *Geosciences* **2014**, *4*, 176–190. [CrossRef]
150. Hussain, S.; Hongxing, S.; Ali, M.; Sajjad, M.M.; Ali, M.; Afzal, Z.; Ali, S. Optimized landslide susceptibility mapping and modelling using PS-InSAR technique: A case study of Chitral valley, Northern Pakistan. *Geocarto Int.* **2021**, *36*, 1–22. [CrossRef]

MDPI

Article

A Novel Model for Landslide Displacement Prediction Based on EDR Selection and Multi-Swarm Intelligence Optimization Algorithm

Junrong Zhang [1], Huiming Tang [1,2,3,*], Dwayne D. Tannant [4], Chengyuan Lin [1], Ding Xia [1], Yankun Wang [5] and Qianyun Wang [2]

[1] Faculty of Engineering, China University of Geosciences, Wuhan 430074, China; zjr@cug.edu.cn (J.Z.); chengyuanlin@cug.edu.cn (C.L.); cug_xia@cug.edu.cn (D.X.)
[2] Three Gorges Research Center for Geohazards of Ministry of Education, China University of Geosciences, Wuhan 430074, China; wangqianyun@cug.edu.cn
[3] Badong National Observation and Research Station of Geohazards, China University of Geosciences, Wuhan 430074, China
[4] School of Engineering, University of British Columbia, Kelowna, BC V1V 1V7, Canada; dwayne.tannant@ubc.ca
[5] School of Geosciences, Yangtze University, Wuhan 430100, China; ykwang@yangtzeu.edu.cn
* Correspondence: tanghm@cug.edu.cn; Tel.: +86-027-6788-3127

Abstract: With the widespread application of machine learning methods, the continuous improvement of forecast accuracy has become an important task, which is especially crucial for landslide displacement predictions. This study aimed to propose a novel prediction model to improve accuracy in landslide prediction, based on the combination of multiple new algorithms. The proposed new method includes three parts: data preparation, multi-swarm intelligence (MSI) optimization, and displacement prediction. In the data preparation, the complete ensemble empirical mode decomposition (CEEMD) is adopted to separate the trend and periodic displacements from the observed cumulative landslide displacement. The frequency component and residual component of reconstructed inducing factors that related to landslide movements are also extracted by the CEEMD and *t*-test, and then picked out with edit distance on real sequence (EDR) as input variables for the support vector regression (SVR) model. MSI optimization algorithms are used to optimize the SVR model in the MSI optimization; thus, six predictions models can be obtained that can be used in the displacement prediction part. Finally, the trend and periodic displacements are predicted by six optimized SVR models, respectively. The trend displacement and periodic displacement with the highest prediction accuracy are added and regarded as the final prediction result. The case study of the Shiliushubao landslide shows that the prediction results match the observed data well with an improvement in the aspect of average relative error, which indicates that the proposed model can predict landslide displacements with high precision, even when the displacements are characterized by stepped curves that under the influence of multiple time-varying factors.

Keywords: landslide displacement prediction; complete ensemble empirical mode decomposition (CEEMD); edit distance for real sequence (EDR); multi-swarm intelligence (MSI); support vector regression (SVR)

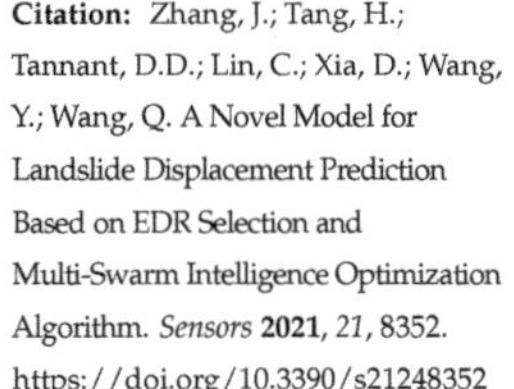

Citation: Zhang, J.; Tang, H.; Tannant, D.D.; Lin, C.; Xia, D.; Wang, Y.; Wang, Q. A Novel Model for Landslide Displacement Prediction Based on EDR Selection and Multi-Swarm Intelligence Optimization Algorithm. *Sensors* **2021**, *21*, 8352. https://doi.org/10.3390/s21248352

Academic Editor: Francesca Cigna

Received: 24 November 2021
Accepted: 12 December 2021
Published: 14 December 2021

1. Introduction

Landslides reactivated by the impoundment of a reservoir or rainfall can cause catastrophic losses such as casualties, road burying, and house damages, which seriously threaten the property and life safety of human society [1,2]. In 2019, there were approximately 6181 geological hazard events in China, causing economic losses of 2.77 billion yuan. Among these, 4220 were landslides, accounting for 68% of the geological hazards [3]. The development of more accurate and effective landslide displacement prediction methods is of great significance for the early warning of catastrophic landslide movements and is an active research area [4–7]. Through the information obtained from the prediction approaches, the landslide status can be evaluated, and the corresponding mitigation measures can be taken in advance to reduce the destructive effects of landslides.

Landslide prediction models can generally be divided into physical–mechanical and phenomenological models [8,9]. The physical–mechanical models are generally recognized as originating from the empirical formula proposed by Saito in 1965 [10], and a series of models have been developed based on creep theory in the following decades [11,12]. Owing to the complexity, strict application conditions, and time-consuming shortcomings of the physical–mechanical models, research on phenomenological models is becoming more and more popular nowadays [13]. By means of mathematical statistics and machine learning, measured landslide displacements are analyzed and modeled while considering the related factors, such as rainfall, the reservoir water level, groundwater level, etc., allowing for the prediction of landslide displacements [12,14].

The support vector machine (SVM) is a frequently used method among all phenomenological models. Nevertheless, when solving regression problems, the performance of the SVM model, also known as support vector regression (SVR), is highly influenced by the determination of penalty parameter C and kernel parameters g [15]. Therefore, research has focused on improving the predictive ability of SVR models for landslide displacements through optimization algorithms. In addition to some classical optimization algorithms such as the genetic algorithm (GA) [16], particle swarm optimization (PSO) [17–19], artificial bee colony (ABC) [20], and ant colony optimization (ACO) [21], recently, studies have advanced with the times, and some newly developed optimization algorithms start to be used [22,23]. Moreover, the continuous enhancement process of the optimization algorithm, as well as the evaluation of the prediction effect after using different frameworks, are also carried out at the same time. Miao et al. [24] adopted a variety of algorithms to optimize the SVR model and achieved a good application effect in the prediction of Baishuihe landslide displacement. Zhang et al. [25] made comparisons of the predictive capability of the SVR model optimized by ACO and GA and found the advantage of ACO-SVR with the consideration of the inducing factors' frequency component. At present, the application of optimization algorithms on SVR-based landslide prediction model parameter optimization is limited. It is still necessary to apply new optimization algorithms to these SVR-based models and compare their performance in landslide prediction.

Although based on the no free lunch (NFL) theorem, any optimization algorithms are equivalent when their performance is averaged across all possible problems; the swarm intelligence optimization algorithms (SIs) still show competitive results in solving optimization problems [26]. Similar to evolutionary algorithms (EA) [27] and artificial neural network algorithms (ANN) [28], the SIs also belong to the nature-inspired metaheuristics method [29]. With its high robustness, the SIs have been applied in many fields, including data clustering, network traffic forecast, data classification, UAV control, etc. Liu et al. [30] proposed a model of a global artificial fish swarm algorithm optimized support vector regression (GAFSA-SVR) for the network traffic forecast; the simulation shows an improvement of forecast precision and is superior to GA and chaos particle swarm optimization (CPSO)-optimized SVR model. Ali et al. [31] adopted the ant lion optimization algorithm (ALOA) in optimal allocation and sizing of renewable distributed generation sources in various distribution networks and results confirmed the effectiveness of the proposed algorithm. Jiang et al. [32] proposed an opposition-based seagull optimization algorithm (OSOA) to overcome the shortage of classification models such as slow computation, instability, and sensitivity to noise. In this paper, six new SIs proposed after 2010, including the bat algorithm (BA) [33], grey wolf optimization (GWO) [34], dragonfly optimization algorithm (DA) [35], whale optimization algorithm (WOA) [36], grasshopper optimization algorithm (GOA) [37], and sparrow search algorithm (SSA) [38], have been tested and compared in the proposed model, and the most suitable optimization algorithm has been identified.

Decomposition of landslide displacement is also a vital step in a prediction model and will directly affect the prediction effect. At present, decomposition method based on signal processing technology, for instance, Fourier transform (FT), discrete wavelet transforms (DWT), wavelet transform (WT), empirical mode decomposition (EMD), variational mode decomposition (VMD), and ensemble empirical mode decomposition (EEMD), are massively used in this field [39–42]. With these methods, the landslide displacement can be decomposed into a trend term and a periodic term, and then these components of the displacement can be predicted by different models. However, when using the CEEMD (complete ensemble empirical mode decomposition), the residual term shows a trend of first decreasing and then increasing, which is difficult to predict as a trend term compared with the residual terms of EMD and EEMD (Figure 1). Hence, a novel prediction model needs to be designed when the CEEMD is adopted in the decomposition of landslide displacement.

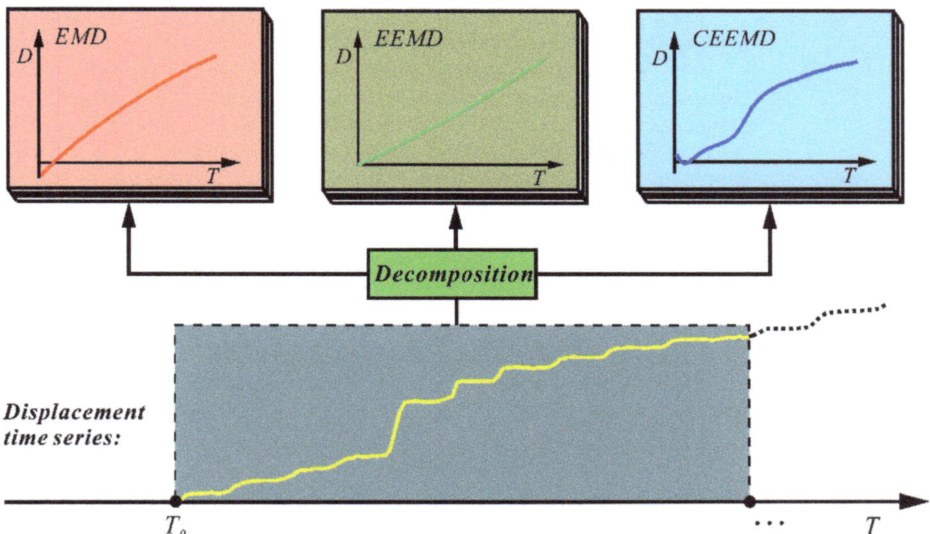

Figure 1. Residual terms of Baishuihe landslide displacements obtained through EMD, EEMD, and CEEMD.

The screening of input parameters for an SVR model from related factors is an important part of prediction model optimization. Grey relational analysis (GRA) is a usual approach for this and has achieved convincing results [43]. Meanwhile, many other statistical methods such as maximal information coefficient (MIC) [23] and mean influence value (MIV) [44] have also been tried for this purpose. Zhang et al. [25] found that, as a similarity measuring method of time series, dynamic time warping (DTW) can be employed and works well in optimal input parameters selection of the SVR model. However, the DTW has the limitation of insensitive to the noise of the time series. To overcome this, the edit distance on real sequence (EDR) has been chosen and utilized in this study [45]. The EDR method is a classic trajectory similarity measurement that calculates the minimal number of editing operations needed for altering one sequence to another. With the advantages of robustness and accuracy, it has been utilized in traffic trajectory classification, physical movement similarity, and fiber segmentation, etc. [46]. It can also be applied in the related components selection for the prediction of landslides. Through calculating the similarity between restructured related factors sequence and periodic displacements sequence after normalization, two restructured related factors with minimum EDR value are the input variables of the SVR model.

This paper aims to improve the accuracy of landslide displacement prediction by constructing a novel model combined with the EDR method and multi-swarm intelligence (MSI). The new method can provide useful predictions of landslide displacements, allowing for the landslide status to be evaluated and the corresponding landslide mitigation measures to be taken before destructive movements occur.

In this paper, the next content is arranged as follows. In Section 2, the CEEMD, EDR, and MSI algorithms are briefly introduced. Section 3 considers the geological conditions and deformation features of the study case, the Shiliushubao landslide. The data preparation and statistical analysis of related factors are shown in Section 4. The predicted results and analysis are shown in Section 5. Section 6 discusses the proposed method, and conclusions are given in Section 7.

2. Methodology

2.1. Data Preprocessing with CEEMD

The CEEMD method is an effective improvement of the EMD method and EEMD method. By adding the white noise in the way of positive and negative pairs to the initial sequence of data, the residual auxiliary noise in the reconstruction signal can be better eliminated. Furthermore, the number of noise sets added can be very low, resulting in higher calculation efficiency. In CEEMD, based on local characteristics, the sequence can be converted to a limited number of intrinsic mode functions (IMF) and a residue. The operation of CEEMD includes three steps [47]:

Step 1: Add white noise consisting of positive and negative pairs to the original sequence data.

$$\begin{bmatrix} \frac{P}{T} \end{bmatrix} = \begin{bmatrix} 1 & 1 \\ 1 & -1 \end{bmatrix} \begin{bmatrix} \frac{\eta(t)}{N} \end{bmatrix} \tag{1}$$

where the original sequence is $\eta(t)$, N is the added white noise, and P and T are two reverse white noise. The number of the decomposed sequences is $2n$, with j as the jth sample.

Step 2: Obtain a series of IMFs by decomposing P and N with the EMD method to generate two sets of IMFs.

$$\begin{cases} P = \sum\limits_{i=1}^{m} IMF_{ji}^{+} \\ T = \sum\limits_{i=1}^{m} IMF_{ji}^{-} \end{cases} \tag{2}$$

where IMF_{ji}^{+} is the ith IMF after adding the positive white noise, IMF_{ji}^{-} is the ith IMF after adding the negative white noise, and m is the number of IMFs.

Step 3: Repeat step 1 and step 2 to get the corresponding IMF terms, and calculate the average of all the IMFs:

$$IMF_j = \frac{\sum_{i=1}^{n} \left(IMF_{ij}^{+} + IMF_{ij}^{-} \right)}{2n} \tag{3}$$

Through this method, the original sequence can be expressed as the sum of some IMFs and a residue $r_n(t)$.

Zhang et al. pointed out that the CEEMD method combined with a t-test can obtain the high-frequency and low-frequency components from related factors such as rainfall and the reservoir water level through a fine-to-coarse reconstruction [25]. Moreover, according to the time series theory, the landslide displacement can be separated into a trend term and a periodic term by methods presented in the Introduction section. In this paper, the CEEMD is adopted as the decomposition method, and the obtained residual term is considered as the trend term. The result after the trend term is subtracted by the cumulative displacement of the landslide is regarded as the period term. Due to the special shape of the trend displacement time series after CEEMD decomposition, the displacement trend term and the period term will be predicted by the SVR model, respectively, later.

2.2. Selection of Optimal Related Factors via EDR

The EDR, which is based on Levenshtein distance, is a traditional and well-established similarity measurement method proposed by Chen et al. [45] and has been used for judging trajectory similarity since [48,49]. The EDR calculates the number of insertions, deletions, or replacement operations required to change the sequence R to T when the threshold is ε. It reduces the effect of noise by quantifying the distance into 0 and 1, and the Levenshtein distance method itself improves the local time-shifting situation (especially when the local time-shifting is not very large). Based on this, the displacement trend term sequence and residue of restructured related factors sequence were set as a reference sample sequence $R = \{r_1, r_2, \ldots, r_n\}$ and a test sample sequence $T = \{t_1, t_2, \ldots, t_m\}$ after normalization. Then, the EDR(R, T) can be calculated as follows:

$$match(r_i, t_j) = true; if \left|r_{ix} - t_{jx}\right| \leq \varepsilon \; and \; \left|r_{iy} - t_{jy}\right| \leq \varepsilon \tag{4}$$

$$D_{EDR}(R,T) = \begin{cases} n; \; if \; m = 0 \\ m; \; if \; n = 0 \\ Min \begin{cases} D_{EDR}(Rest(R), Rest(T)) + subcost, \\ D_{EDR}(Rest(R), T) + 1, \\ D_{EDR}(R, Rest(T)) + 1 \end{cases} & ; \; otherwise \end{cases} \tag{5}$$

$$subcost = \begin{cases} 0, match(r_1, t_1) = true \\ 1, otherwise \end{cases} \tag{6}$$

where the real number $0 < \varepsilon < 1$ is the matching threshold. The cost for a replace, insert, or delete operation is set to 1. Therefore, through calculating the edit distance between two sequences, the smaller the EDR is, the greater the similarity will be. After calculating the EDR between the displacement trend term sequence and residues of original related factors sequence and restructured related factors sequence, three residues with the highest similarity were chosen as the input variable of the SVR model for predicting the displacement trend term.

Similarly, three optimal input variables for predicting the displacement periodic term with an SVR model can be obtained by calculating the EDR between displacement periodic term sequence and original related factors, restructured related factors and related factors frequency sequence.

2.3. Support Vector Regression (SVR)

The support vector regression (SVR) algorithm is a classic landslide displacement prediction model developed from statistical learning theory. With a powerful generalization ability and robust performance, the SVR model can easily solve quadratic programming problems with constraints. The main steps of an SVR model are summarized as follows [50].

Suppose that a nonlinear sample set in low dimensional space is: $\{x_i, y_i\}$, where $x_i = \{x_{i1}, x_{i2}, \ldots, x_{ip}\}$ is the input vector, y_i is the corresponding output vector, i is the number of samples and j is the number of input vectors. Then, the regression estimation function is:

$$f(x) = w^T \varphi(x) + b \tag{7}$$

where w is the weight vector, $\varphi(x)$ is the nonlinear mapping function and b is the offset. Through minimizing the following equation, the value of w and b can be obtained:

$$minJ = \frac{1}{2}\|w\|^2 + C\sum_{i=1}^{n}(\xi_i^+ + \xi_i^-) \tag{8}$$

$$s.t. \begin{cases} y_i - w^T\varphi(x_i) - b \leq \varepsilon + \xi_i^+ \\ w^T\varphi(x_i) + b - y_i \leq \varepsilon + \xi_i^- \\ \xi_i^+, \xi_i^- \geq 0, i = 1, 2, \ldots, n \end{cases} \tag{9}$$

where C and ε are the penalty parameter and the size of the insensitive loss function, respectively. ξ_i^+ and ξ_i^- are the relaxation factors. By solving the quadratic optimization problem, the weight vector w can be expressed as:

$$w = \sum_{i=1}^{n} (\beta_i^* - \beta_i)\varphi(x_i) \tag{10}$$

where β_i^* *and* β_i are Lagrange multipliers. Therefore, the SVR model can be denoted as follows:

$$f(x) = \sum_{i=1}^{n} (\beta_i^* - \beta_i)K(x_i, x_p) + b \tag{11}$$

where $K(x_i, x_p)$ is the kernel function. The SVR kernel function has various forms; in this study, the Gaussian radial basis function (RBF function) is chosen and adopted. Since algorithms for the determination of the penalty factor and the kernel function parameter (C, g) vary, the approach for selecting C and g must be further studied. Different forms of MSI algorithms were explored for the parameter optimization of the SVR model and all of them are briefly described next.

2.4. Multiple Swarm Intelligence

2.4.1. Bat Algorithm (BA)

The bat algorithm (BA), proposed in 2010 by Yang et al., is a novel swarm intelligence optimization technique that simulates the echolocation behavior of microbats [33]. Based on iteration, this algorithm describes the echolocation of microbats and uses it to minimize any objective function and solve optimization problems. In BA, after initializing a group of random solutions, the optimal solution is searched by iteration, and a new local solution is generated by a random flight around the optimal solution, which strengthens the local search. BA is an accurate and effective method of finding the optimal parameter values for an SVR model with few parameters to adjust.

2.4.2. Grey Wolf Optimization (GWO)

The grey wolf optimization (GWO) algorithm is a new swarm intelligent optimization algorithm proposed by Mirjalili et al. [23,51]. Based on the predatory behavior and strict social dominant hierarchy of grey wolves, this algorithm first randomly generates a group of gray wolves in the search space. Then, the wolves are divided into four social hierarchies according to the fitness from high to low, each marked with alpha, beta, delta, and omega. The location and distance between the grey wolves and the prey, which is the possible solution of the optimized SVR model, is obtained through iterative calculation. Finally, through the evolution of the wolf group itself, the distance between them is gradually reduced to realize the optimal hunting of prey. The algorithm has the advantages of strong convergence, few parameters, and easy implementation.

2.4.3. Dragonfly Algorithm (DA)

The dragonfly optimization algorithm (DA) is a swarm intelligent optimization algorithm proposed by Mirjalili et al. [35,52]. The algorithm is based on the dynamic and static swarm behavior of dragonflies in nature, which includes separation behavior, alignment behavior, cohesion behavior, foraging behavior, and distraction from enemy behavior. By establishing a mathematical model of all these behaviors, the dragonfly's latest position vector, which is a possible solution of the objective function, is calculated. This algorithm has the advantages of simple calculation, low complexity, few control parameters, and fast convergence speed.

2.4.4. Whale Optimization Algorithm (WOA)

The WOA algorithm is a new heuristic optimization algorithm. The key idea is to simulate the behavior of humpback whales [36]. The humpback whales hunt in a special way using bubble nets, which can be described as two mechanisms: upward spirals and double loops. The WOA optimization algorithm has three steps: searching and encircling prey, the bubble-net argument attacking method (exploitation phase), and search for prey (exploration phase). Through this, the position vector of humpback whales with the best fitness value can be obtained by satisfying a termination criterion, and the final position vector is chosen as the best solution of the optimized SVR model parameters. The algorithm has the advantages of simple operation, few parameters to adjust, and a strong ability to jump out of a local optimum.

2.4.5. Grasshopper Optimization Algorithm (GOA)

The grasshopper optimization algorithm (GOA), proposed by Saremi et al., in 2017, is a metaheuristic bionic optimization algorithm that mimics the swarming behavior of grasshoppers during population migration (exploration) and foraging behavior (exploitation) [37]. The grasshoppers' position vector is equal to the value of an objective function [53]. When the grasshoppers reach a food source, the parameters reach the optimal variable, and the optimal value of the SVR model parameters is obtained. The algorithm provides a balanced condition between local and global search operators to achieve the final target. Two forces in grasshoppers, attraction and repulsion, provide global search and local search, respectively. To obtained effective solutions, the influence of the grasshopper's current position, its relative position to other grasshoppers, and the position of the target point are regarded as the effective agents to determine the search vector. It has higher search efficiency and faster convergence speed, and its special adaptive mechanism can balance the global and local search processes with better optimization accuracy.

2.4.6. Sparrow Search Algorithm (SSA)

The sparrow search algorithm (SSA), as proposed by Xue et al. [38], was mainly inspired by the foraging behavior and anti-predation behavior of sparrows. Some sparrows are in charge of seeking food and providing locations for the entire population, while the remaining sparrows use the locations to obtain food. Meanwhile, when a sparrow is aware of the danger and alarms, the entire population will immediately take anti-predation behavior. Although idealized, these behaviors are formulated with corresponding rules, and the algorithm classifying the sparrows into producers and scroungers. Their positions are updated according to their own rules, separately. In SSA, the position of each sparrow is equal to a possible solution of the objective function, and the best solution can be obtained when meeting iteration conditions. The algorithm is novel and has the advantages of a strong optimization ability, fast convergence speed, fewer adjustment parameters, and simple calculation.

2.5. Procedure of the Proposed Hybrid Algorithm

The framework of the proposed ensemble prediction model is shown in Figure 2. The entire forecasting process is divided into three steps: data preparation, multi-swarm intelligence (MSI) optimization, and displacement prediction. In the data preparation step, the time-sequences of factors related to the landslide movements, such as rainfall and reservoir water level, are restructured. The frequency component and residual component of all original and restructured sequences are then obtained through the combined application of CEEMD and *t*-test. In the MSI optimization step, MSI optimization algorithms are used to select the optimal C and g for the SVR model. In the displacement prediction step, the trend and periodic displacements are extracted from the observed cumulative landslide displacement through CEEMD. Then EDR is used to select the input variables of the periodic displacement prediction SVR model by calculating the EDR value between the periodic displacement and original related factors, restructured related factors, and frequency related factors after normalization. Similarly, the input variables of the trend displacement prediction SVR model are obtained by calculating the EDR value between the trend term displacement and all residue terms after normalization. Finally, the predictions of the trend and the periodic displacements are performed separately, and the total predicted displacement is obtained by adding them together.

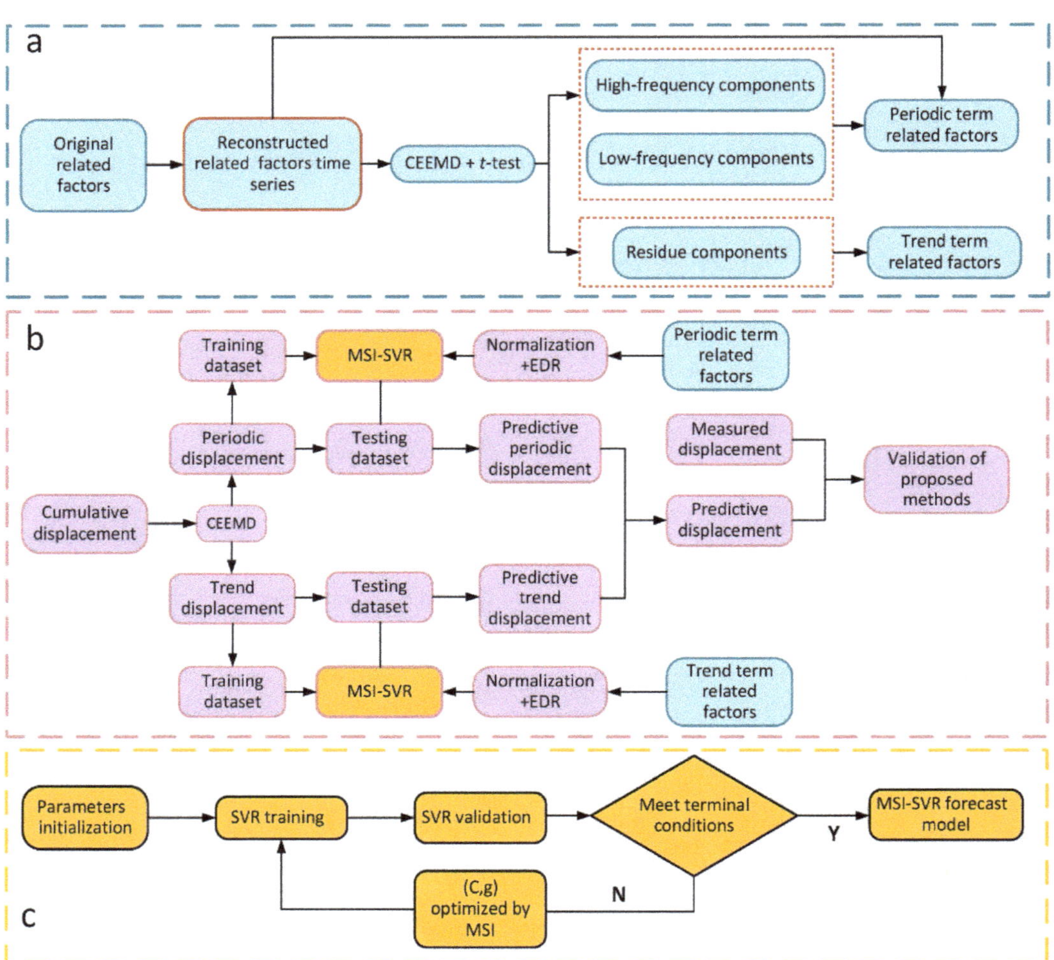

Figure 2. Framework of the proposed ensemble prediction model, (**a**) data preparation step, (**b**) displacement prediction step, and (**c**) MSI optimization step.

2.6. Performance Evaluation Formula

The most commonly used indicators to evaluate the performance of prediction models are coefficient of determination (R^2), root mean square error (RMSE), mean absolute error (MAE), and mean average percentage error (MAPE). These indicators were used in this study and are defined as:

$$R^2 = 1 - \frac{\sum_{i=1}^{N}(y_t - \hat{y}_t)^2}{\sum_{i=1}^{N}(y_t - \overline{y}_t)^2} \tag{12}$$

$$\text{RMSE} = \sqrt{\frac{1}{N}\sum_{i=1}^{N}\left((y_t - \hat{y}_t)^2\right)} \quad (13)$$

$$\text{MAE} = \frac{1}{N}\sum_{i=1}^{N}|y_t - \hat{y}_t| \quad (14)$$

$$\text{MAPE} = \frac{1}{N}\left(\sum_{i=1}^{N}\left|\frac{\hat{y}_t - y_t}{y_t}\right|\right) \times 100\% \quad (15)$$

where y_t is the tth measured value, $\overline{y}_t$ is the mean of the measured value, $\hat{y}_t$ is the tth predicted value, and $\overline{\hat{y}}_t$ is the mean value of the prediction.

3. Cases Study

3.1. Geological Conditions

The Shiliushubao landslide is part of the famous Huanglashi landslide group, one of the large-scale landslides in the Three Gorges Reservoir Area (TGRA). It is located on the north bank of the Yangtze River, 1.5 km east of Badong county, 66 km away from the Three Gorges Dam (TGD) (Figure 3). The landslide's geographical coordinates are 110°26′ east longitude and 31°02′ north latitude. The Shiliushubao landslide is bordered by the Lijiawan valley on the east and the Gan valley on the west, with a tongue-like shape. It is bigger than the well-known Baishuihe landslide with an estimated volume of 11.8 × 10^6 m^3 and covers an area of 0.34 km^2. The top of the landslide is at an elevation of 340 to 358 m with a width of 140 m, and the toe of the landslide is at an elevation of 68 to 80 m with a width of 570 m.

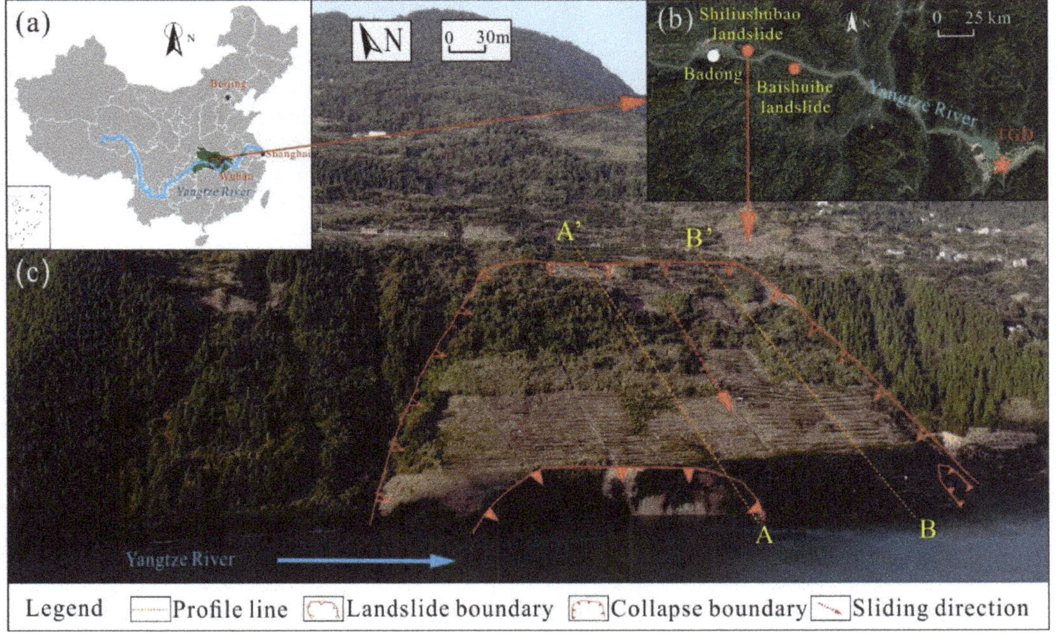

Figure 3. Location (**a**,**b**) and an oblique view (**c**) of Shiliushubao landslide captured by UAV, October 2020.

The cross-section of the ground surface is shown in Figure 4 by the profile B-B′. The average slope angle is 26° along the sliding direction. However, the slope contains a gently sloping bench at an elevation near 200 m, and the slope is much steeper than 26° above and below the bench. The slope angle is up to 40° at elevations below the reservoir level.

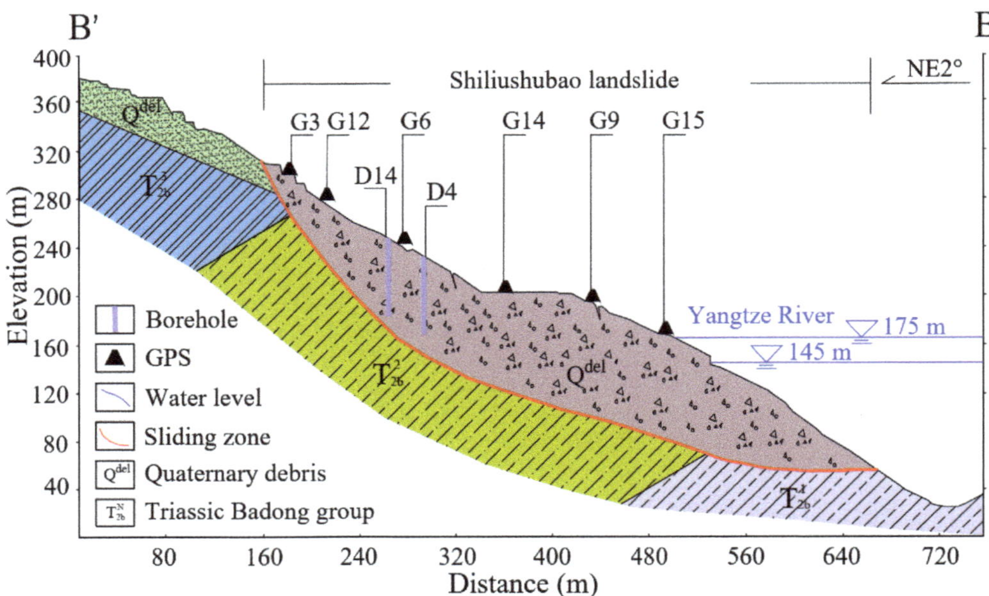

Figure 4. Geological section of the Shiliushubao landslide (B-B′).

The geological profile B-B′ in Figure 4. shows that the Shiliushubao landslide occurs in the Triassic Badong Group consisting of red mudstone, siltstone, gray-green marl, and limestone. These rocks are characterized by high clay mineral content (about 68%). Exposure of the rock to water allows the rock to soften and weaken. The sliding mass also includes near-surface Quaternary soils. The rear edge of the landslide is mainly a loose accumulation of gravel and clay. This soil is weak and is prone to collapses or sliding along the bedrock surface. The sliding zone consists of clay or silty clay with some gravel. The thickness of the sliding zone varies from 1.0 to 4.9 m, with an average thickness of 2.0 m.

The topography of the lower part of the Shiliushubao landslide was mostly altered by the newly formed Hengping landslide (Figure 5), and some landslide materials under 100 m elevation have been removed by erosion. There are some small gullies near the landslide's front edge caused by surface water runoff, which are the main channels for gathering and draining surface water.

3.2. *Rainfall and Reservoir Levels*

The Shiliushubao landslide is located in a subtropical zone, in which rainfall is continuous and concentrated in the summer. The rainy season generally occurs from May to September, which accounts for 70% of the yearly rainfall. Rainfall is one factor that increases the movement of the Shiliushubao landslide. Fluctuation in the reservoir level in the TGRA is another factor influencing the landslide movements, especially the sudden reservoir drops before the flood season.

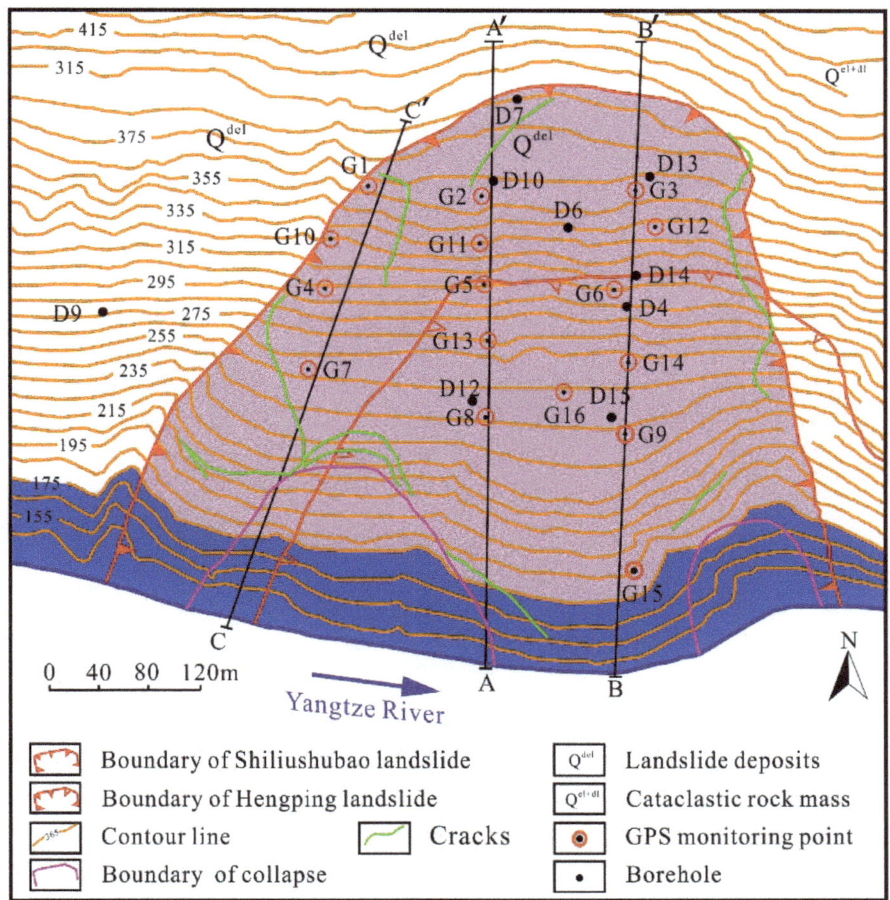

Figure 5. Geological map and monitoring points for the Shiliushubao landslide.

3.3. Deformation Characteristics

Since the reservoir was first impounded in June 2003, the toe area of the slope has experienced repeated small collapses (Figure 6). From 4 to 14 June 2004, four sliding events occurred during a period of rainfall, involving an estimated volume of 6000 m^3. The toe area is very unstable, and slope movements at the toe affect the rest of the slope. At present, the slope's deformation processes are causing small collapses under the influence of rainfall or reservoir level fluctuations.

Slope movements have created ground fissures that have gradually intensified. Areas of subsidence have also occurred. While the existing main cracks continued to expand, a series of new cracks gradually formed at the landslide's rear edge. These cracks have connected and coalesced inside the sliding mass. The maximum crack length obtained by field monitoring is 345 m with opening widths up to 0.5 m and depths over 1 m. Many cracks have occurred in a concrete-lined drainage ditch at the front edge of the landslide. Moreover, some feathery cracks are also scattered along both sides of the landslide.

3.4. Landslide Monitoring

From February 2004 to December 2009, field monitoring was conducted to study the Shiliushubao landslide movements, based on which, the deformation evolution characteristics and development trend of the Shiliushubao landslide can be mastered. A total of sixteen GPS monitoring points and 15 boreholes were arranged on the surface of the sliding mass (Figure 5). Some monitoring points were destroyed due to rainfall, landslide movement, and other reasons. Thus, only monitoring data from February 2004 to December 2009 have been recorded and preserved. The cumulative displacement data from GPS points G1, G2, G4, G8, plus the rainfall and reservoir water levels were selected and shown in Figure 7.

Figure 6. The ground collapses (**a**,**b**) and cracking (**c**,**d**) in the toe area captured by UAV, October 2020.

3.5. Analysis of Monitoring Data

Monitoring data show that the displacement of the sliding mass increases with time in an obvious stepped shape. From February 2004 to December 2009, due to rainfall, the five displacement jumps occurred in the rainy season (May to September). After the rainy season, the landslide resumes movement at a slow, roughly constant speed.

The fluctuation of the reservoir water level is another factor affecting the deformation of the sliding mass. When the reservoir level drops sharply, the movement of the sliding mass accelerates. For example, from January to May 2007, the water level dropped from 155.4 m to 144.7 m, and the landslide displacement rate reached 16.2 mm/month in March when the water level dropped by 5 m. In May, when the water level dropped by 10.7 m, the landslide displacement rate was 44.4 mm/month. Similarly, when the reservoir water level fell in other periods, such as January to July 2009, the landslide displacement rate increased from 1.5 to 47.1 mm/month.

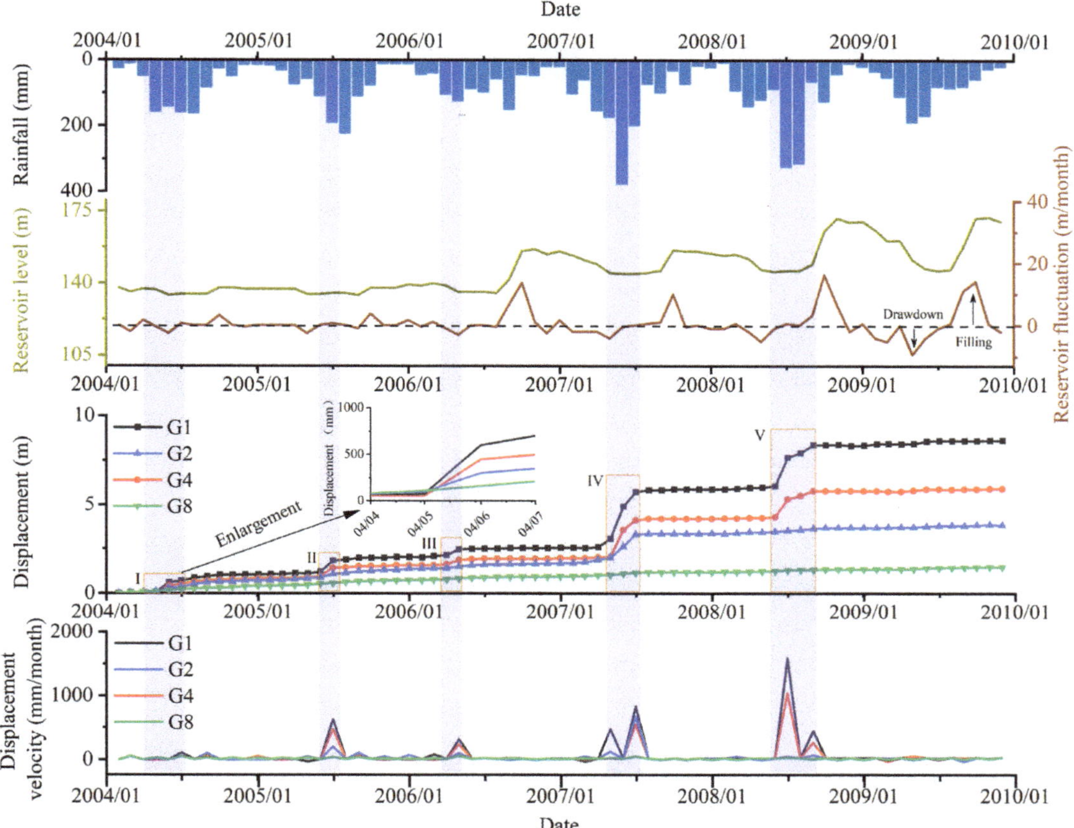

Figure 7. Displacement data from GPS points G1, G2, G4, G8, and rainfall and reservoir level data.

For a better understanding of this seasonal deformation acceleration's related factors, a correlation analysis between displacement velocity at G1 (located at the northeast edge of the landslide) and rainfall, rate of reservoir level change, and reservoir level are shown in Figure 8. The size of the bubbles represents the deformation velocity. The larger bubbles tend to plot where the rainfall is higher. Meanwhile, the large bubbles are mainly concentrated where the reservoir level is between 140 and 150 m and are located where the reservoir level fluctuates slowly (between −4.4 and 9.0 m/month). This indicates that reservoir level fluctuations mainly trigger accelerated landslide movements when the reservoir level is low. The maximum size bubble appears where the rainfall is about 325 mm/month, and the water level rises between 4.5 and 9.0 m/month. The combined effect of heavy rainfall and rising reservoir level on landslide deformation is more significant than low rainfall combined with reservoir level drawdown.

Inclinometer D7 indicates that the main sliding zone is located at a depth of 22 to 26 m (Figure 9). The data show that before June 2003, the shear deformation in the slip zone was slow. Then, with the operation of the TGRA, the displacement in the shear zone increases. Therefore, it can be judged that the Shiliushubao landslide is in the stage of accumulative creep deformation, and the deformation tends to be intensified under the influence of reservoir water.

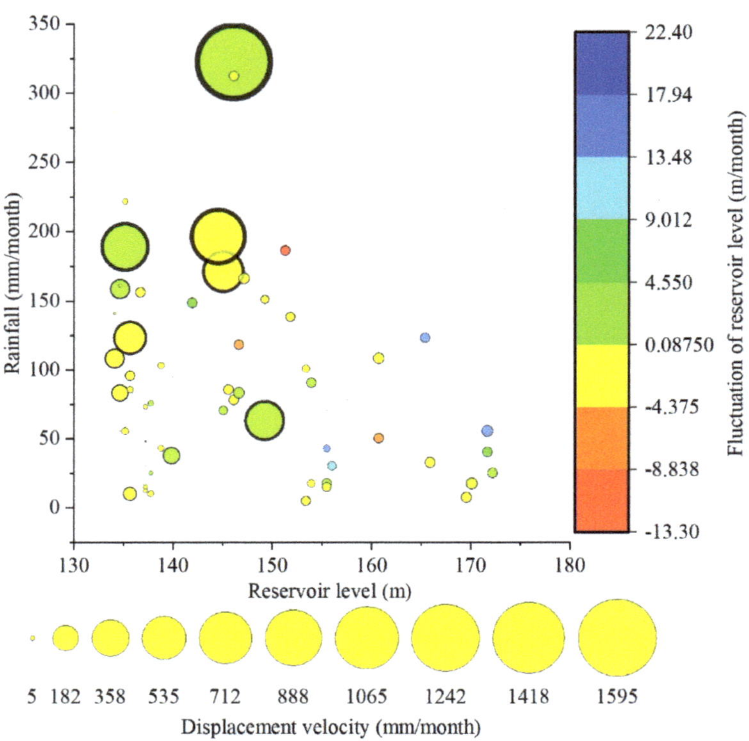

Figure 8. Correlation of displacement velocity at G1 versus the reservoir level, rainfall, and fluctuation of reservoir level.

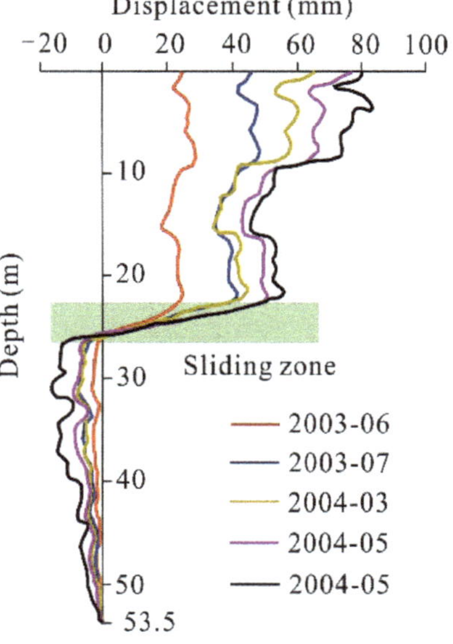

Figure 9. Lateral deformation versus depth in inclinometer D7.

In conclusion, the formation of the Shiliushubao landslide is the result of a series of related factors including internal inducing factors and external inducing factors. Weak rock formations are the inherent cause of deformation. In the Badong Formation, soft rocks characterized by high clay mineral content account for about 68%. The hydrophilicity of the rock determines that the rock has the characteristics of easy softening, muddy and weathering, and lays the material foundation for the deformation and failure of the slope. Water is the external cause of deformation. The impact of concentrated high intensity rainfall and periodic water storage activities in the TGRA, especially the sudden drop before the flood season, are main external inducing factors for the reactive of the Shiliushubao landslide.

4. Data Processing and Statistical Analysis

4.1. CEEMD Decomposition of Landslide Displacement Versus Time Data

Since all displacements at the Shiliushubao landslide show a similar step-like deformation curve, only the displacement data at site G1 is chosen for model validation in this study. The CEEMD method can be used to extract the trend displacements and the periodic displacements. The following parameters were used [2]:

- ensemble member = 200
- standard deviation of added white noise in each ensemble member = 0.2
- threshold variance = 0.2
- threshold for first iteration = 4

The landslide displacement sequence was decomposed into a few IMFs and a residue through CEEMD. The residue is considered to be the trend displacement of the landslide, and the periodic displacement was obtained by adding all the IMFs together.

The results show that the trend displacement component of G1 has local fluctuations and an increasing trend over time, which is consistent with a long-term trend of cumulative displacements. The periodic displacement component shows a cyclical variation in displacements ranging from −800 to 917 mm. The maximum variation range of periodic displacement occurred in the 2007 rainy season when the TGR was first impounded. As the periodic displacement and trend displacement are important components of the cumulative displacement, they will be separately modeled and predicted. Once the best prediction for each component is obtained, the best prediction for cumulative displacement is obtained.

The displacement data are divided into training and testing data sets to establish the SVR prediction model of periodic and trend displacements (Figure 10). The SVR model is organized with the training dataset to establish the regression relationships between displacement and selected variables. The trained SVR model can then be used to predict the current month periodic displacement and compared with the testing dataset to verify the model's accuracy. In this study, the displacement data from February 2004 to September 2008 were selected as the training dataset, and the rest were used as the testing dataset.

4.2. CEEMD Decomposition of Related Factors

Before selecting the input various parameters, the factors related to the landslide deformation are usually restructured first [24]. Original related factors such as the rainfall, reservoir level, and date of displacement were restructured. The current monthly rainfall sequence (L1) was restructured as the accrued precipitation of the previous two months (L2), as were the accrued precipitation of the previous month and the current month (L3), and the accrued precipitation of the previous two and the current month (L4). The current monthly reservoir level data (X1) were restructured as the reservoir level monthly change (X2) and the change of reservoir level between two months (X3). The displacement data (D) were restructured as the previous month displacement (D1) and the accrued displacement of the previous month and the current month (D2).

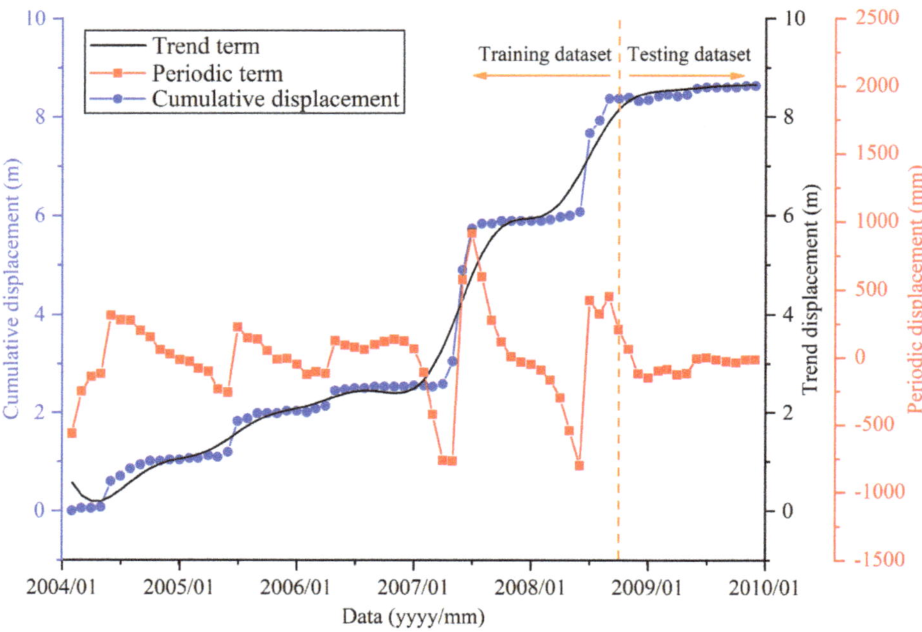

Figure 10. Periodic and trend displacement at site G1 obtained through CEEMD.

Keeping the CEEMD parameters fixed, L1–L4, X1–X3, and D1–D2, can be decomposed into a few IMFs sorted by frequency from highest to lowest and a residue. The mean of IMF_1 was compared to the other IMFs by a paired *t*-test with a significance set at 0.05 (two-tailed) for each decomposed and restructured factor. If the significance values of IMF_i are greater than 0.05, the difference between IMF_1 and IMF_i is not significant. Therefore, the superposition of IMFs from IMF_1 to IMF_i is the high-frequency component, and the superposition of the remaining IMFs is the low-frequency component. The IMFs of each restructured factors are shown in Figure 11, and the results of the paired *t*-test are shown in Table 1.

The results reveal that the IMFs obtained from the decomposition of all factors show a certain periodicity. Their frequency varies, and IMF_1 usually has the highest frequency and fluctuation amplitude. Since there is only one IMF after the CEEMD decomposition of D2, it is considered that there are only high-frequency components in D2. The paired *t*-test results indicate that only IMF_3 in X1 and IMF_4 in X3 has a significance value that is less than 0.05, which denotes that the low-frequency components only exist in X1 and X3. Taking these two as the low-frequency components of X1 and X3, the high-frequency components of the other factors will be the sum of the remaining IMFs. Therefore, in addition to the variables mentioned above, new variables can also be chosen as input to an SVR model of the periodic displacements after reconstruction: high-frequency current monthly rainfall sequence ($L1^H$), high-frequency accrued precipitation of the previous two months ($L2^H$), high-frequency accrued precipitation of the previous month and the current month ($L3^H$), high-frequency accrued precipitation of the previous two and the current months ($L4^H$), high-frequency current monthly reservoir level data ($X1^H$), low-frequency current monthly reservoir level data ($X1^L$), high-frequency reservoir level monthly change ($X2^H$), high-frequency change of reservoir level between two months ($X3^H$), low-frequency change of reservoir level between two months ($X3^L$), high-frequency previous month displacement ($D1^H$), and high-frequency accrued displacement of the previous month and the current month ($D2^H$).

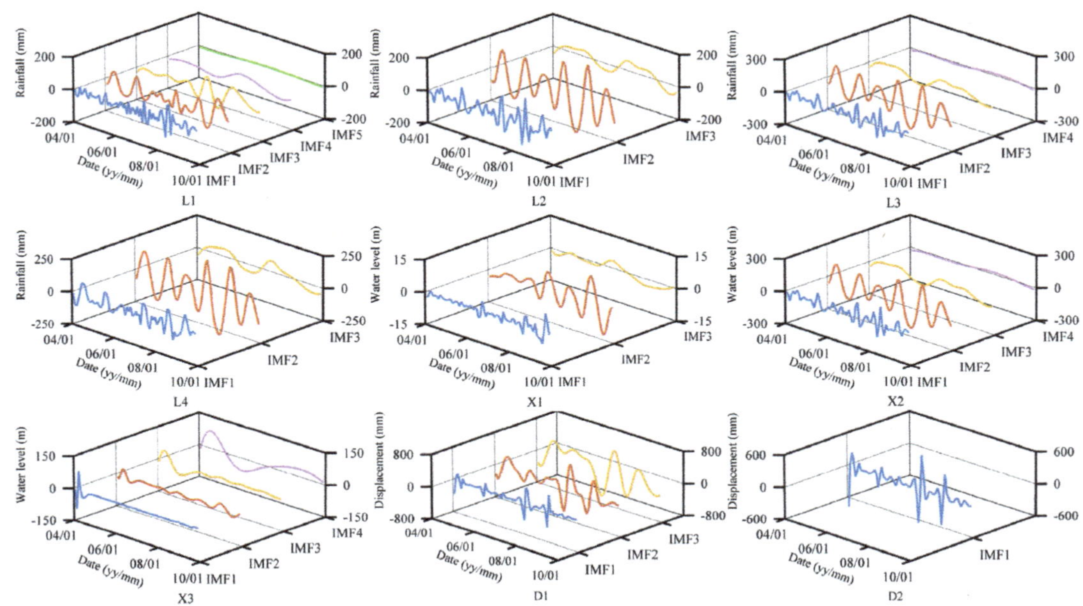

Figure 11. IMFs of restructured factors derived through CEEMD.

Table 1. Paired *t*-test results of all decomposed IMF.

Groups	Restructured Factor	Component	t	Sig.	Mean (mm)	Std. Deviation (mm)
Rainfall	L1	IMF_2	0.22	0.83	1.50	56.92
		IMF_3	−0.20	0.84	−1.23	50.75
		IMF_4	0.60	0.55	3.02	42.24
		IMF_5	0.10	0.92	0.41	36.49
	L2	IMF_2	0.47	0.64	6.16	110.1
		IMF_3	−1.34	0.18	−10.09	63.28
	L3	IMF_2	0.23	0.82	3.05	111.1
		IMF_3	−1.70	0.09	−12.42	61.56
		IMF_4	−1.08	0.28	−6.75	52.48
	L4	IMF_2	0.38	0.70	5.43	120.1
		IMF_3	−0.61	0.54	−5.02	69.11
Reservoir water level	X1	IMF_2	0.47	0.64	0.26	4.73
		IMF_3	2.07	0.04	0.66	2.70
	X2	IMF_2	0.22	0.83	2.91	111.6
		IMF_3	−1.58	0.12	−11.65	62.23
		IMF_4	−0.98	0.33	−6.13	52.64
	X3	IMF_2	−0.17	0.86	−0.37	18.16
		IMF_3	−0.52	0.61	−1.49	24.30
		IMF_4	−2.19	0.03	−10.94	42.05
Displacement	D1	IMF_2	−0.04	0.97	−1.16	229.4
		IMF_3	−1.07	0.29	−40.82	320.5

The residue terms of restructured factors derived through CEEMD are shown in Figure 12. The results demonstrate that, except for L1, all the residue terms show a roughly increasing trend that is similar to the trend displacement term. This suggests that the residual terms roughly reflect the trend of the cumulative displacement, allowing the residue terms to be used as input parameters for the SVR to predict the displacement trend term.

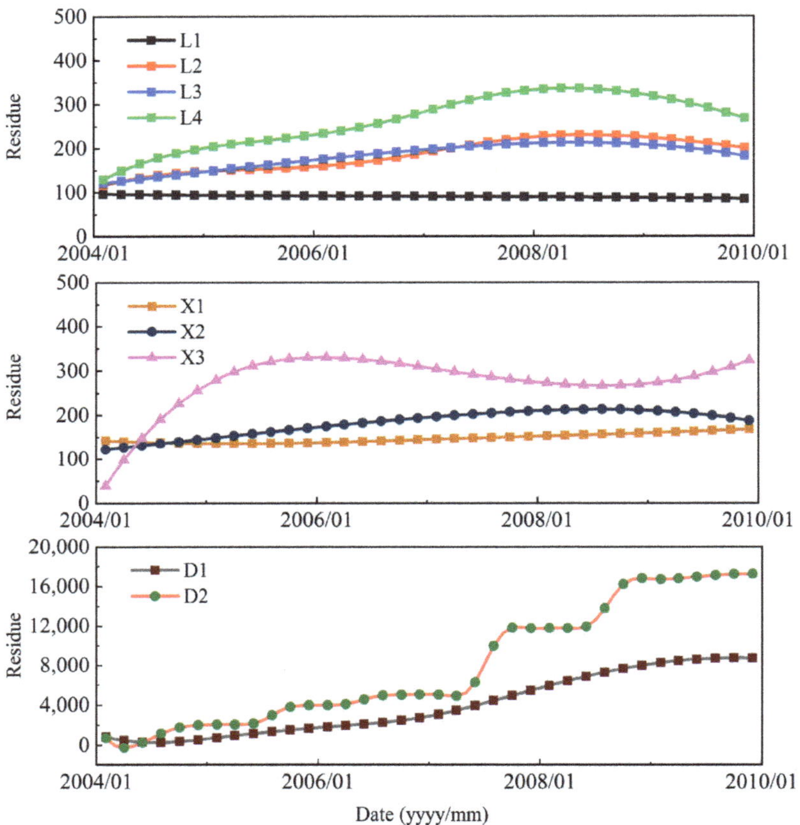

Figure 12. Residue term of restructured factors derived through CEEMD.

4.3. Factors Affecting Landslide Displacement Selected by EDR

Previous analyzes demonstrated a strong association between the landslide displacement and the aforementioned factors. Thus, it is vital to determine which factors that have the greatest influence on landslide displacement. The EDR distance was determined between each factor and the displacements to determine the specific factors most closely related to the landslide's periodic displacement and trend displacement, respectively. This helps to identify the best factors to use the SVR model. The original restructured factors and their frequency components were chosen to compute the EDR distance with the periodic displacement. Simultaneously, the residue term for each factor and restructured factors were utilized to compute the EDR distance with the trend displacement. Normalization can be used to eliminate the influence of the numerical magnitude on analysis results due to the dimension difference between the displacement time series and the related factors. The calculated EDR distances are shown in Table 2.

Table 2. EDR distance between periodic displacements and related factors.

Groups	Component	Periodic Displacement			Trend Displacement
		Origin	High	Low	
Rainfall	L1	61	60	/	68
	L2	56	53	/	24
	L3	56	53	/	42
	L4	54	49	/	42
Reservoir level	X1	53	33	44	22
	X2	56	53	/	41
	X3	69	41	58	60
Displacement	D1	67	21	/	3
	D2	66	32	/	2

After dividing all the related factors into rainfall, reservoir water level, and displacement groups, the factors with a smaller EDR distance can be regarded as more interrelated with the landslide displacement component in each group. The results show that, for periodic displacement, the high-frequency accrued precipitation of the previous two and current months ($L4^H$), the high-frequency current monthly reservoir level data ($X1^H$), and the high-frequency previous month displacement ($D1^H$) are the most relevant factors in each group. Thus, when predicting periodic displacement, $L4^H$, $X1^H$, and $D1^H$ are the input variables for the periodic displacement SVR model. Similarly, related factors for predicting trend displacement are the residual terms of L2, X1, and D2 according to the EDR results in each group, and these were chosen as the input parameters for the trend displacement SVR model.

To verify the effectiveness of the EDR method, grey relational analysis (GRA), a common method for selecting input variables in landslide displacement prediction, was used to calculate the grey relational degree (GRD) between the selected factors and the displacement component. The periodic displacement component is chosen as the research object, and the factor's GRD and periodic displacement velocity are shown and compared in Figure 13. The factors with a GRD value higher than 0.6 are regarded as closely interrelated with the periodic displacement. Therefore, the high-frequency accrued precipitation of the previous two and current months ($L4^H$), the high-frequency current monthly reservoir level data ($X1^H$), and the high-frequency previous month displacement ($D1^H$) are the most relevant related factors in each group, which is consistent with the results selected by EDR.

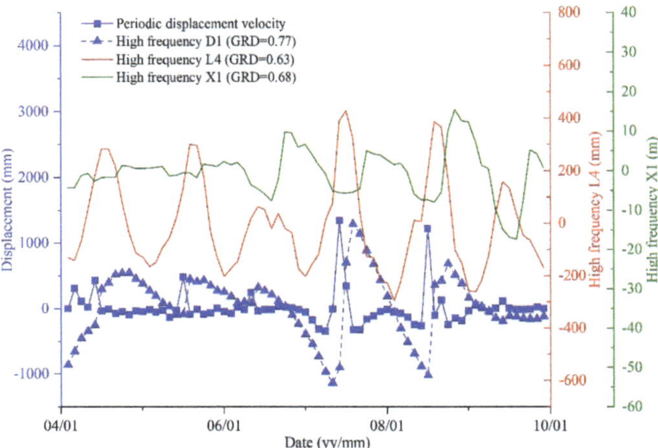

Figure 13. Landslide periodic displacement compared with selected factors affecting landslide movement.

5. Prediction Results and Comparison

5.1. Parameter Optimization

For quantitatively measuring the optimization performance of the six SIs adopted in this study, three selected benchmark functions (Table 3) with different features are employed as test functions and results are shown in Figure 14. Different from $F_2(x)$ and $F_3(x)$, the $F_1(x)$ is smoother and has a unique extreme point in the solution space of x_1 and x_2. The calculation results and process show that the slopes of the convergence curves of SSA and GWO are close, indicating that the convergence performance of the two is close and is the best among the six algorithms. The solutions obtained by each SI in $F_1(x)$ and $F_3(x)$ are relatively scattered, and some algorithms (such as BA) will fall into a local optimum.

Table 3. Three benchmark functions.

Function	Range	Theoretical Minimum Value
$F_1(x) = \sum_{i=1}^{n} x_i^2$	$x_i \in [-100, 100],\ i = 1,\ 2$	0
$F_2(x) = \sum_{i-1}^{n} ix_i^4 + random(0,1)$	$x_i \in [-1.28, 1.28],\ i = 1,\ 2$	0
$F_3(x) = \sum_{i=1}^{n} [x_i^2 - 10\cos(2\pi x_i) + 10]$	$x_i \in [-5.12, 5.12],\ i = 1,\ 2$	0

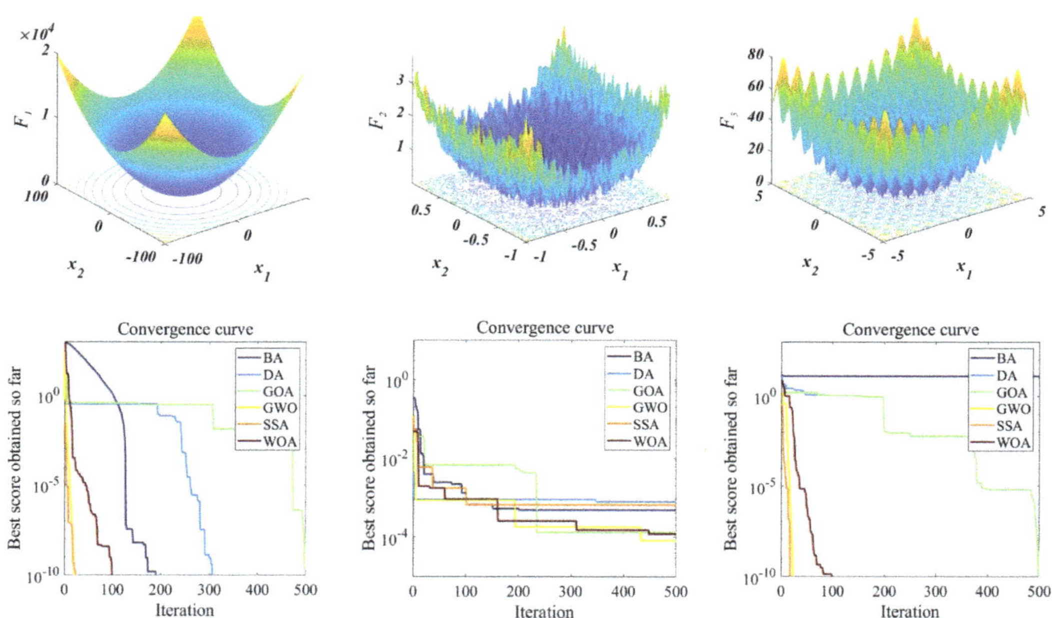

Figure 14. Iterative curves of three benchmark functions solved using multiple-SI.

Determining the optimal value of the penalty factor C and the kernel function parameter g of the SVR model is a vital procedure dominating the accuracy of a displacement prediction. The parameters C and g in this study are optimized with MSI algorithms and are conducted independently for periodic and trend terms. For each MSI algorithm, the parameters C and g make a two-dimensional searching space. A population of simple agents communicate locally with each other and with their environment and move in specific patterns to search for the best result. The parameter settings and initial conditions in the MSI algorithm jointly affect the result. The parameter settings are iteratively adjusted and recalculated according to the optimal prediction effect. The results of the optimization are shown in Table 4. The optimized C and g are later used in the SVR-based model to predict the periodic and trend displacements.

Table 4. Parameter and results of each optimization algorithm.

Algorithm	Parameters			Periodic		Trend	
				C	g	C	g
BA-SVR	Sizepop = 20	Max_iter. = 200	A = 0.2	220.67	0.00109	657.16	0.00106
	$Lb = 1 \times 10^{-2}$	$Ub = 1 \times 10^{2}$	r = 0.5				
	Freq_min = 0.1	Freq_min = 0.2	Alpha = 0.2				
DA-SVR	Sizepop = 30	Max_iter. = 200	e = f = 0.1	66506	0.00001	83702	0.00001
	$lb = 1 \times 10^{-5}$	$ub = 1 \times 10^{5}$	c = 0.7				
	w = 0.5	s = 0.1	a = 0.1				
GOA-SVR	Sizepop = 30	Max_iter. = 200	l = 1.5	16.13	0.00100	29.68	0.01000
	$lb = 1 \times 10^{-3}$	$ub = 1 \times 10^{3}$	f = 0.5				
GWO-SVR	Sizepop = 30	Max_iter. = 200	dim = 2	474.94	0.00100	706.29	0.00100
	$lb = 1 \times 10^{-3}$	$ub = 1 \times 10^{3}$	/				
SSA-SVR	Sizepop = 30	Max_iter. = 200	pNum = 20%	16.17	0.00100	9677.9	0.00014
	$lb = 1 \times 10^{-4}$	$ub = 1 \times 10^{4}$	sNum = 20%				
OA-SVR	Sizepop = 20	Max_iter. = 200	dim = 2	1.74	0.01000	48277.4	0.00001
	$lb = 1 \times 10^{-5}$	$ub = 1 \times 10^{5}$	b = 1				

5.2. Prediction of Periodic and Trend Displacements

An MSI-based SVR prediction model was developed with the optimized input factors to predict the periodic displacements and the trend displacements separately, as shown in Figure 15. The prediction accuracy and error of each model are shown and compared in Figure 16. For the periodic displacements, the prediction accuracy with the largest R^2 and smallest MAPE, RMSE, and MAE was obtained using the DA algorithm among all of the given models. The corresponding result of MAPE, RMSE, MAE, and R^2 is 3.654173, 63.0435, 119.2786, 0.824217, respectively. Meanwhile, the GWO-based SVR model gave the best prediction for the trend displacements compared to the other optimization algorithms, with the result of MAPE, RMSE, MAE, and R^2 being 0.010273, 95.9178, 184.4194, and 0.99473, respectively. Overall, the prediction results provided by the SVR model optimized by MSI matched well with the observation results.

5.3. Prediction of Cumulative Displacements

The predicted cumulative displacements of the Shiliushubao landslide can be obtained by adding the predicted periodic and trend displacements. The predicted cumulative displacements are shown in Figure 17, and these are in good agreement with the observed displacements. The maximum relative error of monthly displacement is generally less than 3% and the average relative error of less than 1%. The results show the usefulness of the proposed model. The most appropriate optimization algorithm and the most relevant landslide related factors were selected and applied.

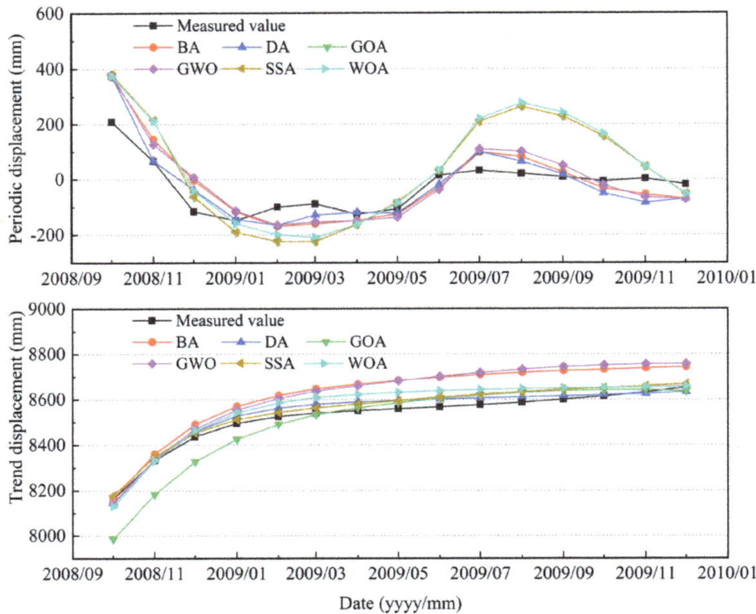

Figure 15. Comparison of prediction results by MSI-SVR model with monitoring data.

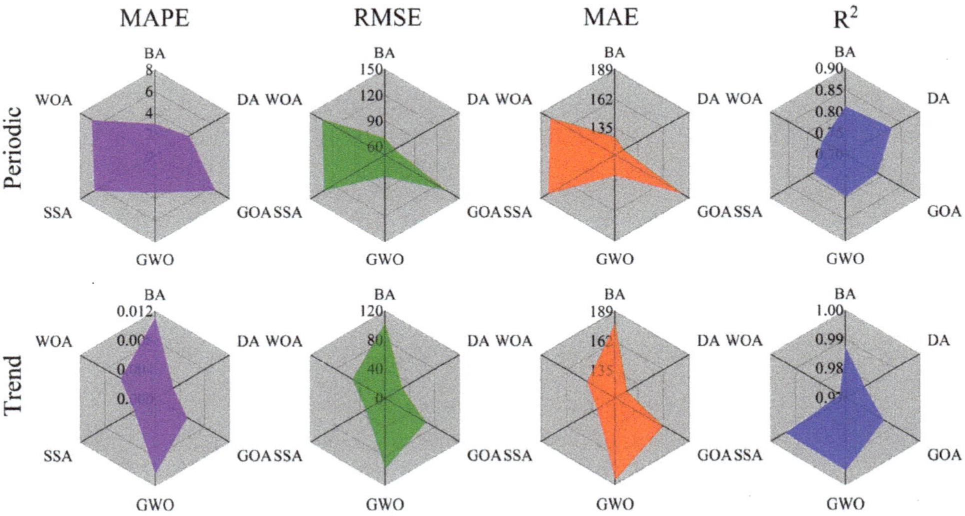

Figure 16. Rose diagram for each model's performance.

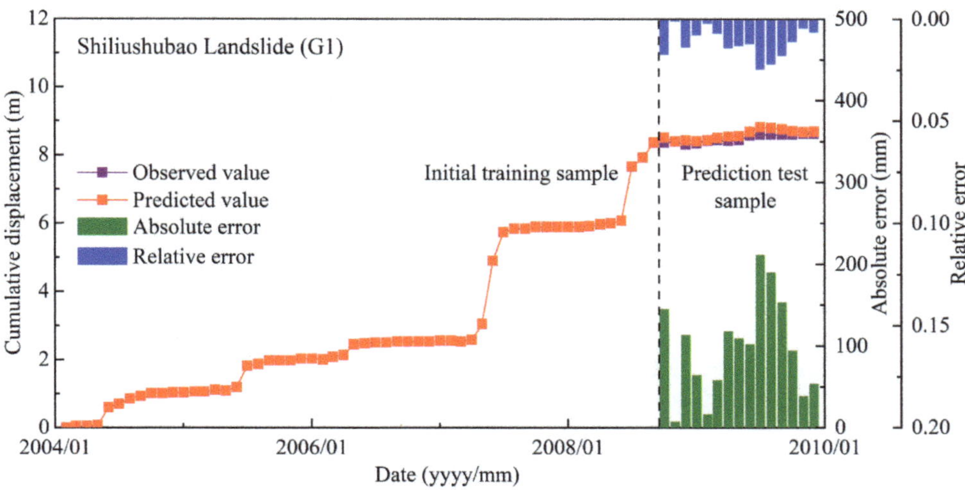

Figure 17. Comparison of predicted displacement and observed displacement.

To further verify the effectiveness of the proposed prediction model, the displacement at ZG93 of the well-known Baishuihe landslide is chosen as another case and predicted. The prediction accuracy of each SI and cumulative displacement prediction result are shown and compared in Table 5 and Figure 18.

Table 5. Prediction accuracy of each SI in Baishuihe landslide.

Optimization Algorithm	Periodic Displacement				Trend Displacement			
	MAPE	RMSE	MAE	R^2	MAPE	RMSE	MAE	R^2
BA	0.688	13.691	30.118	0.757	0.395	1065.132	926.683	0.8621
DA	0.788	13.652	30.367	0.761	0.008	20.448	66.336	0.9997
GOA	0.692	13.663	30.110	0.758	0.008	19.649	66.214	0.9997
GWO	0.680	13.592	29.558	0.751	0.008	20.448	66.336	0.9997
SSA	0.786	13.589	30.307	0.762	0.009	22.766	64.733	0.9998
WOA	0.788	13.629	30.329	0.761	0.008	20.448	66.336	0.9997

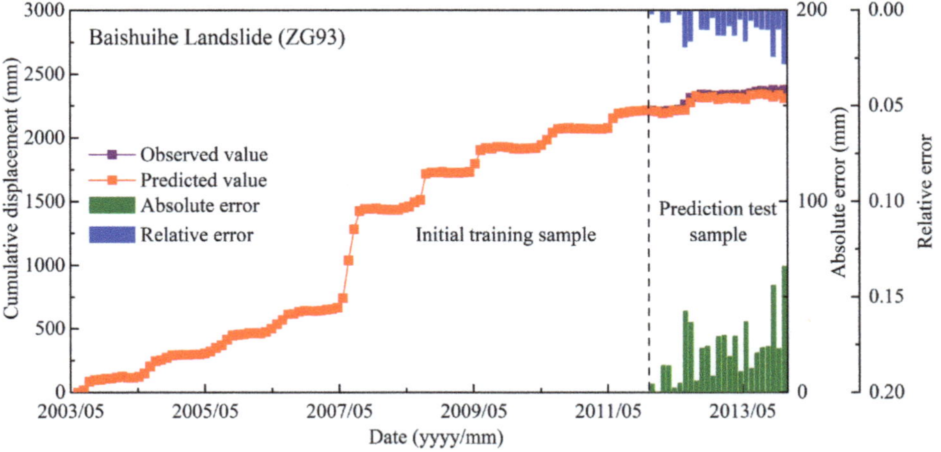

Figure 18. Comparison of predicted displacement and measured displacement.

The SSA method has achieved the best results in predicting both periodic displacement and trend displacement, with the largest value of R^2, which is 0.762 and 0.9998, respectively. The cumulative displacement prediction results are in good agreement with the measured displacement, with an absolute error of monthly displacement that is generally less than 67mm and the maximum relative error of monthly displacement that less than 3%. The average relative error of the proposed prediction model is 0.898%, which is slightly smaller than the result obtained by the prediction model of Deng et al. [54]. The comparative study shows the effective improvement of the proposed model in terms of prediction performance and the universality of it to predict the displacement of slow-moving landslides all around the world.

6. Discussion

This paper aims to improve the accuracy of landslide displacement prediction by constructing a novel prediction model combined with the CEEMD method, EDR method, and multi-swarm-intelligence (MSI) algorithm. The new prediction model can forecast landslide movements so that the landslide status can be evaluated, and appropriate stabilization measures can be implemented in advance to reduce the destructive effects of landslide movements. The CEEMD method was first employed for the landslide displacement decomposition, and a new prediction based on this was proposed to overcome its defect by optimizing the model's framework. The trend displacement obtained from CEEMD decomposition can reflect the long-term trend of landslide deformation. The periodic displacement obtained from CEEMD decomposition shows a cyclical variation in displacements consistent with periodic changes of related factors such as rainfall and reservoir levels. The frequency components of related factors that change periodically can be decomposed by the CEEMD. Combining with the *t*-test, the high-frequency and low-frequency components of related factors can be separated. With the EDR method, the most relevant factors related to the landslide displacements among the original related factors, reconstruction related factors, and frequency related factors can be selected by calculating the distance between all related factors and the extracted displacement component. The relevant factors that were identified are consistent with the results obtained by GRA.

The factors related to landslide displacement prediction can be separated into three groups: rainfall, reservoir level, and previous displacement. The most relevant factors for the Shiliushubao landslide's periodic displacement are $L4^H$, $X1^H$, and $D1^H$, and the most relevant factors for the trend displacements are the residual terms of L2, X1, and D2. MSI (BA, DA, GOA, GWO, SSA, and WOA) was used to optimize the proposed prediction model. For the Shiliushubao landslide, the DA-based SVR model performs best to predict periodic displacements, and the GWO-based SVR model works best for predicting trend displacements. The prediction of cumulative displacements is in good agreement with the measured displacements with a maximum relative error of monthly displacement of less than 3%. The trail of the proposed model on the Baishuihe landslide, another landslide in the reservoir area, is also satisfied with the average relative error of 0.898%, which performs slightly better than that from the previous study.

While the proposed methodology yielded satisfactory results, there are also some limitations. First, the CEEMD method has limits in the decomposition of measured displacements and related factors when the time series does not have enough extreme points, which limits the applicability of this method. When there is only one IMF sequence after CEEMD decomposition, a *t*-test cannot be carried out, and the IMF itself is high frequency. Second, the trend displacement and residue term for related factors after CEEMD decomposition may still have local fluctuations. It might contain some periodic fluctuation information, which can lead to prediction error, which needs to be further studied in the future. Third, the values of thrsh, sthresh, N, and alpha used in CEEMD will have an indirect impact on the prediction results. The appropriate range of these parameters and their influence on the results are still unclear. Fourth, when using MSI to optimize the parameters of the SVR model, the search for the g value is usually close to the lowest value of the search interval, and the result does not gradually increase as the search boundary continues to widen. Different SI optimization algorithms may perform differently for different landslides, so for new landslide data, the trial of different optimization algorithms for the best results is needed.

The deformation and failure of landslides are usually closely bonded with the groundwater effect [55]. The evaluation of the landslide stability with groundwater nowadays has developed into several hotspot branches, which includes analytical methods, such as the Limit Equilibrium Analysis with the Reliability Analysis and the Intelligent Algorithms on sliding zone searching, and numerical methods, such as the Fast Lagrange Analysis, the Finite Element Method, and the Discrete Element Method coupled with hydraulic calculations. These advanced evaluation methods have their status in the practical industry on the slope stability and deformation assessment, based on the current state and data gathered in the field and laboratory; however, these mechanism-based methods took insufficient account of the history state and data of the slope. The novel prediction model proposed in this paper can consider the historical influence of rainfall and reservoir fluctuation that precisely related to the displacement periodic component and displacement trend component with the help of the CEEMD method and EDR, thus improving the accuracy of landslide displacement prediction. It is a profitable attempt and a good way to improve the accuracy of landslide movement prediction. Although some in-depth research in consideration of historical factors of inducing factors has been carried out in this study, the predictive capability of the proposed model is still flawed in the sense that they cannot say anything about changes that are caused by external factors not captured by the available data series. Therefore, it is very important to develop multi-field (displacement field, seepage field, stress field, etc.) monitoring technology for the landslides, and the innovative prediction models based on this can more reflect the evolution process of the sliding mass.

Landslides in complex water environments could develop different deformation patterns, both categorized by history data and potential failure mechanism [4]. The pattern is highly related to the interaction between soil and water in a certain engineering geological condition. In the proposed novel displacement prediction model, the interaction mechanism is still not included, which limits the adaptability and comparability among different landslide cases. A better insight into the landslide development patterns is to be developed, combining the failure-mechanism-based evaluation method, in the future model for displacement prediction. Other than from the pure displacement prediction based on displacement, rainfall, and water level data sequence, an evaluation of the critical rainfall intensity and critical water level fluctuation rate is needed to be conducted under certain landslide development patterns in the further study.

In addition, landslide displacement is a noisy and non-stationary process that varies with time, which is highly affected by internal factors such as formation lithology and geological structure and external factors such as the rainfall, reservoir water level, and snow melting. Due to the complex nonlinear relationship between all these various inducing factors and landslide displacement, the landslide displacement prediction is subject to considerable uncertainties [56]. The limitations of the machine learning model, parameter selection, and data noise will increase the uncertainty of prediction [57]. The prediction model proposed in this paper is a deterministic point prediction model which cannot estimate the variability and uncertainty related to a given landslide displacement prediction, which limits its reliability under uncertain conditions. This should be addressed in the future study.

7. Conclusions

A reservoir landslide's movement is closely associated with the related factors including reservoir level fluctuations, rainfall intensity, and previous deformations. The complex nonlinear relationship between all these various inducing factors and landslide displacement increased the challenge of forecasting in the form of considerable uncertainties. In this study, a novel prediction model for landslide displacement prediction was proposed to improve the accuracy by the combination of multiple algorithms. The EDR method can identify the most relevant factors influencing a landslide's movements to use as input variables for an SVR model. The CEEMD method is suitable for the decomposition of various time series and can be used to extract the trend displacement of slow-moving landslide displacement. The CEEMD method can also highlight local fluctuations in the time series of related factors, and the frequency components of these time series can be extracted by combining the t-test method. With the help of MSI optimization algorithms, the optimal value of the penalty factor C and the kernel function parameter g for an SVR model can be obtained. This paper proposes an SVR model based on the CEEMD method, EDR selection, and MSI optimization algorithm that can capture the deformation characteristics of the landslide before failure.

Measurements of landslide displacements for the Shiliushubao landslide in the TGRA were used to demonstrate the novel displacement prediction model. The predicted displacements, including season fluctuations and the long-term trend, were found to be consistent with the observed data, which indicates that the proposed model has good predictive performance, even when the displacement characteristics are cyclic and complex. The DA- and GWO-based SVR model provided the best prediction of periodic displacement and trend displacement, respectively. The prediction model proposed in this paper has wider applicability. It can enhance the prediction of landslide displacements characterized by slow-moving, step-like displacements that are influenced by multiple related factors with frequency conversion characteristics.

Author Contributions: Conceptualization, J.Z.; methodology, J.Z.; supervision, H.T.; writing—original draft preparation, J.Z. and D.D.T.; funding acquisition, H.T.; resources, C.L.; software, C.L.; investigation, Y.W. and D.X.; visualization, D.X.; writing—review and editing, Q.W. All authors have read and agreed to the published version of the manuscript.

Funding: This work was supported by the Major Program of the National Natural Science Foundation of China (No. 42090055), the National Key Research and Development Program of China (No. 2017YFC1501305), the National Major Scientific Instruments and Equipment Development Projects of China (No. 41827808), the National Natural Sciences Foundation of China under Grant Nos. 42177147, 41807263 and China Postdoctoral Science Foundation (Grant No. 2021M703002).

Acknowledgments: The authors gratefully appreciate the editor and anonymous reviewers for their constructive criticism and comments on the earlier version of this paper and for offering valuable suggestions that have helped to improve this paper.

Conflicts of Interest: The authors declare no conflict of interest.

References

1. Tang, H.; Wasowski, J.; Juang, C.H. Geohazards in the three Gorges Reservoir Area, China–Lessons learned from decades of research. *Eng. Geol.* **2019**, *261*, 105267. [CrossRef]
2. Niu, M.; Wang, Y.; Sun, S.; Li, Y. A novel hybrid decomposition-and-ensemble model based on CEEMD and GWO for short-term PM2.5 concentration forecasting. *Atmos. Environ.* **2016**, *134*, 168–180. [CrossRef]
3. Wang, J.; Schweizer, D.; Liu, Q.; Su, A.; Hu, X.; Blum, P. Three-dimensional landslide evolution model at the Yangtze River. *Eng. Geol.* **2021**, *292*, 106275. [CrossRef]
4. Wang, Y.; Tang, H.; Wen, T.; Ma, J. A hybrid intelligent approach for constructing landslide displacement prediction intervals. *Appl. Soft Comput.* **2019**, *81*, 105506. [CrossRef]
5. Zhang, Y.G.; Tang, J.; He, Z.Y.; Tan, J.; Li, C. A novel displacement prediction method using gated recurrent unit model with time series analysis in the Erdaohe landslide. *Nat. Hazards* **2020**, *105*, 783–813. [CrossRef]
6. Zhang, J.; Tang, H.; Tannant, D.D.; Lin, C.; Xia, D.; Liu, X.; Zhang, Y.; Ma, J. Combined forecasting model with CEEMD-LCSS reconstruction and the ABC-SVR method for landslide displacement prediction. *J. Clean. Prod.* **2021**, *293*, 126205. [CrossRef]
7. Yin, Y.P.; Huang, B.L.; Wang, W.P.; Wei, Y.J.; Ma, X.H.; Ma, F.; Zhao, C.J. Reservoir-induced landslides and risk control in Three Gorges Project on Yangtze River, China. *J. Rock Mech. Geotech. Eng.* **2016**, *8*, 577–595. [CrossRef]
8. Zhou, C.; Yin, K.; Cao, Y.; Intrieri, E.; Ahmed, B.; Catani, F. Displacement prediction of step-like landslide by applying a novel kernel extreme learning machine method. *Landslides* **2018**, *15*, 2211–2225. [CrossRef]
9. Ma, J.; Niu, X.; Tang, H.; Wang, Y.; Wen, T.; Zhang, J. Displacement Prediction of a Complex Landslide in the Three Gorges Reservoir Area (China) Using a Hybrid Computational Intelligence Approach. *Complexity* **2020**, *2020*, 2624547. [CrossRef]

10. Saito, M. Forecasting the time of occurrence of a slope failure. In Proceedings of the 6th International Congress on Soil Mechanics and Foundation Engineering, Montreal, QC, Canada, 8–15 September 1965; pp. 537–541.
11. Crosta, G.B.; Agliardi, F. Failure forecast for large rock slides by surface displacement measurements. *Can. Geotech. J.* **2003**, *40*, 176–191. [CrossRef]
12. Ma, J.W.; Tang, H.M.; Liu, X.; Hu, X.L.; Sun, M.J.; Song, Y.J. Establishment of a deformation forecasting model for a step-like landslide based on decision tree C5.0 and two-step cluster algorithms: A case study in the Three Gorges Reservoir area, China. *Landslides* **2017**, *14*, 1275–1281. [CrossRef]
13. Zou, Z.; Yang, Y.; Fan, Z.; Tang, H.; Zou, M.; Hu, X.; Xiong, C.; Ma, J. Suitability of data preprocessing methods for landslide displacement forecasting. *Stoch. Environ. Res. Risk Assess.* **2020**, *34*, 1105–1119. [CrossRef]
14. Ma, J.; Liu, X.; Niu, X.; Wang, Y.; Wen, T.; Zhang, J.; Zou, Z. Forecasting of landslide displacement using a probability-scheme combination ensemble prediction technique. *Int. J. Environ. Res. Public Health* **2020**, *17*, 4788. [CrossRef]
15. Tharwat, A.; Gabel, T. Parameters optimization of support vector machines for imbalanced data using social ski driver algorithm. *Neural. Comput. Appl.* **2019**, *32*, 6925–6938. [CrossRef]
16. Cai, Z.; Xu, W.; Meng, Y.; Shi, C.; Wang, R. Prediction of landslide displacement based on GA-LSSVM with multiple factors. *Bull. Eng. Geol. Environ.* **2015**, *75*, 637–646. [CrossRef]
17. Zhou, C.; Yin, K.; Cao, Y.; Ahmed, B. Application of time series analysis and PSO–SVM model in predicting the Bazimen landslide in the Three Gorges Reservoir, China. *Eng. Geol.* **2016**, *204*, 108–120. [CrossRef]
18. Zhang, J.; Yin, K.; Wang, J.; Huang, F. Displacement prediction of Baishuihe landslide based on time series and PSO-SVR model. *Chin. J. Rock Mech. Eng.* **2015**, *34*, 382–391.
19. Peng, L.; Niu, R.; Wu, T. Time series analysis and support vector machine for landslide displacement prediction. *J. Zhejiang Univ. (Eng. Sci.)* **2013**, *47*, 1672–1679.
20. Zhou, C.; Yin, K.; Cao, Y.; Ahmed, B.; Fu, X. A novel method for landslide displacement prediction by integrating advanced computational intelligence algorithms. *Sci. Rep.* **2018**, *8*, 7287. [CrossRef]
21. Ding, L.; Lv, J.; Li, X.; Li, L. Support vector regression and ant colony optimization for HVAC cooling load prediction. In Proceedings of the 2010 International Symposium on Computer, Communication, Control and Automation (3CA), IEEE, Tainan, Taiwan, 5–7 May 2010; Volume 1, pp. 537–541.
22. Balogun, A.L.; Rezaie, F.; Pham, Q.B.; Gigović, L.; Drobnjak, S.; Aina, Y.A.; Panahi, M.; Yekeen, S.T.; Lee, S. Spatial prediction of landslide susceptibility in western Serbia using hybrid support vector regression (SVR) with GWO, BAT and COA algorithms. *Geosci. Front.* **2021**, *12*, 101104. [CrossRef]
23. Li, L.W.; Wu, Y.P.; Miao, F.S.; Liao, K.; Zhang, F.L. Displacement prediction of landslides based on variational mode decomposition and GWO-MIC-SVR model. *Chin. J. Rock Mech. Eng.* **2018**, *37*, 1395–1406. (In Chinese)
24. Miao, F.; Wu, Y.; Xie, Y.; Li, Y. Prediction of landslide displacement with step-like behavior based on multialgorithm optimization and a support vector regression model. *Landslides* **2018**, *15*, 475–488. [CrossRef]
25. Zhang, J.; Tang, H.; Wen, T.; Ma, J.; Tan, Q.; Xia, D.; Liu, X.; Zhang, Y. A Hybrid Landslide Displacement Prediction Method Based on CEEMD and DTW-ACO-SVR—Cases Studied in the Three Gorges Reservoir Area. *Sensors* **2020**, *20*, 4287. [CrossRef]
26. Zhang, Y.; Agarwal, P.; Bhatnagar, V.; Balochian, S.; Yan, J. Swarm Intelligence and Its Applications. *Sci. World J.* **2013**, *2013*, 528069. [CrossRef]
27. Chen, W.; Tsangaratos, P.; Ilia, I.; Duan, Z.; Chen, X. Groundwater spring potential mapping using population-based evolutionary algorithms and data mining methods. *Sci. Total Environ.* **2019**, *684*, 31–49. [CrossRef]
28. Kişi, Ö. Streamflow Forecasting Using Different Artificial Neural Network Algorithms. *J. Hydrol. Eng.* **2007**, *12*, 532–539. [CrossRef]
29. Beni, G. From Swarm Intelligence to Swarm Robotics. In *Swarm Robotics*; Springer: Berlin/Heidelberg, Germany, 2005; pp. 1–9.
30. Liu, Y.; Wang, R. Study on network traffic forecast model of SVR optimized by GAFSA. *Chaos Solitons Fractals* **2016**, *89*, 153–159. [CrossRef]
31. Ali, E.S.; Abd Elazim, S.M.; Abdelaziz, A.Y. Ant Lion Optimization Algorithm for Renewable Distributed Generations. *Energy* **2016**, *116*, 445–458. [CrossRef]
32. Jiang, H.; Yang, Y.; Ping, W.; Dong, Y. A Novel Hybrid Classification Method Based on the Opposition-Based Seagull Optimization Algorithm. *IEEE Access* **2020**, *8*, 100778–100790. [CrossRef]
33. Yang, X.S.; He, X. Bat algorithm: Literature review and applications. *Int. J. Bio-Inspir. Comput.* **2013**, *5*, 141. [CrossRef]
34. Emary, E.; Zawbaa, H.M.; Hassanien, A.E. Binary grey wolf optimization approaches for feature selection. *Neurocomputing* **2016**, *172*, 371–381. [CrossRef]
35. Mirjalili, S. Dragonfly algorithm: A new meta-heuristic optimization technique for solving single-objective, discrete, and multi-objective problems. *Neural Comput. Appl.* **2015**, *27*, 1053–1073. [CrossRef]
36. Mirjalili, S.; Lewis, A. The Whale Optimization Algorithm. *Adv. Eng. Softw.* **2016**, *95*, 51–67. [CrossRef]
37. Mirjalili, S.Z.; Mirjalili, S.; Saremi, S.; Faris, H.; Aljarah, I. Grasshopper optimization algorithm for multi-objective optimization problems. *Appl. Intell.* **2017**, *48*, 805–820. [CrossRef]
38. Xue, J.; Shen, B. A novel swarm intelligence optimization approach: Sparrow search algorithm. *Syst. Sci. Control* **2020**, *8*, 22–34. [CrossRef]

39. Du, H.; Song, D.; Chen, Z.; Shu, H.; Guo, Z. Prediction model oriented for landslide displacement with step-like curve by applying ensemble empirical mode decomposition and the PSO-ELM method. *J. Clean. Prod.* **2020**, *270*, 122248. [CrossRef]
40. Li, Y.; Sun, R.; Yin, K.; Xu, Y.; Chai, B.; Xiao, L. Forecasting of landslide displacements using a chaos theory based wavelet analysis-Volterra filter model. *Sci. Rep.* **2019**, *9*, 19853. [CrossRef]
41. Xu, S.; Niu, R. Displacement prediction of Baijiabao landslide based on empirical mode decomposition and long short-term memory neural network in Three Gorges area, China. *Comput. Geosci.* **2018**, *111*, 87–96. [CrossRef]
42. Ren, F.; Wu, X.; Zhang, K.; Niu, R. Application of wavelet analysis and a particle swarm-optimized support vector machine to predict the displacement of the Shuping landslide in the Three Gorges, China. *Environ. Earth Sci.* **2014**, *73*, 4791–4804. [CrossRef]
43. Yang, B.; Yin, K.; Lacasse, S.; Liu, Z. Time series analysis and long short-term memory neural network to predict landslide displacement. *Landslides* **2019**, *16*, 677–694. [CrossRef]
44. Huang, H.F.; Wu, Y.I.; Yi-Liang, L. Study on variables selection using SVR-MIV method in displacement prediction of landslides. *Chin. J. Undergr. Space Eng.* **2016**, *12*, 213–219. (In Chinese)
45. Chen, L.; Özsu, M.T.; Oria, V. Robust and fast similarity search for moving object trajectories. In *SIGMOD '05, Proceedings of the 24th ACM International Conference on Management of Data, New York, NY, USA, 13–15 June 2005*; ACM Press: New York, NY, USA, 2005; pp. 491–502. [CrossRef]
46. Mai, S.T.; Goebl, S.; Plant, C. A Similarity Model and Segmentation Algorithm for White Matter Fiber Tracts. In Proceedings of the 2012 IEEE 12th International Conference on Data Mining, IEEE, Washington, DC, USA, 10–13 December 2012.
47. Xu, Y.; Zhang, M.; Zhu, Q.; He, Y. An improved multi-kernel RVM integrated with CEEMD for high-quality intervals prediction construction and its intelligent modeling application. *Chemometr. Intell. Lab. Syst.* **2017**, *171*, 151–160. [CrossRef]
48. Ranacher, P.; Tzavella, K. How to compare movement? A review of physical movement similarity measures in geographic information science and beyond. Cartogr. *Geogr. Inf. Sci.* **2014**, *41*, 286–307. [CrossRef]
49. Moayedi, A.; Abbaspour, R.A.; Chehreghan, A. An evaluation of the efficiency of similarity functions in density-based clustering of spatial trajectories. *Ann. GIS* **2019**, *25*, 313–327. [CrossRef]
50. Liu, H.; Mi, X.; Li, Y.; Duan, Z.; Xu, Y. Smart wind speed deep learning based multi-step forecasting model using singular spectrum analysis, convolutional Gated Recurrent Unit network and Support Vector Regression. *Renew. Energy* **2019**, *143*, 842–854. [CrossRef]
51. Mirjalili, S.; Mirjalili, S.M.; Lewis, A. Grey Wolf Optimizer. *Adv. Eng. Softw.* **2014**, *69*, 46–61. [CrossRef]
52. Li, Z.; Xie, Y.; Li, X.; Zhao, W. Prediction and application of porosity based on support vector regression model optimized by adaptive dragonfly algorithm. *Energy Sources Part A Recovery Util. Environ. Eff.* **2019**, *43*, 1073–1086. [CrossRef]
53. Barman, M.; Dev Choudhury, N.B. Hybrid GOA-SVR technique for short term load forecasting during periods with substantial weather changes in North-East India. *Procedia Comput. Sci.* **2018**, *143*, 124–132. [CrossRef]
54. Deng, D.; Liang, Y.; Wang, L.; Wang, C.-S.; Sun, Z.-H.; Wang, C.; Dong, M.-M. Displacement prediction method based on ensemble empirical mode decomposition and support vector machine regression—A case of landslides in Three Gorges Reservoir area. *Rock Soil Mech.* **2017**, *38*, 3660–3669.
55. Hongtao, N. Smart safety early warning model of landslide geological hazard based on BP neural network. *Saf. Sci.* **2020**, *123*, 104572. [CrossRef]
56. Adnan, M.S.G.; Rahman, M.S.; Ahmed, N.; Ahmed, B.; Rabbi, M.F.; Rahman, R.M. Improving Spatial Agreement in Machine Learning-Based Landslide Susceptibility Mapping. *Remote. Sens.* **2020**, *12*, 3347. [CrossRef]
57. Ahmad, H.; Ningsheng, C.; Rahman, M.; Islam, M.M.; Pourghasemi, H.R.; Hussain, S.F.; Habumugisha, J.M.; Liu, E.; Zheng, H.; Ni, H.; et al. Geohazards Susceptibility Assessment along the Upper Indus Basin Using Four Machine Learning and Statistical Models. *ISPRS Int. J. Geo-Inf.* **2021**, *10*, 315. [CrossRef]

 sensors

Article

Using Complementary Ensemble Empirical Mode Decomposition and Gated Recurrent Unit to Predict Landslide Displacements in Dam Reservoir

Beibei Yang [1], Ting Xiao [2,*], Luqi Wang [3] and Wei Huang [4]

[1] School of Civil Engineering, Yantai University, Yantai 264005, China; yangbeibei@ytu.edu.cn
[2] School of Geosciences and Info-Physics, Central South University, Changsha 410083, China
[3] School of Civil Engineering, Chongqing University, Chongqing 400044, China; wlq93@cqu.edu.cn
[4] The Seventh Geological Brigade of Hubei Geological Bureau, Yichang 443000, China; huangwei@cug.edu.cn
* Correspondence: shouting1993@163.com; Tel.: +86-180-8649-8957

Abstract: It is crucial to predict landslide displacement accurately for establishing a reliable early warning system. Such a requirement is more urgent for landslides in the reservoir area. The main reason is that an inaccurate prediction can lead to riverine disasters and secondary surge disasters. Machine learning (ML) methods have been developed and commonly applied in landslide displacement prediction because of their powerful nonlinear processing ability. Recently, deep ML methods have become popular, as they can deal with more complicated problems than conventional ML methods. However, it is usually not easy to obtain a well-trained deep ML model, as many hyperparameters need to be trained. In this paper, a deep ML method—the gated recurrent unit (GRU)—with the advantages of a powerful prediction ability and fewer hyperparameters, was applied to forecast landslide displacement in the dam reservoir. The accumulated displacement was firstly decomposed into a trend term, a periodic term, and a stochastic term by complementary ensemble empirical mode decomposition (CEEMD). A univariate GRU model and a multivariable GRU model were employed to forecast trend and stochastic displacements, respectively. A multivariable GRU model was applied to predict periodic displacement, and another two popular ML methods—long short-term memory neural networks (LSTM) and random forest (RF)—were used for comparison. Precipitation, reservoir level, and previous displacement were considered to be candidate-triggering factors for inputs of the models. The Baijiabao landslide, located in the Three Gorges Reservoir Area (TGRA), was taken as a case study to test the prediction ability of the model. The results demonstrated that the GRU algorithm provided the most encouraging results. Such a satisfactory prediction accuracy of the GRU algorithm depends on its ability to fully use the historical information while having fewer hyperparameters to train. It is concluded that the proposed model can be a valuable tool for predicting the displacements of landslides in the TGRA and other dam reservoirs.

Keywords: reservoir landslide; displacement prediction; time series analysis; complementary ensemble empirical mode decomposition; gated recurrent unit

Citation: Yang, B.; Xiao, T.; Wang, L.; Huang, W. Using Complementary Ensemble Empirical Mode Decomposition and Gated Recurrent Unit to Predict Landslide Displacements in Dam Reservoir. *Sensors* **2022**, *22*, 1320. https://doi.org/10.3390/s22041320

Academic Editor: Francesca Cigna

Received: 16 December 2021
Accepted: 3 February 2022
Published: 9 February 2022

1. Introduction

Landslides are one of the most catastrophic disasters and are widely distributed in numerous parts of the world [1–4]. In China, annual reports from China Institute of Geo-Environment Monitoring (IGEM) show that landslides account for more than 50% of all geological hazards in recent years [5]. In 2020, for instance, 7840 geology-related hazards occurred in China, resulting in 139 deaths or people missing, 58 people injured, and a direct economic loss of CNY 5.02 billion. Among these geological disasters, 4810 were landslides, accounting for 61.3% of the total. Other types of hazards in 2020 included 1797 avalanches, 899 debris flows, 183 ground collapses, 143 ground fissures, and 8 cases of ground subsidence.

As one of the most landslide-prone areas in China, the Three Gorges Reservoir Area (TGRA) has been given much attention concerning severe landslides [6]. One main reason is that the construction of the Three Gorges Dam (TGD) has significantly changed the regional hydrogeological conditions [7,8]. Some landslides in the TGRA (e.g., Bazimen landslide) have deformed continuously for several decades, whereas some landslides (e.g., Woshaxi landslide) have achieved a displacement of 28,065.9 mm, and the deformation is still increasing [9,10]. Once landslides in dam reservoirs occur, they can cause severe damage along both sides of the reservoir area. In addition, these reservoir landslides can induce secondary surge disasters, endangering the shipping and bridges along the river and its tributaries [11]. The Honyanzi landslide, which occurred on 24 June 2015, was such an example, initiating a reservoir tsunami that resulted in two deaths and severe damage to shipping facilities (Figure 1) [12]. These risks can be mitigated if one can establish reliable early warning systems. As landslide displacement can represent its evolution intuitively, accurate landslide displacement prediction is an effective means of establishing such reliable early warning systems [10,13,14].

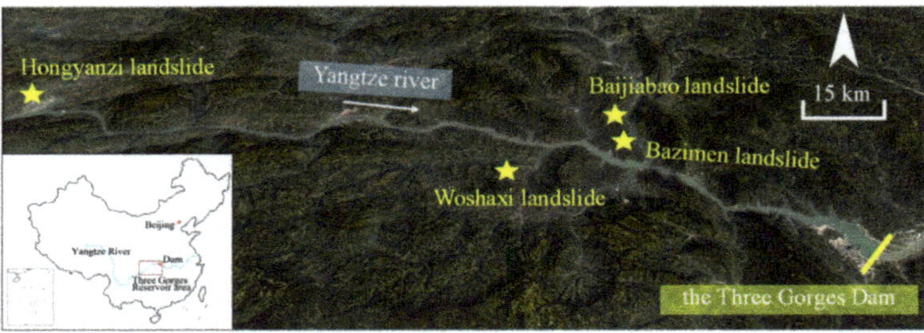

Figure 1. Location map of landslides in TGRA mentioned in the paper.

In situ displacement monitoring techniques have been available since the 1940s, especially the global positioning system (GPS) technique [15–17]. These techniques make it possible to acquire real-time monitoring information. These monitoring data have been applied extensively in landslide displacement prediction (LDP). The research of LDP dates back to the 1960s with the presentation of the Saito model. Subsequently, numerous LDP theories and models have been successively proposed [18]. The development of LDP research can be summarized into three stages [14,19,20]. The first stage (from the 1960s to 1970s) is the phenomenological and empirical prediction, mainly based on the macroscopic deformation phenomenon before landslide failure. The prediction accuracy is usually unsatisfied because of a high dependence on the gained experience. The second stage (during the 1980s) is the displacement-time statistical analysis prediction, leading qualitative prediction to quantitative prediction. Benefiting from the development of mathematical sciences, various statistical mathematical models have been proposed and applied to the LDP (e.g., grey system theory) [21]. Without considering influencing factors, these models are built from statistics and mathematics. Hence, these approaches are primarily valid for landslides with similar deformation characteristics [22]. The third stage (from the 1990s to the present) is the nonlinear prediction and intelligent integrated prediction. Numerous nonlinear and intelligent LDP models have been proposed and applied in cases. These models can build relationships between landslide displacement and multiple triggering factors. Their prediction performance has shown encouraging improvement.

As intelligent algorithms, machine learning (ML) models have been extensively utilized to predict landslide displacements because of their nonlinear processing ability. These models, such as the back-propagation (BP) neural network [23,24], extreme learning machine (ELM) [25–29], random forest (RF) [30,31], and support vector machine (SVM) [32–34], have become popular and have been adopted in some landslide cases in the TGRA. Influencing factors and displacement are set as the input and output of the models, respectively. The trained models have achieved encouraging performances. Zhou et al. [27] selected an artificial bees colony (ABC) to optimize the parameters of a kernel-based extreme learning machine (KELM) for LDP. Li et al. [28] proposed an ensemble-based ELM and copula model to predict the displacement of the Baishuihe landslide in the TGRA. Hu et al. [30] developed an integrated LDP model by combining the Verhulst inverse function (VIF) and RF algorithm, which provided a practical approach for predicting the long-term deformation of landslides. Bui et al. [34] adopted ABC optimization to model the least squares support vector regression (LSSVR). These forecasting models belong to static models, whereas the evolution of landslides is a complex nonlinear dynamic process [35]. The deformation conditions of landslides at one time can be affected by that of the former time [36]. A dynamic model—long short-term memory (LSTM) neural networks—was applied to LDP [9]. Jiang et al. [37] combined the support vector regression (SVR) algorithm and LSTM model to forecast the displacement of the Shengjibao landslide in the TGRA. As a deep ML method, LSTM can deal with more complicated time series predictions. With the increment of the number of available monitoring data and the improvements in computer hardware and software, the LSTM model has become a priority choice to deal with more complicated time series prediction [38,39]. One drawback of LSTM is that it has more parameters to be trained than classical ML methods, which makes it challenging to obtain the optimum of all parameters simultaneously [10]. An improved version of the LSTM—the gated recurrent unit (GRU)—is proposed and adopted in LDP. GRU replaced the three gates (input gate, forget gate, and output gate) of LSTM with two new gates (reset gate and update gate). This structure of GRU makes it possible to reduce the number of hyperparameters required for training. Thus, it can be easier for GRU to obtain a well-trained model than the LSTM [31].

In general, the LDP in the dam reservoir involves decomposing the total displacement into several components (trend term, periodic term, and stochastic term) according to time series analysis and then through predicting each component by different methods. Each displacement component has clear mathematical and physical significance. This treatment of LDP has been proven to be effective in previous studies [10,23,31,33,36,40–42]. Several decomposition methods have been adopted, such as the average moving method [10,33], double exponential smoothing [10], variational mode decomposition (VMD) [40], empirical mode decomposition (EMD) [37], ensemble empirical mode decomposition (EEMD) [40], and wavelet transform (WT) [41]. It is critical to forecast periodic displacement accurately to ensure the good prediction performance of accumulated displacement for landslides [23]. The prediction of periodic displacement is a heated topic, and the predictive models are summarized as mentioned above. The trend displacement is usually modeled and predicted by fitting the curve of displacement–time with polynomial functions [23,31,33]. A piecewise curve may need several polynomial functions [10]. Another displacement component—the stochastic term—is usually ignored [10,32,37,43]. The main reason is that stochastic displacement is influenced by varied, ever-present, and unquantifiable stochastic factors.

This paper decomposed accumulated displacement into a trend term, periodic term, and stochastic term by CEEMD. A univariate and a multivariable GRU model were used to predict the trend and stochastic displacements, respectively. A multivariable GRU model was adopted to predict periodic term displacement, and another two popular ML methods—LSTM and RF—were used for comparison. The proposed model was applied in the displacement prediction of the Baijiabao landslide in the TGRA. The deep dynamic model has the advantages of a powerful prediction ability with a simpler structure and fewer trained hyperparameters. In addition, the stochastic displacement, neglected in most exiting prediction models, was considered in the proposed model.

2. Approach to Model Displacements in Three Gorges Dam Reservoir

2.1. Time Series Decomposition

The change in landslide accumulated displacement is determined by geological conditions, triggering factors, and stochastic factors [10,33]. Geological conditions involve internal factors, such as the geological structure, topography, lithology, etc. Triggering factors for landslides in the TGRA are mainly the seasonal rainfall and reservoir level fluctuation. Stochastic factors appear with uncertainties, including earthquakes, traffic load, wind load, etc. The displacement components induced by the above three factors can be represented as trend displacement, periodic displacement, and stochastic displacement, respectively. Consequently, the accumulated displacement can be expressed as Equation (1):

$$\mathrm{A} = \mathrm{T} + \mathrm{P} + \mathrm{S} \tag{1}$$

where A is accumulated displacement, T is trend displacement, P is periodic displacement, and S is stochastic displacement.

2.2. Complementary Ensemble Empirical Mode Decomposition

Empirical mode decomposition (EMD) was firstly proposed by Huang et al. [44]. They implemented EMD by converting a nonlinear sequence into a set of stationary sequences that consisted of several intrinsic mode functions (IMFs) and a residual. EMD, however, has the disadvantage of mode mixing, and thus ensemble empirical mode decomposition (EEMD) was presented by Wu et al. [45]. In EEMD, uncorrelated finite white noise is added into the original signal, and the final IMF is obtained by averaging all the IMFs. Due to the dependence of the added noise in EEMD, Yeh et al. [46] presented a modified algorithm of EEMD named complete ensemble empirical mode decomposition (CEEMD) to decompose the signal into different scale IMFs. By adding opposite random white noise into the decomposition results of EEMD, CEEMD realized the advantages of an improved decomposition, better denoising, and higher computational efficiency. The following steps settle the process of CEEMD decomposing the original time series.

The first step is to add positive and negative white noise pairs to the original time series.

$$\begin{bmatrix} B_i(t) \\ C_i(t) \end{bmatrix} = \begin{bmatrix} 1 & 1 \\ 1 & -1 \end{bmatrix} \begin{bmatrix} S_i(t) \\ a_i(t) \end{bmatrix} \tag{2}$$

where $B_i(t)$ and $C_i(t)$ are the time series after adding positive and negative white noise, respectively, $S_i(t)$ is the original time series, and $a_i(t)$ is the added white noise.

Subsequently, the EMD algorithm is used to decompose $B_i(t)$ and $C_i(t)$.

$$\begin{cases} B_i(t) = \sum\limits_{j}^{J} IMF_{ij}^{+} \\ C_i(t) = \sum\limits_{j}^{J} IMF_{ij}^{-} \end{cases} \tag{3}$$

where J is the number of *IMF* after decomposing, and IMF_{ij}^{+} and IMF_{ij}^{-} are the jth components of IMF after adding positive and negative white noise, respectively.

N sets of *IMFs* can be obtained after repeating the above two steps.

$$\left\{ \begin{array}{c} \left\{ \left\{ IMF_{1j}^{+}, IMF_{2j}^{+}, \cdots, IMF_{Nj}^{+} \right\} \right\} \\ \left\{ \left\{ IMF_{1j}^{-}, IMF_{2j}^{-}, \cdots, IMF_{Nj}^{-} \right\} \right\} \end{array} \right\} \tag{4}$$

We can obtain the final jth *IMF* by averaging its positive and negative components.

$$IMF_j = \frac{1}{2N} \sum_{i=1}^{N} \left(IMF_{ij}^{+} + IMF_{ij}^{-} \right) \tag{5}$$

Finally, the time series $S_i(t)$ is decomposed as Equation (6):

$$S(t) = \sum_{j=1}^{N} IMF_j \quad (6)$$

2.3. Machine Learning Methods

2.3.1. Long Short-Term Memory Neural Network

Long short-term memory (LSTM) neural networks are in the category of dynamic recurrent neural networks (RNN). Due to the issues of gradient vanishing and gradient exploding in conventional RNN, they cannot handle the dependency of a long time series. To avoid such disadvantages of conventional RNN, Hochreite and Schmidhuber [47] proposed LSTM in 1997. In LSTM, a memory block is used as the basic unit of its hidden layer, consisting of a memory cell and three gates, named the input gate, forget gate, and output gate (Figure 2) [48].

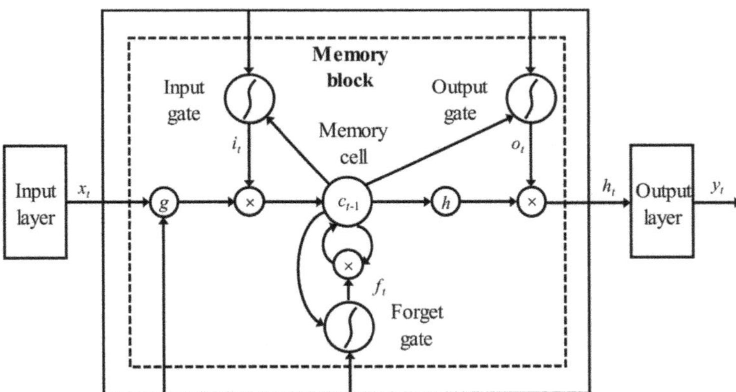

Figure 2. Architecture of LSTM neural network.

The input gate controls the flow of input activations into the memory cell. The information from the hidden state at step t − 1 (h_{t-1}) and the current input value (x_t) is firstly passed along to the sigmoid function (σ). Then, the information of input data from the current step and previous data from the last step is used to update and generate a new vector. The forget gate is responsible for filtering information by means of passing along useful information to the next step and abandoning useless information. The output gate controls the transfer of useful information into other memory blocks.

We recorded the input sequence as $x = (x_1, x_2, \ldots, x_T)$, and can obtain the output sequence $y = (y_1, y_2, \ldots, y_T)$ by treating Equation (7) to Equation (12).

$$i_t = \sigma(W_{xi}x_t + W_{hi}h_{t-1} + W_{ci}c_{t-1} + b_i) \quad (7)$$

$$f_t = \sigma\left(W_{xf}x_t + W_{hf}h_{t-1} + W_{cf}c_{t-1} + b_f\right) \quad (8)$$

$$c_t = f_t c_{t-1} + i_t \tan h(W_{xc}x_t + W_{hc}h_{t-1} + b_c) \quad (9)$$

$$O_t = \sigma(W_{xo}x_t + W_{ho}h_{t-1} + W_{co}c_{t-1} + b_o) \quad (10)$$

$$h_t = o_t \tan h(c_t) \quad (11)$$

$$y_t = W_{hy}h_t + b_y \quad (12)$$

where i_t, f_t, o_t, and c_t are the values of the input gate, forget gate, output gate, and a memory cell at time t; b_i, b_f, b_o, and b_c are their corresponding bias values; W_x are the weights between input nodes and hidden nodes; W_h are the weights between hidden nodes and cell memory; W_c are the weights connecting the memory cell to output nodes; σ is the sigmoid activation function; $\tan h$ is the hyperbolic tangent function mapping data to $[-1, 1]$; and h_t is the hidden state, containing information about the history of earlier elements in the series.

2.3.2. Gated Recurrent Unit

The gated recurrent unit (GRU) is an improved version of LSTM. Compared with LTSM, GRU has the advantages of fewer hyperparameters and faster training by using two new gates (update gate and reset gate) (Figure 3). These two gates are utilized to store as much information as possible for a long time series [49,50]. The reset gate is responsible for determining how much information at the previous moment is passed along, and resets the information at the current moment. The update gate controls the extent of information from both the previous time step and the current time step that will be passed along to the memory cell. The equations in GRU are given as follows:

$$u_t = \sigma(W_{xu}x_t + W_{hu}h_{t-1} + b_u) \tag{13}$$

$$r_t = \sigma(W_{xr}x_t + W_{hr}h_{t-1} + b_r) \tag{14}$$

$$h' = \tan\mathrm{h}(W_{xh}x_t + (r_t \odot h_{t-1})W_{hh} + b_h) \tag{15}$$

$$h_t = (1 - u_t) \odot h' + u_t \odot h_{t-1} \tag{16}$$

where u_t and r_t are the values of the upset gate and reset gate, respectively; h' is the value after resetting; W and b are the weights and deviations, respectively; $\odot$ represents pointwise multiplication between tensors. Other parameters indicate the same meaning as those in LSTM.

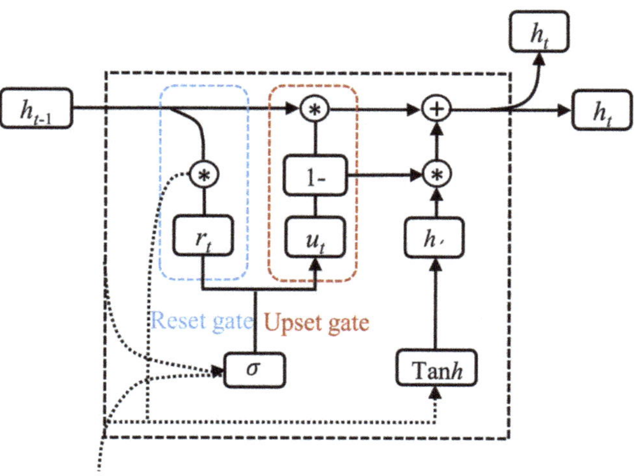

Figure 3. Structure chart of GRU.

2.3.3. Random Forest

Random Forest (RF) is an ensemble ML method that has been well-developed for classification, regression, and other tasks [51]. This method has some advantages, including great robustness, data adaptability, and low overfitting [52]. The RF algorithm is realized based on multiple decision trees by sampling from the original dataset (both samples and their features) [53].

To build a decision tree, we divide the predictor space into the number of J regions that are distinct and non-overlapping and represented as $R_1, \ldots, R_j$. The division is implemented by minimizing the root of the sum of squares.

$$\sum_{j=1}^{J} \sum_{i \in R_j} \left(y_i - \hat{y}_{R_j} \right)^2 \tag{17}$$

where y_i is the observation belonging to R_j, and $\hat{y}_{R_j}$ is the mean response for the training observations within the jth region.

Bagging is used to select training sets from the original dataset, and each training set is utilized for building a decision tree. The final prediction result $\hat{y}_{bag}$ can be achieved by averaging the results of all decision trees (Equation (18)), which can improve the prediction accuracy by doing so.

$$\hat{y}_{bag} = \frac{1}{M} \sum_{i=1}^{M} \hat{y}_i \tag{18}$$

where $\hat{y}_i$ is the prediction result of the ith decision tree and M is the number of decision trees.

2.4. Prediction Process with the Proposed Model

In the establishment of the proposed model (Figure 4), we adopted CEEMD to decompose the monitored accumulated displacement into a trend component and a periodic component. Subsequently, we used a univariate GRU model and a multivariate GRU model to predict the trend term and periodic term, respectively. The univariate GRU model described the trend displacement versus time, whereas the multivariate GRU model described the relationships between periodic displacement and influencing factors. A multivariate LSTM model and a multivariate RF model were also utilized for forecasting periodic displacement to verify the prediction performance of the GRU model. We adopted a multivariate GRU model to predict stochastic displacement.

The error analysis introduces the root mean square error (*RMSE*), mean absolute percentage error (*MAPE*), and the goodness of fit (R^2) for validations. Smaller values of *RMSE* and *MAPE* and a larger value of R^2 reflect a better prediction performance.

$$RMSE = \sqrt{\frac{1}{N} \sum_{i=1}^{N} (x_i - \hat{x}_i)^2} \tag{19}$$

$$MAPE = 100\% \times \frac{1}{N} \sum_{i=1}^{N} \left| \frac{x_i - \hat{x}_i}{x_i} \right| \tag{20}$$

$$R^2 = 1 - \frac{N \sum (x_i - \hat{x}_i)^2}{N \sum x_i^2 - \sum \hat{x}_i{}^2} \tag{21}$$

where x_i and $\hat{x}_i$ represent the ith observed displacement and predicted displacement, respectively, and N is the record number of displacement.

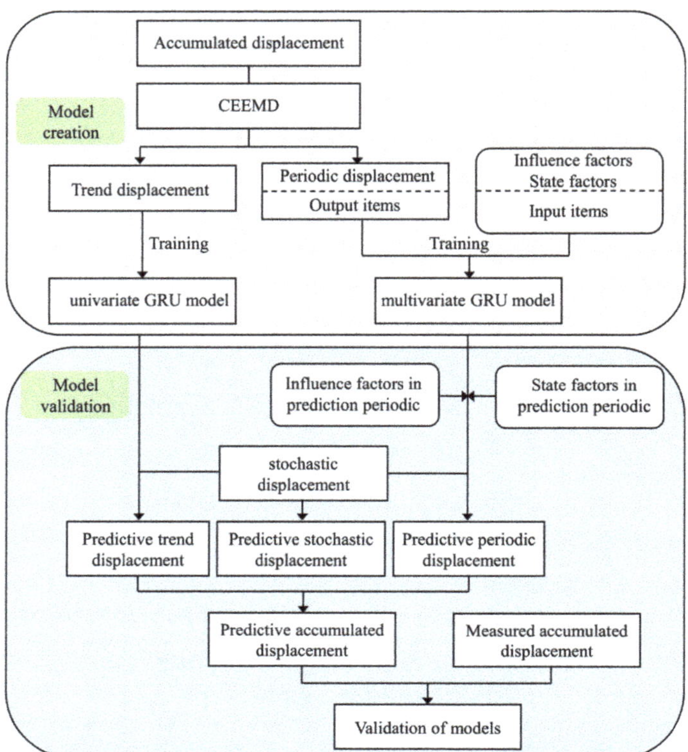

Figure 4. Flowchart of the proposed predictive model.

3. Baijiabao Landslide Case Study

3.1. Overview of the Baijiabao Landslide

3.1.1. Geological Conditions

The Baijiabao landslide is located on the west bank of the Xiangxi River and belongs to Zigui County, Hubei Province, China (Figure 5). The Xiangxi River is a major tributary of the Yangtze River, approximately 2.5 km upstream from the estuary. The main sliding direction of the landslide is perpendicular to the Xiangxi River and orientated at N 82° E. The front part of the landslide is submerged in the Xiangxi River, whereas the interface between bedrock and soil bounds the upper edge. The left and right boundaries are defined by seasonal homologous gullies (Figure 6). The landslide has a leading-edge elevation of 160–175 m, a trailing-edge elevation of 265 m, a width of approximately 550 m, a length of approximately 400 m, an average thickness of 45 m, and an estimated volume of 9.9×10^6 m^3 [25].

The sliding mass is mainly composed of silty clay and fragmented rubble. These sliding materials form a loose and disordered structure of the slope. The slip bed is silty mudstones and muddy siltstones of the Jurassic Xiangxi group, which dig into the hill by a direction of 260° with an angle of 30° [9]. The sliding surface is defined by the interface between colluvial materials and subjacent bedrock. The sliding zone is mainly composed of silty clay (Figure 7).

The Baijiabao landslide experienced large deformations since the impoundment of the Three Gorges Dam (TGD) in 2003 and kept deforming in the following years. In June 2007, tensile cracks with a length of 160 m and depth of 10 cm occurred at both side boundaries of the landslide close to the trailing edge. In May 2009, tensile cracks were observed on the road in the front and right parts of the landslide. A similar road deformation appeared in the middle of the landslide. In June 2012, cracks of the trailing edge showed a connecting tendency. Besides, cracks of the boundaries extended to the front part of the landslide. In June 2015, several tensile cracks, both on the right boundary and Zi-Xing road, became larger. Before the impoundment of the TGD, 165 residents used to live in the landslide area, whereas now, only 20 residents live there.

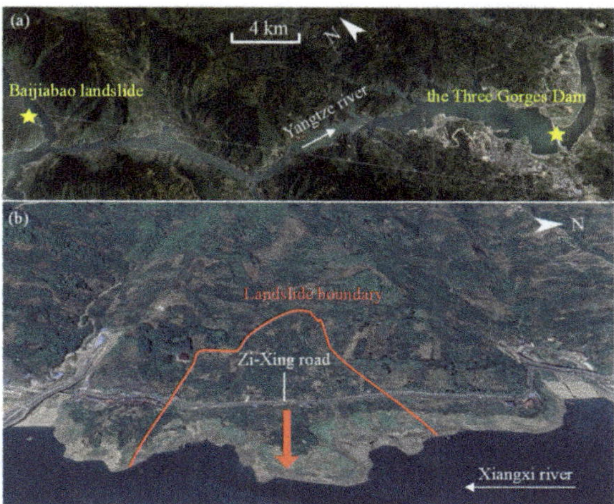

Figure 5. (**a**) Location of the Baijiabao landslide; (**b**) overall view of the Baijiabao landslide.

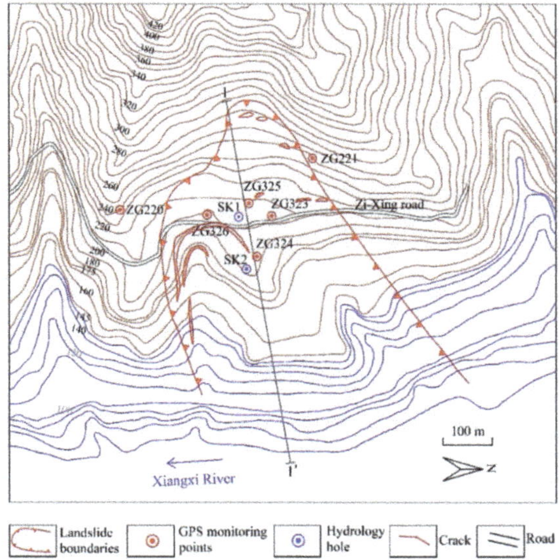

Figure 6. Monitoring arrangement in the Baijiabao landslide.

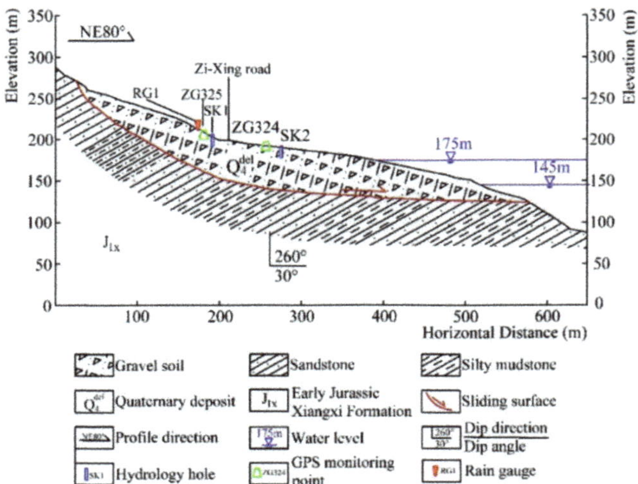

Figure 7. Schematic geological cross-section A–1′ of the Baijiabao landslide.

3.1.2. Monitoring Data and Deformation Characteristics of the Landslide

Four GPS stations numbered ZG323, ZG324, ZG325, and ZG326 were installed in the landslide area to monitor the surface displacements at one time per month since late 2006. Another two stations numbered ZG320 and ZG321 were established as the datum stations. Monitoring data from January 2007 to July 2018 were acquired (Figure 8). The displacements of the four monitoring stations showed a similar trend of step-wise, which meant that the landslide deformed distinctly in steps during April and September (especially from May to July) and became unremarkable in other times of the year.

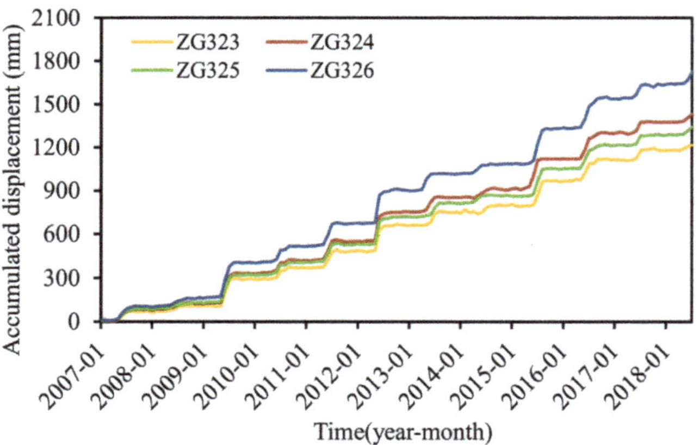

Figure 8. Accumulated displacement in the Baijiabao landslide.

Cao et al. [25] analyzed the deformation characteristics and evolution of the Baijiabao landslide. The analysis showed that the Baijiabao landslide deformed as an entity. Station ZG324, located in the central position of the landslide, was chosen as a representative for establishing the displacement forecasting model. Figure 9 displayed the accumulated displacements at station ZG324, monthly rainfall, and reservoir water level, and all the data were obtained by measurement. The annual displacement, displacement during step-wise deformation period (from May to September), and the maximum monthly displacement were summarized in Figure 10.

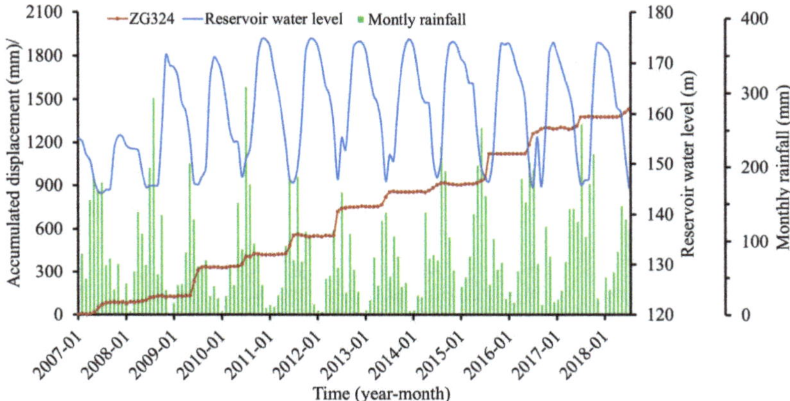

Figure 9. Rainfall, reservoir water level, and accumulated displacement at ZG324, Baijiabao landslide.

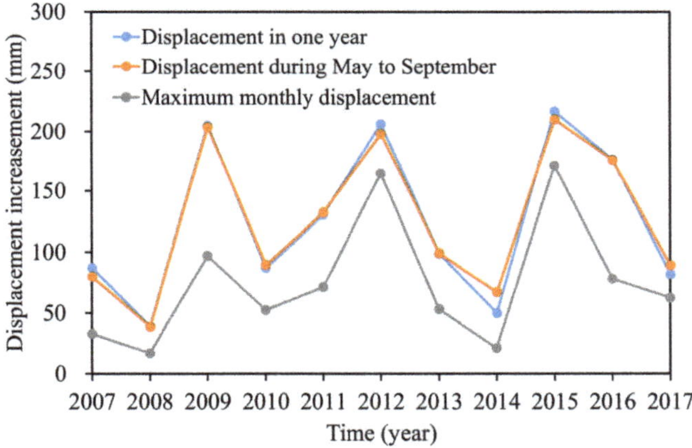

Figure 10. Annual displacement increment, displacement during step-wise deformation period, and the maximum monthly displacement at ZG324, Baijiabao landslide.

It can be seen that a sharp displacement increment occurred every few years (2009, 2012, and 2015) that was more than 200 mm (204.81 mm, 206.18 mm, and 216.92 mm, respectively). The displacement in other years increased by less than 100 mm. Another phenomenon was that the displacement during the step-wise deformation period (from May to September) contributed to the majority of the displacement in the whole year, especially from May to July, which contributed to more than 70% of the annual displacement. The maximum monthly displacement occurred in June or July each year, except 2015 (occurred in August). For example, the yearly displacement in 2012 was 206.18 mm; the displacement increment between May and July was 187.55 mm and occupied 91% of the whole year displacement. The maximum monthly rainfall occurred in June and was up to 164 mm. The reservoir level dropped between May and July 2012, and the cumulative rainfall rose to 349.73 mm. Thus, the time from May to July can be the critical early warning period for step-wise landslides. The deformation during this period was mainly controlled by reservoir water level decline and heavy rainfall.

3.2. Accumulated Displacement Decomposition

The monitored data of station ZG324 from January 2007 to July 2017 and from August 2017 to July 2018 were selected as training and testing data sets, respectively. An appropriate decomposition method is crucial in establishing a landslide displacement prediction model. Several methods have been used in accumulated displacement decomposition, as mentioned in the introduction, and each has advantages and disadvantages. Zhu et al. [54] and Fu et al. [55] have demonstrated that CEEMD is an effective method for reconstructing landslide displacement, with the advantages of a high stability and complete decomposition. Therefore, the CEEMD method was adopted here to decompose accumulated displacement into trend term and periodic term displacements.

In the training of the forecast model, we tested 200 trials and set the standard deviation of the added white noise in each ensemble to 0.25. We used the CEEMD to decompose the accumulated displacement into several IMFs and a residual, while the residual represented a trend component. Subsequently, we can obtain the periodic displacement by summing up all of the IMFs or subtracting the trend term from the accumulated displacement. Figure 11 displayed the trend and periodic components of ZG324 in the Baijiabao landslide.

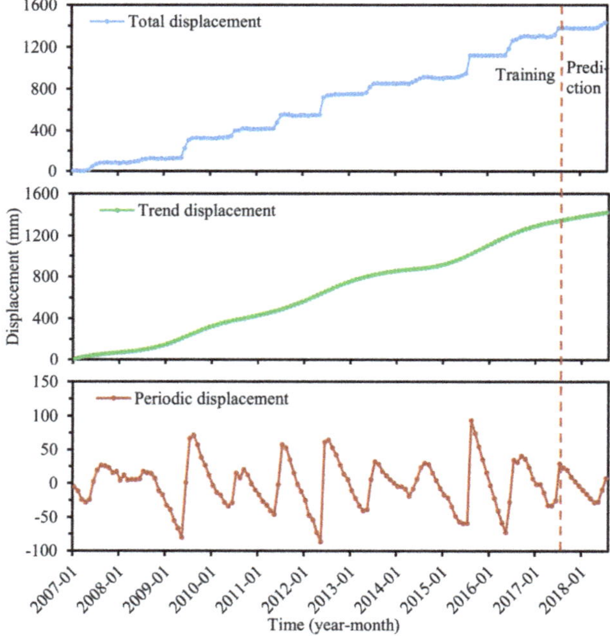

Figure 11. Displacement decomposition at ZG324.

3.3. Trend Displacement Prediction

Controlled by "internal" conditions, the trend displacement increases monotonically with time [23]. Some researchers forecasted trend displacement by fitting the displacement–time curve, and a polynomial was commonly used [33,37]. However, a single function can be insufficient to fit the curve properly [10]. A univariate GRU model was adopted to forecast the trend displacement in this study, and the established model achieved an excellent prediction performance (Figure 12). The prediction results of RMSE, MAPE, and R^2-values were 2.09 mm, 0.14%, and 0.9984, respectively.

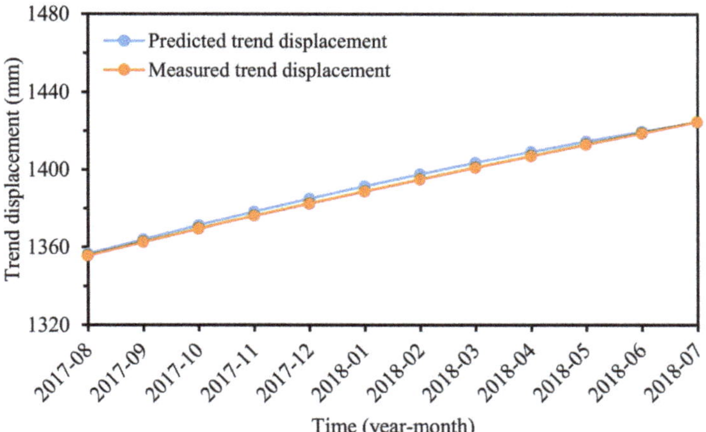

Figure 12. Predicted and measured trend displacement.

3.4. Periodic Displacement Prediction

3.4.1. Triggering Factors Selection

Triggering factors selection is essential to guarantee the accuracy of a displacement predictor. According to the monitoring data of the Baijiabao landslide (Figures 9 and 10), rainfall and reservoir water level fluctuation are two major factors triggering its step-wise deformation. Selby [56] proposed that the evolutionary state of landslides was also an influential factor in the dependence of the movement on external factors. By referring to the research [9,25,31,36] and our previous work [42], seven candidate triggering factors were considered here.

Gray relational analysis (GRA) was used to check the degree of correlation between the periodic displacement and candidate triggering factors [57]. In GRA, we chose periodic displacement and candidate triggering factors as primary sequence and sub-sequences, respectively. All the sequences were normalized in the following way:

$$X_k(i)' = X_k(i) / \frac{1}{n} \sum_{i=0}^{n} X_k(i) \tag{22}$$

where $i = 0, 1, \cdots, n$; $k = 0, 1, \cdots, m$; n is the number of data points; m is the number of candidate triggering factors. The correlation coefficients were thus obtained by Equation (23):

$$\delta\big((x_0(i)', x_k(i)'\big) = \frac{p + \rho q}{|X_k(i)' - X_0(i)'| + \rho q} \tag{23}$$

$$p = \min_k \min_i \big(X_k(i)' - X_0(i)'\big) \tag{24}$$

$$q = \max_k \max_i \big(X_k(i)' - X_0(i)'\big) \tag{25}$$

where ρ is the resolution coefficient and is usually set to 0.5.

The grey relational grade (GRG) was adopted to evaluate the correlation between variables, and was calculated by Equation (26):

$$r(x_0, x_i) = \frac{1}{n} \sum_{k=1}^{n} \delta\big((x_0(i)', x_k(i)'\big) \tag{26}$$

The GRG values vary from 0 to 1, with GRG values above 0.6 indicating a strong correlation between variables. The results were summarized in Table 1. GRG values between all the variables were above 0.6, suggesting that the candidate triggering factors can be used as the input of the prediction model.

Table 1. Candidate factors for the periodic displacement of Baijiabao landslide.

Inputs 1–7	Grey Relational Grade (GRG)
Input 1: the 1-month antecedent rainfall	0.68
Input 2: the 2-month antecedent rainfall	0.68
Input 3: average reservoir elevation in the current month	0.69
Input 4: change in reservoir level over the last month	0.72
Input 5: the displacement over the past month	0.71
Input 6: the displacement over the past two months	0.70
Input 7: the displacement over the past three months	0.69

3.4.2. Establishment of the Prediction Model

The training dataset was divided into training and validation sections, and they accounted for 70% and 30% of the total [9,35]. The triggering factors and periodic displacement were normalized to [−1, 1], and they were used as the input sequence and output sequence of the models, respectively. In this experiment, all the models used in the paper were implemented on MATLAB R2021a software, where the ML toolbox and deep ML toolbox were used. The GRU model had three layers: two were GRU layers, and the other one was a hidden layer. In the established GRU model, the number of hidden units was 200. The values of maximum epochs, minimum batch size, and initial learning rate were 250, 10, and 0.05, respectively. Those parameters of LSTM were 250, 1, and 0.01, respectively. In the RF model, the number of predictors and trees were 5 and 10, respectively.

The predicted values of GRU, LSTM, and RF models in the training process were shown in Figure 13. The prediction accuracy of the trained models was shown in Table 2. It indicated that the predicted displacements fitted well with the measured displacement in the trained LSTM and GRU models and were more satisfied than the RF model.

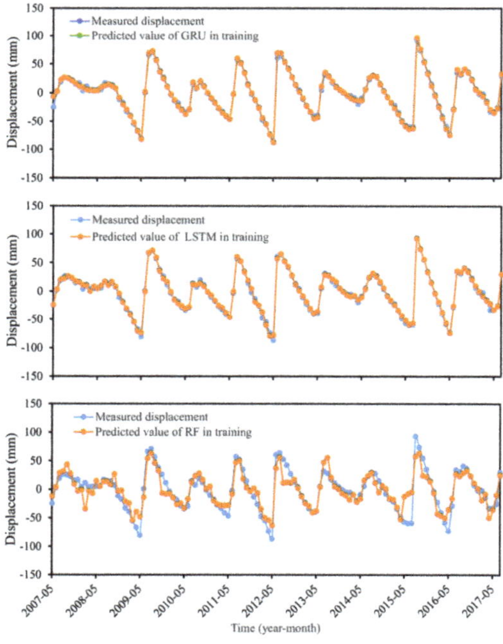

Figure 13. Measured and predicted displacements of GRU, LSTM, and RF models in training.

Table 2. Prediction accuracy of the trained models.

Model	RMSE (mm)	MAPE (%)	R^2
GRU	3.12	21.22	0.9929
LSTM	3.67	30.04	0.9916
RF	15.95	109.21	0.8009

3.4.3. Predicted Periodic Displacement

Figures 14 and 15 compared the measured and predicted periodic displacement at locations ZG324 using the GRU, LSTM, and RF models. The prediction accuracy of each model was summarized in Table 3. The GRU model gave the best agreement with the measured values in the three models, with RMSE, MAPE, and R^2 values of 1.21 mm, 11.87%, and 0.9952. Another deep ML method—LSTM—showed a lower prediction accuracy than the GRU model. Its RMSE, MAPE, and R^2 were 3.67 mm, 26.67%, and 0.9672, respectively. Compared with the two deep ML methods—LSTM and GRU—the ensemble model RF did not demonstrate a satisfied prediction performance, and the accuracy factors were 7.35 mm, 69.84%, and 0.8517.

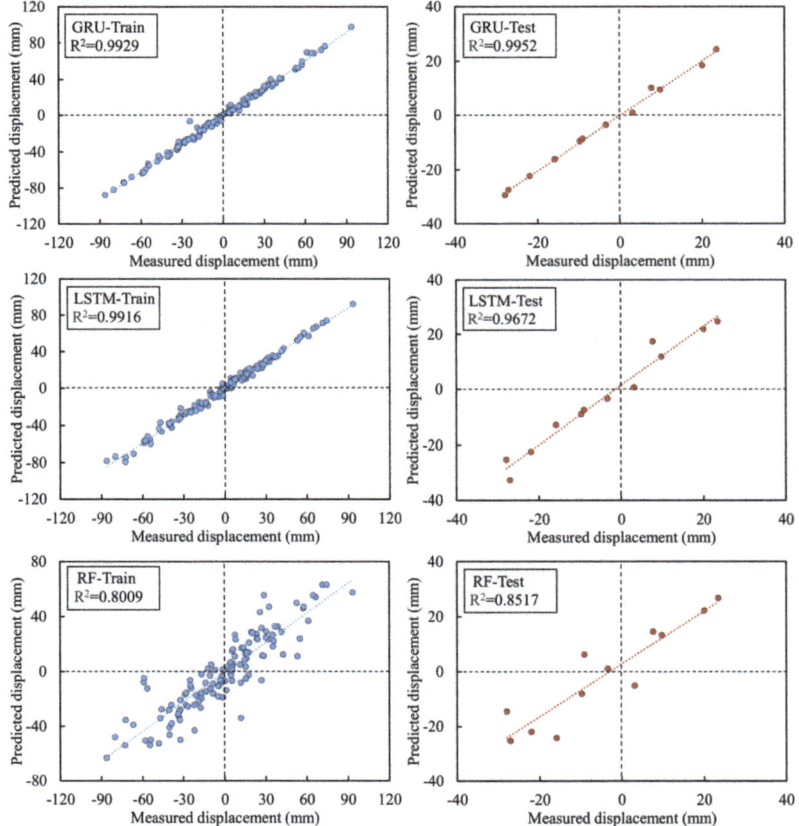

Figure 14. Training and prediction process of each model.

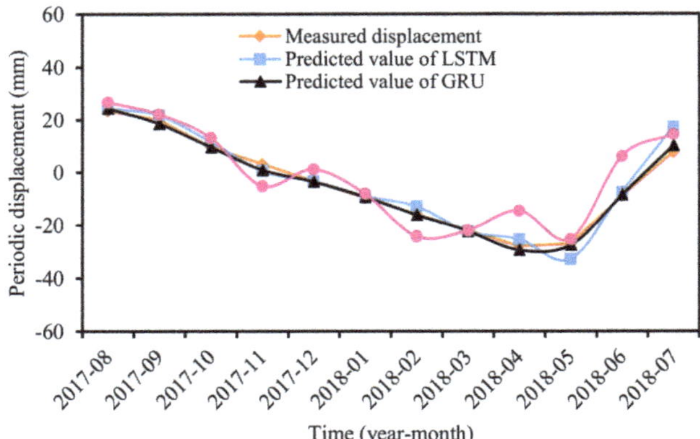

Figure 15. Predicted and measured periodic displacement.

Table 3. Prediction accuracy of periodic displacement.

Model	RMSE (mm)	MAPE (%)	R^2
GRU	1.21	11.87	0.9952
LSTM	3.67	26.67	0.9672
RF	7.35	69.84	0.8517

The predicted displacements of GRU and LSTM aligned well with the measured displacement, including in the critical early warning period of the step-wise landslides (from May to July). During May to July 2018, the reservoir water level decreased from 160.39 m to 145.33 m, and the cumulative precipitation rose to 397.83 mm. The above two influencing factors caused the displacement to increase sharply. Several local peaks existed in the curve of the predicted results for the RF model. The error of each prediction time point (each month) was distributed disorderly.

It should be noted that the GRU model showed a better prediction performance than the LSTM and RF models on the whole rather than at every time point. For example, for the displacement prediction of March, 2018, the absolute error (AE) and relative error (RE) of the GRU model were 0.38 mm and 1.72%, whereas the indicators of the RF model were 0.27 mm and 1.25%.

3.5. Stochastic Displacement Prediction

According to displacement component composition, stochastic displacement can be obtained by removing the trend term and the periodic term from the accumulated displacement series. The results were shown in Figure 16, which indicated that stochastic displacement varied with time disorderly.

In this paper, the stochastic displacement of the Baijiabao landslide was trained and predicted by a multivariate GRU model. All of the impact factors and stochastic displacements were converted to a $[-1, 1]$ format in sample data preprocessing. The prediction results were shown in Figure 17. The RMSE, MAPE, and R^2 values were 1.48 mm, 94.36%, and 0.0793, respectively. The prediction accuracy was not satisfied, whereas the whole variant trend between the predicted value and measured stochastic displacement was identical.

3.6. Accumulated Displacement Prediction

According to the accumulated displacement composition, the total displacement can be obtained by making the sum of the predicted trend and periodic and stochastic displacements. Figure 18 showed that the predicted accumulated displacements compared well with the measured displacement. The RMSE, MAPE, and R^2 values were 1.48 mm, 0.09%, and 0.9936.

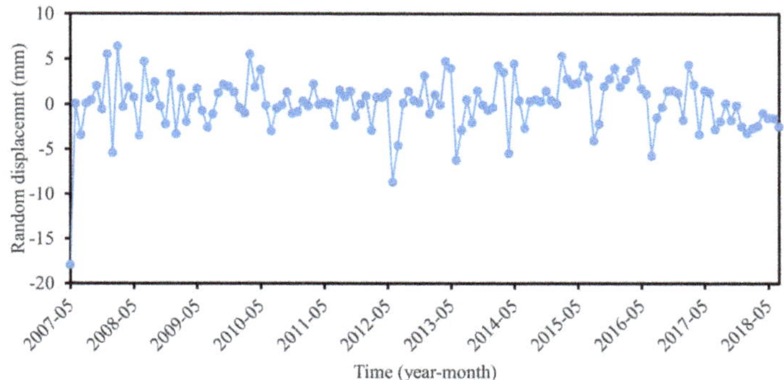

Figure 16. Stochastic displacement at ZG324.

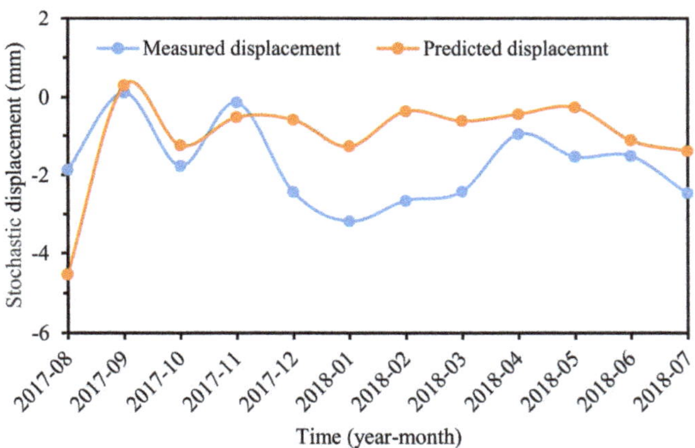

Figure 17. Predicted and measured stochastic displacement.

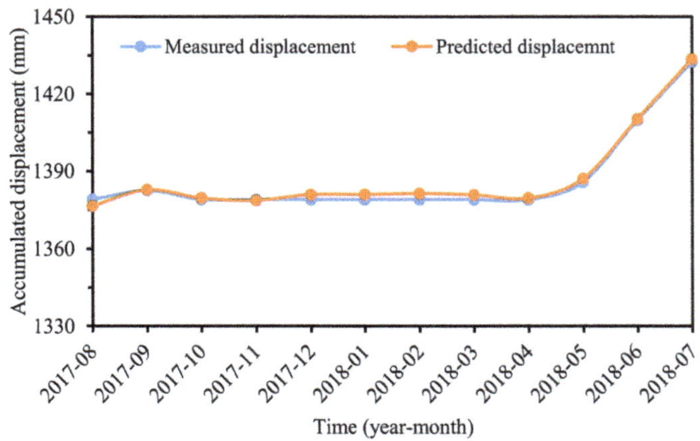

Figure 18. Predicted and measured accumulated displacement.

4. Discussion

It is critical to forecast periodic displacement accurately in the prediction of accumulated displacement for landslides with step-wise deformation [23]. Multiple ML methods have been proposed and adopted in the periodic displacement prediction, such as BPNN, ElM, SVM, RF, etc. The evolution process of landslides is a dynamic, complex, and nonlinear system. With the advantages of handling complex nonlinear problems and considering the dynamic evolution, a deep dynamic model—GRU—is thus selected to predict landslide periodic displacement.

The performance of the model was validated with the observations of the Baijiabao landslide. Another two popular models, LSTM and RF, were adopted for comparison. The results showed that GRU achieved the best prediction accuracy in the three models. Compared with RF, GRU has the ability to establish connections between adjacent time steps, and this structure contributes to improving the prediction performance of the models. Compared with LSTM, GRU has a simpler structure and fewer hyperparameters. Thus, it can be easier to establish a well-trained GRU model and achieve a better prediction accuracy. It should be noted that though GRU indicated a higher prediction accuracy for one monitoring point in the Baijiabao landslide, this does not mean that the model applies to all landslides. The limitation of generalization inherent in the GRU model makes it difficult to predict all cases accurately. Such a limitation exists in all models [37]. To deal with this problem, ensemble models can be established by combining several models with different weights of the individual model [58]. In addition, switched prediction methods can be adopted to select the appropriate individual prediction model from several candidate models for a landslide [59].

Although the GRU model achieved an encouraging prediction accuracy, it has some drawbacks. One drawback is that the GRU uses the stochastic gradient descent optimization algorithm to update weights, which risks falling into local optimization [60]. Another drawback is that the deep GRU model demands a larger dataset size than conventional ML models [10]. The monitoring frequency is one time per month for the GPS data used in the Baijiabao case. It may take years to obtain enough data for the prediction model. If not enough training samples are available, the neural network cannot be fully trained, and therefore the prediction accuracy of the model will be affected. This drawback of GRU places a higher requirement on the monitored data of landslide deformation.

The stochastic displacement is induced by some stochastic factors, including earthquakes, wind load, and vehicle load, which make it a disordered series (Figure 14). This feature contributes to the difficulty in stochastic displacement accurate prediction. Little research on stochastic displacement prediction has been reported [33]. If a slope is marginally stable or even unstable, a slight stochastic "load" can lead to disequilibrium and intense deformation. The ignorance or underestimation of stochastic displacement may make landslide planners carry out nothing, thus increasing the possibility of landslide accidents. In this paper, stochastic component displacement was considered in accumulated displacement prediction. The stochastic displacement was determined by deducting the trend and periodic displacements from accumulated displacement, and was predicted by a multivariable GRU model. The prediction performance was unsatisfactory due to the varied, ever-present, and unquantifiable stochastic factors. The work is still a helpful experiment for understanding landslide displacement components and serves as an early warning for landslides. One should consider methods to develop optimal models for predicting stochastic displacement in the future [37].

The temporal prediction of landslides is one of the main components of early warning systems [61]. Empirical methods based on the trend of landslide rate and semi-empirical practices based on the displacement rate and acceleration can provide an estimation of landslide failure time [62]. In addition, multiple parameters relating to displacement, such as the displacement rate, displacement acceleration, and tangential angle, have been proposed as thresholds to suggest a probable failure, although these approaches cannot provide a time frame for such an occurrence [63]. Realizing the temporal prediction of landslides at slope-scale based on relating the displacement would require a deeper dissertation in future work.

5. Conclusions

Displacement prediction is a vital and economic measure for landslide risk reduction and always emphasizes landslide research. This paper decomposed accumulated displacement into different displacement components by CEEMD. A univariate GRU model and a multivariable GRU model were used to predict the trend and stochastic displacements. A multivariable GRU model was used to establish a predictor for periodic displacement prediction, and two other popular ML models—LSTM and RF—were adopted for comparison. The predicted accumulated displacement was gained by the superposition of the three predicted displacement components. The results showed that predictors of deep ML methods—GRU and LSTM—had a higher prediction accuracy than the RF model in the studied case, which revealed the superiority of deep ML methods in long time series prediction. Both as deep ML methods, the GRU model achieved a better prediction performance than the LSTM model. One main reason is that the GRU algorithm has fewer hyperparameters to be trained in the model establishment than the LSTM algorithm. A prediction model with the structure of CEEMD—univariate GRU (trend displacement), multivariable GRU (periodic displacement), and multivariable GRU (stochastic displacement)—was proposed and achieved an encouraging prediction performance. The proposed model can be a potential tool for landslide risk reduction in the dam reservoir.

Author Contributions: Conceptualization, T.X. and B.Y.; analysis, B.Y. and T.X.; investigation, L.W. and W.H.; writing—original draft preparation, B.Y.; writing—review and editing, B.Y. and T.X.; supervision, B.Y. All authors have read and agreed to the published version of the manuscript.

Funding: This research was supported by the Natural Science Foundation of Shandong Provincial, China (No. ZR2021QD032).

Institutional Review Board Statement: Not applicable.

Informed Consent Statement: Not applicable.

Data Availability Statement: Some or all data, models, or code that support the findings of this study are available from the corresponding author upon reasonable request.

Acknowledgments: We wish to thank the Seventh Geological Brigade of Hubei Geological Bureau for their landslide investigation.

Conflicts of Interest: The authors declare no conflict of interest.

References

1. Mirus, B.B.; Jones, E.S.; Baum, R.L.; Godt, J.W.; Slaughter, S.; Crawford, M.M.; Lancaster, J.; Stanley, T.; Kirschbaum, D.B.; Burns, W.J. Landslides across the USA: Occurrence, susceptibility, and data limitations. *Landslides* **2020**, *17*, 2271–2285. [CrossRef]
2. Pereira, S.; Zêzere, J.L.; Quaresma, I.D.; Bateira, C. Landslide incidence in the North of Portugal analysis of a historical landslide database based on press releases and technical reports. *Geomorphology* **2014**, *214*, 514–525. [CrossRef]
3. Wen, T.; Tang, H.; Huang, L.; Wang, Y.; Ma, J. Energy evolution: A new perspective on the failure mechanism of purplish-red mudstones from the Three Gorges Reservoir area, China. *Eng. Geol.* **2020**, *264*, 105350. [CrossRef]
4. Roodposhti, M.S.; Aryal, J.; Pradhan, B. A novel rule-based approach in mapping landslide susceptibility. *Sensors* **2019**, *19*, 2274. [CrossRef] [PubMed]
5. Yang, B.B. Deformation Characteristics and Displacement Prediction of Colluvial Landslides in Wanzhou County, Three Georges Reservoir. Ph.D. Thesis, China University of Geosciences, Wuhan, China, 2019.
6. Xiao, T.; Yu, L.; Tian, W.; Zhou, C.; Wang, L. Reducing local correlations among causal factor classifications as a strategy to improve landslide susceptibility mapping. *Front. Earth Sci.* **2021**, 997. [CrossRef]
7. Yang, B.; Yin, K.; Xiao, T.; Chen, L.; Du, J. Annual variation of landslide stability under the effect of water level fluctuation and rainfall in the Three Gorges Reservoir, China. *Environ. Earth Sci.* **2017**, *76*, 564. [CrossRef]
8. Xiao, T.; Segoni, S.; Chen, L.; Yin, K.; Casagli, N. A step beyond landslide susceptibility maps: A simple method to investigate and explain the different outcomes obtained by different approaches. *Landslides* **2020**, *17*, 627–640. [CrossRef]
9. Wen, T.; Tang, H.; Wang, Y.; Lin, C.; Xiong, C. Landslide displacement prediction using the GA-LSSVM model and time series analysis: A case study of Three Gorges Reservoir, China. *Nat. Hazards Earth Syst. Sci.* **2017**, *17*, 2181–2198. [CrossRef]

10. Yang, B.; Yin, K.; Lacasse, S.; Liu, Z. Time series analysis and long short-term memory neural network to predict landslide displacement. *Landslides* **2019**, *16*, 677–694. [CrossRef]
11. Xing, Y.; Yue, J.; Chen, C.; Qin, Y.; Hu, J. A hybrid prediction model of landslide displacement with risk-averse adaptation. *Comput. Geosci.* **2020**, *141*, 104527. [CrossRef]
12. Xiao, L.; Wang, J.; Ward, S.N.; Chen, L. Numerical modeling of the June 24, 2015, Hongyanzi Landslide generated impulse waves in Three Gorges Reservoir, China. *Landslides* **2018**, *15*, 2385–2398. [CrossRef]
13. Yao, W.; Zeng, Z.; Lian, C.; Tang, H. Training enhanced reservoir computing predictor for landslide displacement. *Eng. Geol.* **2015**, *188*, 101–109. [CrossRef]
14. Xing, Y.; Yue, J.; Chen, C.; Cai, D.; Hu, J.; Xiang, Y. Prediction interval estimation of landslide displacement using adaptive chicken swarm optimization-tuned support vector machines. *Appl. Intell.* **2021**, *51*, 8466–8483. [CrossRef]
15. Zhu, Z.W.; Liu, D.Y.; Yuan, Q.Y.; Liu, B.; Liu, J.C. A novel distributed optic fiber transduser for landslides monitoring. *Opt. Lasers Eng.* **2011**, *49*, 1019–1024. [CrossRef]
16. Tagliavini, F.; Mantovani, M.; Marcato, G.; Pasuto, A.; Silvano, S. Validation of landslide hazard assessment by means of GPS monitoring technique-a case study in the Dolomites (Eastern Alps, Italy). *Nat. Hazards Earth Syst. Sci.* **2007**, *7*, 185–193. [CrossRef]
17. Wang, G.Q. Kinematics of the Cerca del Cielo, Puerto Rico landslide derived from GPS observations. *Landslides* **2012**, *9*, 117–130. [CrossRef]
18. Saito, M. Forecasting the time of occurrence of a slope failure. In Proceedings of the 6th International Mechanics and Foundation Engineering, Montreal, QC, Canada, 8–15 September 1965; pp. 537–541.
19. Wu, X.; Zhan, F.B.; Zhang, K.; Deng, Q. Application of a two-step cluster analysis and the Apriori algorithm to classify the deformation states of two typical colluvial landslides in the Three Gorges, China. *Environ. Earth Sci.* **2016**, *75*, 146. [CrossRef]
20. Pradhan, B. A comparative study on the predictive ability of the decision tree, support vector machine and neuro-fuzzy models in landslide susceptibility mapping using GIS. *Comput. Geosci.* **2013**, *51*, 350–365. [CrossRef]
21. Deng, J.L. Control problems of grey systems. *Syst. Control Lett.* **1982**, *1*, 288–294. [CrossRef]
22. Jibson, R.W. Regression models for estimating coseismic landslide displacement. *Eng. Geol.* **2007**, *91*, 209–218. [CrossRef]
23. Du, J.; Yin, K.; Lacasse, S. Displacement prediction in colluvial landslides, Three Gorges Reservoir, China. *Landslides* **2013**, *10*, 203–218. [CrossRef]
24. Mayoraz, F.; Vulliet, L. Neural networks for slope movement prediction. *Int. J. Geomech.* **2002**, *2*, 153–173. [CrossRef]
25. Cao, Y.; Yin, K.; Alexander, D.E.; Zhou, C. Using an extreme learning machine to predict the displacement of step-like landslides in relation to controlling factors. *Landslides* **2016**, *13*, 725–736. [CrossRef]
26. Zhang, Y.; Chen, X.; Liao, R.; Wan, J.; He, Z.; Zhao, Z.; Zhang, Y.; Su, Z. Research on displacement prediction of step-type landslide under the influence of various environmental factors based on intelligent WCA-ELM in the Three Gorges Reservoir area. *Nat. Hazards* **2021**, *107*, 1709–1729. [CrossRef]
27. Zhou, C.; Yin, K.; Cao, Y.; Ahmed, B.; Fu, X. A novel method for landslide displacement prediction by integrating advanced computational intelligence algorithms. *Sci. Rep.* **2018**, *8*, 7287. [CrossRef]
28. Li, H.; Xu, Q.; He, Y.; Deng, J. Prediction of landslide displacement with an ensemble-based extreme learning machine and copula models. *Landslides* **2018**, *15*, 2047–2059. [CrossRef]
29. Jiang, Y.; Xu, Q.; Lu, Z.; Luo, H.; Liao, L.; Dong, X. Modelling and predicting landslide displacements and uncertainties by multiple machine-learning algorithms: Application to Baishuihe landslide in Three Gorges Reservoir, China. *Geomat. Nat. Hazards Risk* **2021**, *12*, 741–762. [CrossRef]
30. Hu, X.; Wu, S.; Zhang, G.; Zheng, W.; Liu, C.; He, C.; Liu, Z.; Guo, X.; Zhang, H. Landslide displacement prediction using kinematics-based random forests method: A case study in Jinping Reservoir area, China. *Eng. Geol.* **2021**, *283*, 105975. [CrossRef]
31. Liu, Z.Q.; Guo, D.; Lacasse, S.; Li, J.H.; Yang, B.B.; Choi, J.C. Algorithms for intelligent prediction of landslide displacements. *J. Zhejiang Univ.—SCIENCE A* **2020**, *21*, 412–429. [CrossRef]
32. Zhou, C.; Yin, K.; Cao, Y.; Ahmed, B. Application of time series analysis and PSO-SVM model in predicting the Bazimen landslide in the Three Gorges Reservoir, China. *Eng. Geol.* **2016**, *204*, 108–120. [CrossRef]
33. Miao, F.; Wu, Y.; Xie, Y.; Li, Y. Prediction of landslide displacement with step-like behavior based on multialgorithm optimization and a support vector regression model. *Landslides* **2017**, *15*, 475–488. [CrossRef]
34. Bui, D.; Bui, K.; Bui, Q.; Doan, C.; Hoang, N. Model based on least squares support vector regression and artificial bee colony optimization for time-series modeling and forecasting horizontal displacement of hydropower dam. *Handb. Neural Comput.* **2017**, 279–293. [CrossRef]
35. Qin, S.; Jiao, J.; Wang, S. A nonlinear dynamical model of landslide evolution. *Geomorphology* **2002**, *43*, 77–85. [CrossRef]
36. Xu, S.; Niu, R. Displacement prediction of Baijiabao landslide based on empirical mode decomposition and long short term memory neural network in Three Gorges area, China. *Comput. Geosci.* **2018**, *111*, 87–96. [CrossRef]
37. Jiang, H.; Li, Y.; Zhou, C.; Hong, H.; Glade, T.; Yin, K. Landslide displacement prediction combining LSTM and SVR algorithms: A case study of Shengjibao Landslide from the Three Gorges Reservoir Area. *Appl. Sci.* **2020**, *10*, 7830. [CrossRef]
38. Ghorbanzadeh, O.; Blaschke, K.; Gholamnia, K.; Meena, S.; Tiede, D.; Aryal, J. Evaluation of different machine learning methods and deep-learning convolutional neural networks for landslide detection. *Remote Sens.* **2019**, *11*, 196. [CrossRef]
39. Wang, H.; Zhang, L.; Yin, K.; Luo, H.; Li, J. Landslide identification using machine learning. *Geosci. Front.* **2021**, *12*, 351–364. [CrossRef]

40. Guo, Z.; Chen, L.; Gui, L.; Du, J.; Yin, K.; Do, H.M. Landslide displacement prediction based on variational mode decomposition and WA-GWO-BP model. *Landslides* **2019**, *17*, 567–583. [CrossRef]
41. Niu, X.; Ma, J.; Wang, Y.; Zhang, J.; Chen, H.; Tang, H. A novel decomposition-ensemble learning model based on ensemble empirical mode decomposition and recurrent neural network for landslide displacement prediction. *Appl. Sci.* **2021**, *11*, 4684. [CrossRef]
42. Yang, B.; Liu, Z.; Lacasse, S.; Nadim, F. Landslide displacement prediction based on wavelet transform and long short-term memory neural network. In Proceedings of the XVII European Conference on Soil Mechanics and Geotechnical Engineering, Reykjavik, Iceland, 1–6 September 2019.
43. Zhang, J.; Tang, H.; Wen, T.; Ma, J.; Tan, Q.; Xia, D.; Xiu, X.; Zhang, Y. A hybrid landslide displacement prediction method based on CEEMD and DTW-ACO-SVR-Cases studied in the Three Gorges Reservoir Area. *Sensors* **2020**, *20*, 4287. [CrossRef] [PubMed]
44. Huang, N.E.; Shen, Z.; Long, S.R.; Wu, M.C.; Shih, H.H.; Zheng, Q.; Yen, N.; Tung, C.; Liu, H.H. The empirical mode decomposition and the Hilbert spectrum for nonlinear and non-stationary time series analysis. *Proc. R. Soc.* **1998**, *454*, 903–995. [CrossRef]
45. Wu, Z.; Huang, N.E. Ensemble empirical mode decomposition a noise assisted data analysis method. *Adv. Adapt. Data Anal.* **2009**, *1*, 1–41. [CrossRef]
46. Yeh, J.R.; Shieh, J.S.; Huang, N.E. Complementary ensemble empirical mode decomposition a novel noise enhanced data analysis method. *Adv. Adapt. Data Anal.* **2010**, *2*, 135–156. [CrossRef]
47. Hochreiter, S.; Schmidhuber, J. Long short-term memory. *Neural Comput.* **1997**, *9*, 1735–1780. [CrossRef]
48. Gers, F.A.; Schmidhuber, J. Recurrent nets that time and count. In Proceedings of the IEEE-INNS-ENNS International Joint Conference on Neural Networks, Como, Italy, 27 July 2000; pp. 189–194. [CrossRef]
49. Fan, Y.; Qian, Y.; Xie, F.L.; Soong, F.K. TTS Synthesis with bidirectional LSTM based recurrent neural networks. In Proceedings of the Fifteenth Annual Conference of the International Speech Communication QAssociation, Singapore, 14–18 September 2014.
50. Ma, Z.; Mei, G.; Prezioso, E.; Zhang, Z.; Xu, N. A deep learning approach using graph convolutional networks for slope deformation prediction based on time-series displacement data. *Neural Comput. Appl.* **2021**, *33*, 14441–14457. [CrossRef]
51. Rajbhandari, S.; Aryal, J.; Osborn, J.; Musk, R.; Lucieer, A. Benchmarking the applicability of ontology in geographic object-based image analysis. *ISPRS Int. J. Geo-Inf.* **2017**, *6*, 386. [CrossRef]
52. Breiman, L. Random forests. *Mach. Learn.* **2001**, *45*, 5–32. [CrossRef]
53. Krkač, M.; Gazibara, S.B.; Arbanas, Ž.; Sečanj, M.; Arbanas, S.M. A comparative study of random forests and multiple linear regression in the prediction of landslide velocity. *Landslides* **2020**, *17*, 2515–2531. [CrossRef]
54. Zhu, S.; Lian, X.; Wei, L.; Che, J.; Shen, X.; Yang, L.; Qiu, X.; Liu, X.; Gao, W.; Ren, X.; et al. PM2.5 forecasting using SVR with PSOGSA algorithm based on CEEMD, GRNN and GCA considering meteorological factors. *Atmos. Environ.* **2018**, *183*, 20–32. [CrossRef]
55. Fu, Z.; Long, J.; Chen, W.; Li, C.; Zhang, H.; Yao, W. Reliability of the prediction model for landslide displacement with step-like behavior. *Stoch. Environ. Res. Risk Assess.* **2021**, *35*, 2335–2353. [CrossRef]
56. Selby, M.J. Landslides causes, consequences and environment. *J. R. Soc. N. Zealand* **1988**, *18*, 343. [CrossRef]
57. Tan, F.; Hu, X.; He, C.; Zhang, Y.; Zhang, H.; Zhou, C.; Wang, Q. Identifying the main control factors for different deformation stages of landslide. *Geotech. Geol. Eng.* **2018**, *36*, 469–482. [CrossRef]
58. Li, J.; Wang, W.; Han, Z. A variable weight combination model for prediction on landslide displacement using AR model, LSTM model, and SVM model: A case study of the Xinming landslide in China. *Environ. Earth Sci.* **2021**, *80*, 386. [CrossRef]
59. Lian, C.; Zeng, Z.; Yao, W.; Tang, H. Multiple neural networks switched prediction for landslide displacement. *Eng. Geol.* **2015**, *186*, 91–99. [CrossRef]
60. Saud, A.S.; Shakya, S. Analysis of gradient descent optimization techniques with gated recurrent unit for stock price prediction: A case study on banking sector of Nepal stock exchange. *J. Inst. Sci. Technol.* **2019**, *24*, 17–21. [CrossRef]
61. Intrieri, E.; Gigli, G.; Gigli, N.; Nadim, F. Brief communication "Landslide Early Warning System: Toolbox and general concepts". *Nat. Hazards Earth Syst. Sci.* **2013**, *13*, 85–90. [CrossRef]
62. Intrieri, E.; Carlà, T.; Gigli, G. Forecasting the time of failure of landslides at slope-scale: A literature review. *Earth-Sci. Rev.* **2019**, *193*, 333–349. [CrossRef]
63. Xu, Q.; Yuan, Y.; Zeng, Y.; Hack, R. Some new pre-warning criteria for creep slope failure. *Sci. China Technol. Sci.* **2011**, *54*, 210–220. [CrossRef]

 sensors

Article

Data Mining and Deep Learning for Predicting the Displacement of "Step-like" Landslides

Fasheng Miao [1,2], Xiaoxu Xie [1], Yiping Wu [1,*] and Fancheng Zhao [1]

[1] Faculty of Engineering, China University of Geosciences, Wuhan 430074, China; fsmiao@cug.edu.cn (F.M.); xiexx@cug.edu.cn (X.X.); zhaofancheng@cug.edu.cn (F.Z.)

[2] Engineering Research Center of Rock-Soil Drilling & Excavation and Protection, Ministry of Education, Wuhan 430074, China

* Correspondence: ypwu1971@163.com or ypwu@cug.edu.cn; Tel.: +86-027-6788-3124

Abstract: Landslide displacement prediction is one of the unsolved challenges in the field of geological hazards, especially in reservoir areas. Affected by rainfall and cyclic fluctuations in reservoir water levels, a large number of landslide disasters have developed in the Three Gorges Reservoir Area. In this article, the Baishuihe landslide was taken as the research object. Firstly, based on time series theory, the landslide displacement was decomposed into three parts (trend term, periodic term, and random term) by Variational Mode Decomposition (VMD). Next, the landslide was divided into three deformation states according to the deformation rate. A data mining algorithm was introduced for selecting the triggering factors of periodic displacement, and the Fruit Fly Optimization Algorithm–Back Propagation Neural Network (FOA-BPNN) was applied to the training and prediction of periodic and random displacements. The results show that the displacement monitoring curve of the Baishuihe landslide has a "step-like" trend. Using VMD to decompose the displacement of a landslide can indicate the triggering factors, which has clear physical significance. In the proposed model, the R^2 values between the measured and predicted displacements of ZG118 and XD01 were 0.977 and 0.978 respectively. Compared with previous studies, the prediction model proposed in this article not only ensures the calculation efficiency but also further improves the accuracy of the prediction results, which could provide guidance for the prediction and prevention of geological disasters.

Keywords: Three Gorges Reservoir; Baishuihe landslide; data mining; displacement prediction; VMD-FOA-BPNN

Citation: Miao, F.; Xie, X.; Wu, Y.; Zhao, F. Data Mining and Deep Learning for Predicting the Displacement of "Step-like" Landslides. *Sensors* **2022**, 22, 481. https://doi.org/10.3390/s22020481

Academic Editor: Domenico Calcaterra

Received: 5 December 2021
Accepted: 8 January 2022
Published: 9 January 2022

Publisher's Note: MDPI stays neutral with regard to jurisdictional claims in published maps and institutional affiliations.

1. Introduction

Landslides occur frequently around the world and are one of the most destructive geological disasters in the world [1,2]. Landslide displacement prediction is one of the geological engineering problems that at present has not been solved, especially for mountain and reservoir areas. Reservoir impoundment usually affects the surrounding geological environment, resulting in landslide disasters. As the largest power station in terms of installed capacity in the world since 2012, the water level of the Three Gorges Reservoir fluctuates between 145 and 175 m all year round. Hence, a large number of landslide disasters have developed in the Three Gorges reservoir [3,4]. Because the Three Gorges reservoir plays an important role in flood control and power generation, it is of great significance to study geological landslides in the Three Gorges Reservoir area [5,6].

Landslide displacement prediction is a hot topic at the forefront of natural hazard research [7]. Displacement prediction is the basis of early warning systems for landslide disasters. Accurate landslide displacement prediction can reduce the losses caused by such disasters as much as possible, so as to ensure the safety of people's lives and property. Due to the complex geological environment, the accuracy of current methods for directly predicting total displacement is not sufficient [8]. Hence, landslide displacement should be divided into several parts by the decomposition technique. At present, landslide displacement decomposition mainly adopts two methods. The first is the time series and simple moving average method [9,10]. This method is simple and practical, and the displacement component obtained has a clear physical meaning. However, due to defects of the decomposition method itself, the random displacement cannot be obtained. The second is empirical mode decomposition (EMD), wavelet analysis, and ensemble empirical mode decomposition (EEMD), which can divide the total displacement into a specific number of components, so it has clear physical significance [11–13].

The landslide displacement prediction model has experienced rapid development in the past 50 years, which was from the initial empirical model to the mathematical statistical model, and then to the non-linear theoretical model and the comprehensive model [14]. Nowadays, with the development of high-speed computers, various machine learning models including deep learning have been widely used for predicting landslide displacement, such as ELM (Extreme Learning Machine) [15], EML (Evaluating Machine Learning) [16], BPNN (Back Propagation Neural Network) [17,18], SVR (Support Vector Regression) [19], KELM (Kernel Extreme Learning Machine) [20,21], LSTM (Long Short-Term Memory) [9,22], and so on. Many algorithms have been used to optimize the parameters for the prediction models, including GS (Grid Search algorithm) [10], PSO (Particle Swarm Optimization) [23], GA (Genetic Algorithm) [24], FOA (Fruit Fly Optimization Algorithm) [25], GWO (Grey Wolf Optimizer) [26], and so on. Therefore, selection of the influencing factors plays a crucial role in the development of landslide prediction. Besides, for landslides in a reservoir area, the fluctuation of the reservoir level and rainfall are usually used as the hydrologic triggering factors of landslide deformation and failure [27]. However, the increase of input factors does not necessarily lead to higher prediction accuracy in the model of landslide displacement prediction. Based on the above facts, for different types of displacement, it is necessary to select the appropriate inducing factor as the input layer to establish the model. At present, data mining technology has been widely used in the field of geological hazards. Nevertheless, the research on data mining technology in landslides mostly focuses on association criterions and thresholds of triggering factors, while there are few publications on the joint use of data mining technology and deep learning. In order to optimize the triggering factors to find the most suitable factors for displacement prediction, data mining technology could be used.

In this paper, the Baishuihe landslide was taken as an example, which was in the east of Three Gorges Reservoir area. Data mining and deep learning were used for predicting the displacement. Based on the time series analysis of landslides, the displacement and triggering factors are decomposed by VMD. Periodic and random terms were predicted by FOA-BPNN. A flow chart of this work is shown in Figure 1.

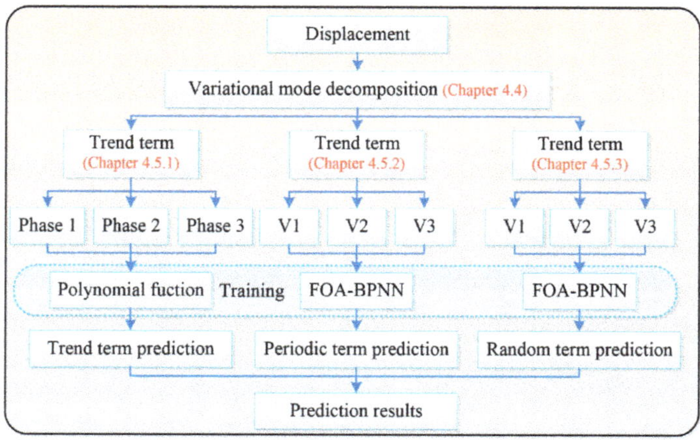

Figure 1. Flow chart of the displacement prediction.

2. Methodology

2.1. Two-Step Clustering

The two-step clustering algorithm is usually applied to deal with large-scale types of data, which divides and integrates data through a two-step process of pre-clustering and clustering to complete the data classification [28]. For sample data including both numerical and subtype variables, the two-step clustering algorithm usually uses a log-likelihood function. If clustered into j classes, it is defined as:

$$l = \sum_{j=1}^{J} \sum_{i \in I_j} \log p(X_i \mid \theta_i) = \sum_{j=1}^{J} l_j \quad (1)$$

where, p is the likelihood function; I_j is the set of samples of jth class; θ_j is the parameter vector of jth class; J is the number of clusters. For all samples, the log-likelihood clusters are obtained as the aggregation of the log-likelihood clusters for each category.

For the certain ith class and jth class, the combination is noted as $<i, j>$, and then their distance can be defined as:

$$d(i,j) = \xi_i + \xi_j - \xi_{\langle i,j \rangle} \quad (2)$$

where ξ_i and ξ_j are the log-likelihood distance of ith class and jth class, respectively. $\xi_{<i,j>}$ is the log-likelihood distance of the combination of $<i, j>$. ξ is the specific form of the log-likelihood function:

$$\xi_v = -N_V \left(\sum_{k=1}^{K^A} \frac{1}{2} \log\left(\hat{\sigma}_k^2 + \hat{\sigma}_{vk}^2 \right) + \sum_{k=1}^{K^B} E_{vk} \right) \quad (3)$$

where,

$$\hat{E}_{vk} = -\sum_{l=1}^{L_k} \frac{N_{vkl}}{N_v} \log \frac{N_{vkl}}{N_v} \quad (4)$$

where K^A is the number of numerical variables; K^B is the number of categorical variable; $\hat{\sigma}_k^2$ and $\hat{\sigma}_{vk}^2$ denote the total variance of the kth numerical variable and the variance in vth class respectively; N_v and N_{vkl} are the sample size of category v and the first category in the kth subtype variable; L_k is the category of the kth subtype variable.

After ith class and jth class are combined, $-\xi_{<i,j>}$ is greater than ξ_i + ξ_j, and hence $d(i,j)$ is less than 0. Moreover, the smaller $d(i,j)$ is, the more it means that the merging of ith class and jth class will not cause a significant increase in intra-class differences. Specially, when $d(i,j)$ is less than the threshold C, ith class and jth class can be merged. Conversely, when $d(i,j)$ is greater than the threshold C, indicating that merging will cause a significant increase in variability within the clustered clusters, and the ith class and jth class cannot be merged.

The threshold value C is given by:

$$C = \log(V) \quad (5)$$

$$V = \prod_k R_k \prod_m L_m \quad (6)$$

where R_k is the range of values of the kth numeric variable; L_m is the sample size of the mth subtype variable.

2.2. Apriori Algorithm

The a priori algorithm was proposed by Agrawal [29]. This algorithm can deal only with categorical variables rather than numeric variables. The algorithm mainly includes two steps: (1) generating frequent item sets that meet the minimum support values, and (2) generating association rules that satisfy the minimum credibility in the frequent item set generated in the first step.

The frequent item set T contains item a (frequent item set). If its support is equal to or greater than the support threshold specified by the user, as shown in Equation (1), the a priori algorithm uses the iterative method of layer-by-layer searching to generate frequent item sets.

$$\frac{|T(a)|}{|T|} \geq \min \text{supp} \tag{7}$$

Frequent k-item sets are used to explore and generate (k + 1)-item sets. The algorithm implementation process is shown in Figure 2.

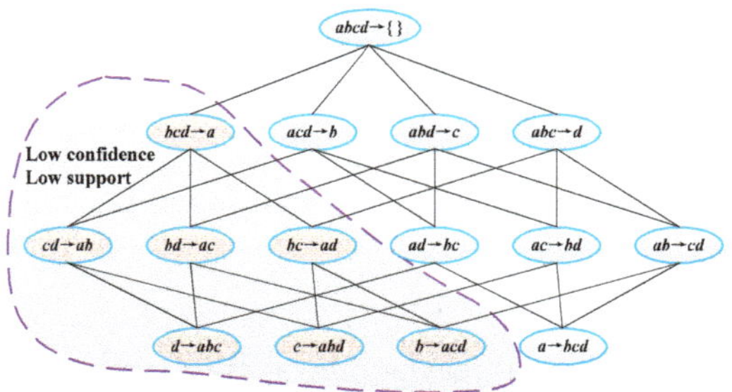

Figure 2. The implementation process of the a priori algorithm.

Simple association rules are generated from the frequent item sets, and association rules with confidence levels greater than the threshold value are selected to form an effective rule set. If $C_{L' \to (L-L')}$ is greater than the confidence threshold specified by the user (see Equation (2)), then the association rule can be generated.

$$C_{L' \to (L-L')} = \frac{|T(L)|}{|T(L')|} \geq \min conf \tag{8}$$

2.3. VMD

Based on the EMD model, the VMD model was proposed in 2013, which is an adaptive method for signal processing and modal variation [30]. The constraint variation can be expressed as:

$$\left\{ \begin{array}{l} \min\limits_{\{u_k\}\{\omega_k\}} \left\{ \sum\limits_{k=1}^{K} \| \partial_t \left[\left(\sigma(t) + \frac{j}{\pi t} \right) * u_k(t) \right] \mathrm{e}^{-j\omega_k t} \|_2^2 \right\} \\ \text{s.t. } \sum\limits_{k=1}^{K} u_k = f(t) \end{array} \right\} \tag{9}$$

where $f(t)$ is the original signal, K is the number of components, ∂_t denotes the Dirac function, $\{\omega_k\}$ denotes the actual central frequency, $\{u_k\}$ denotes the component obtained after decomposition, $\left(\sigma(t) + \frac{j}{\pi t}\right) * u_k(t)$ denotes the analytical signal of each component, $e^{-j\omega_k t}$ denotes the estimated central frequency of each analytical signal, and * denotes the convolution operator. We then obtain the following:

$$L(\{u_k\}, \{\omega_k\}, \lambda) = \alpha \sum_k \| \partial_t \left[\left(\sigma(t) + \frac{j}{\pi t} \right) * u_k(t) \right] \mathrm{e}^{-j\omega_k t} \|_2^2 + \| f(t) - \sum_k u_k(t) \|_2^2 + \left\langle \lambda(t), f(t) - \sum_k u_k(t) \right\rangle \tag{10}$$

where λ denotes the Lagrange multiplier.

By using the alternative direction method of multipliers (ADMM), the saddle point of the model without an upper constraint can be obtained, which is the optimal solution of the constrained variational model, so that the original signal can be decomposed into IMF components.

2.4. FOA-BPNN

The BPNN is a multilayer feedforward neural network based on error back propagation algorithm training, which was first proposed by Rumelhart and McClelland [31]. BPNNs have arbitrary complex pattern classification and good multidimensional function mapping ability. In addition, it can solve XOR and other problems that simple perceptrons cannot solve. Structurally, BPNNs are composed of an input layer, a hidden layer, and an output layer. The BP algorithm takes the network's square error as the objective function and uses the gradient descent method to calculate the minimum objective function.

In addition, as proposed by Wen-Tsao Pan [32], the FOA is a new method of global optimization, which is based on the foraging behavior of *Drosophila melanogaster*. Because the fruit fly is superior to other species in terms of smell and vision, the olfactory organ of *Drosophila* can collect all kinds of smells floating in the air, even the smells of food sources 40 km away. Then, after flying to the vicinity of the food location, they can use their sharp vision to find the food or observe the gathering position of their companions, and fly in that direction. The optimization procedure of the FOA-BPNN is shown in Figure 3.

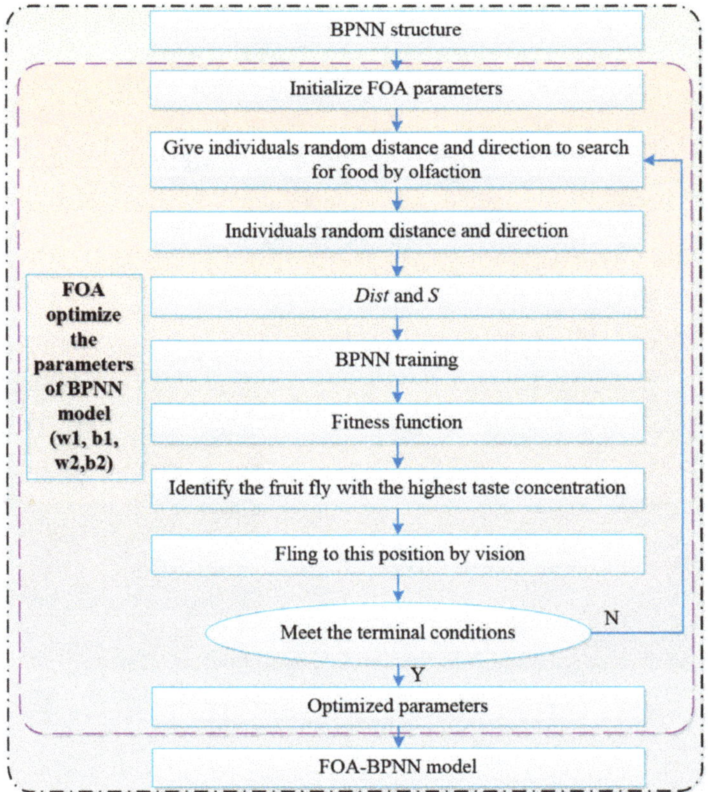

Figure 3. Optimization procedure of an FOA-BPNN.

3. Case Study

3.1. Geological Settings of Three Gorges Reservoir Area

The Three Gorges reservoir is an artificial lake formed after the completion of the Three Gorges hydropower station, situated in the middle part of China. The total lengths of the Yangtze River and surrounding area are 660 km and 1084 km^2 respectively. The altitude drops from the highest part to the west and east, forming a hilly landform and medium altitude mountains, respectively. The trend of the mountains is controlled by the main geological structures. The strata in the Three Gorges Reservoir area are from pre Sinian to Quaternary. Jurassic red strata are dominant in the Three Gorges Reservoir area, mainly exposed in the west of Zigui county east of Fengjie county (the red strata refer to sandstone, mudstone, and sandstone interbedded with mudstone layers). In addition, other sedimentary rocks (limestone, marl, and dolomite) also exist in the area between Fengjie and Zigui. These hard rocks form a steep canyon in Fengjie–Zigui area. Metamorphic complexes and magmatic rocks appear in the area near the dam site on a relatively small scale. Controlled by the complex geological conditions, coupled with seasonal rainfall and periodic fluctuation of reservoir water level, a large number of geological disasters have developed in the Three Gorges Reservoir area. A total of 4429 geological disasters have been found up to the present time, most of which are landslides, rock falls, and debris flows [4]. A geological map of the Three Gorges Reservoir area is shown in Figure 4.

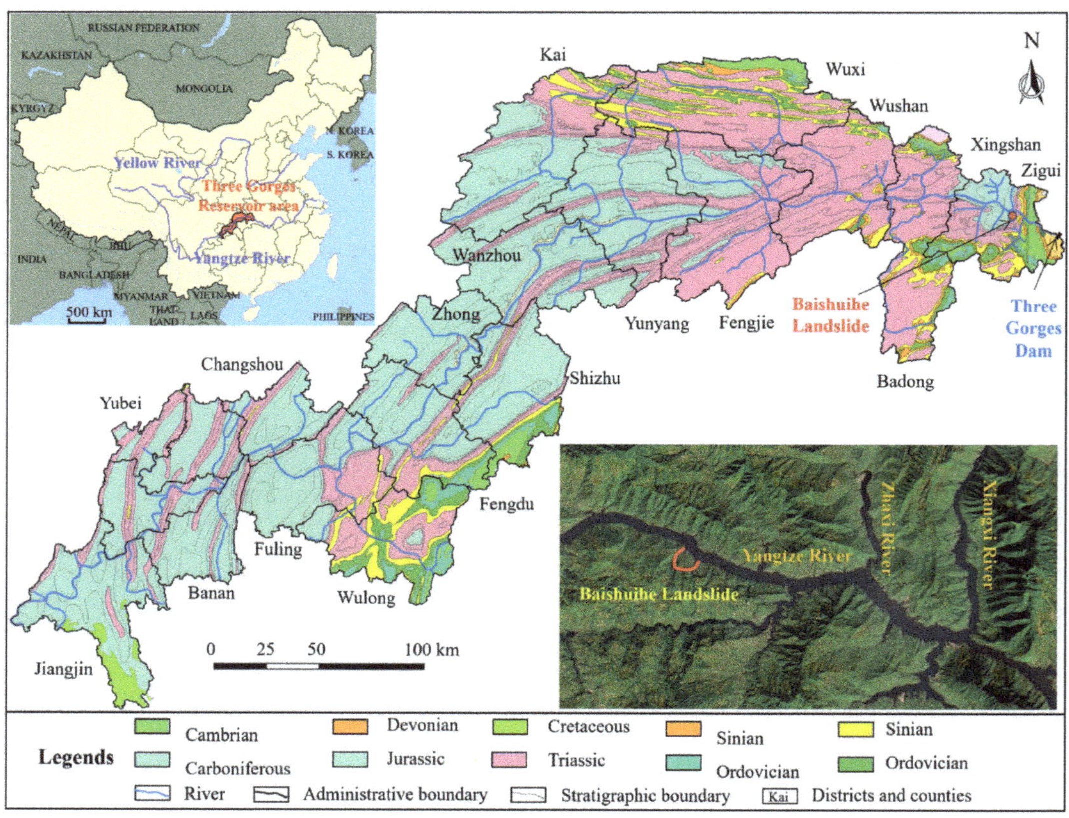

Figure 4. Geological map of the Three Gorges Reservoir area.

3.2. Local Environmental Conditions

The Baishuihe landslide is in the Zigui County area of the Three Gorges Reservoir area, located in the middle latitude, belonging to a subtropical continental monsoon climate zone, with a warm and humid climate, sufficient light, abundant rainfall, and distinct seasons. The average annual rainfall of Zigui County is 1493.2 mm. Rainfall is generally concentrated in the flood season in this area, and the maximum daily rainfall has historically reached up to 358 mm. The monsoon is mainly southerly. Limited by the terrain, the wind speed is generally low. The Yangtze River is the lowest erosion base level in this area and flows through the front edge of the landslide from west to east. The cross section of the river valley is a "V" shape, steep to the north and gentle to the south. There are several gullies short in length and depth in the landslide area, all of which are trunk gullies. Only temporary flood flows are formed after rainstorms, which constitute the primary discharge channel of surface water in the area.

The Baishuihe landslide is located on the south bank (convex bank) of the Yangtze River. The elevation of the landslide gradually decreases from south to north. The elevation of the toe and rear edges is about 70 m and 400 m, and the bedrock ridge is the boundary between the eastern and western sides. The deformation of the middle and front part of the landslide is relatively strong. Pinnate fissures are continuously distributed on both sides of the boundary, and the boundary between the east side and the rear edge is basically connected. The slope of the landslide is 30° to 35°, and the landslide has an average thickness of 30 m with a volume of 1.26×10^7 m^3. The topographic map and a schematic geological profile of the Baishuihe landslide are shown in Figures 5 and 6.

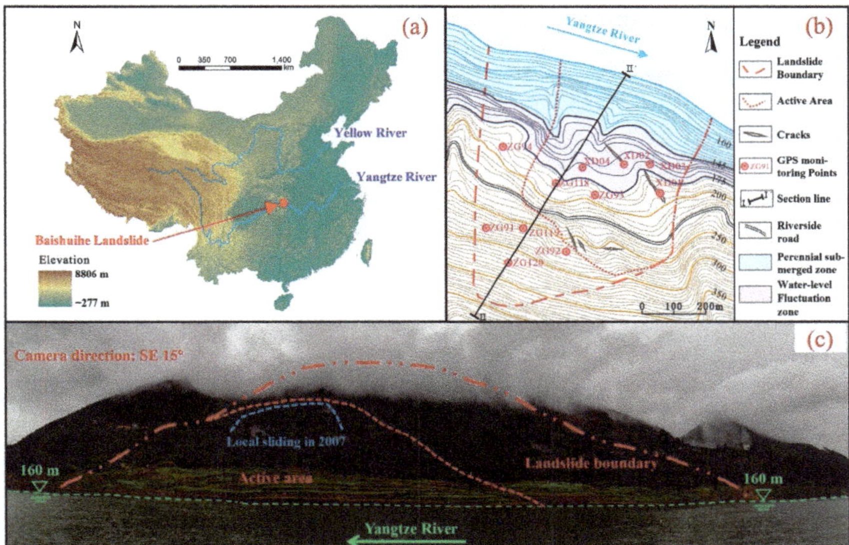

Figure 5. (**a**) Location of the Baishuihe landslide; (**b**) Topographic map of the Baishuihe landslide; (**c**) Overall view of the Baishuihe landslide © 2022 Springer Nature [33].

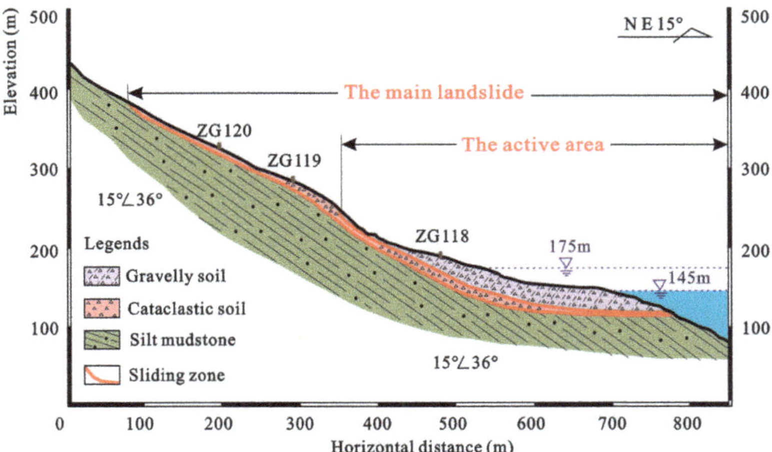

Figure 6. Schematic geological profile of the Baishuihe landslide (II–II′) © 2022 Springer Nature [10].

3.3. Deformation of the Landslide

The Baishuihe landslide has been monitored since June 2003, and the layout of the monitoring surface layout is shown in Figure 5b. According to the characteristics of the surface monitoring displacement and surface macro-deformations, the Baishuihe landslide can be divided into two areas; however, the active area (area A) is the middle and front part of the landslide, which has strong deformation. After the completion of the Three Gorges Dam, the landslide has produced obvious displacement due to the impoundment of the reservoir. Several transverse tension cracks have appeared in the east of the landslide. Specifically, the eastern and posterior boundaries are basically connected, and the western boundary's cracks are in the shape of pinnately distributed cracks. From August 2005 to August 2006, there were many landslides on the inner side slope of the riverside highway with an elevation of about 220 m, and many subsidence and tension cracks appeared on the surface of the landslide. Approximately 100,000 m^3 of landslide debris piled on the road towards the rear of the active area in June 2007 (Figure 7).

Figure 7. Macroscopic deformation of the Baishuihe landslide © 2022 Springer Nature [33].

3.4. Analysis of the Monitoring Data

There are three monitoring sections and six GPS monitoring points in the active Baishuihe landslide area. Among them, the monitoring points ZG93 and ZG118 have been in place since June 2003; XD01 and XD02 were added in May 2005, and XD03, XD04 were added in October 2005. Note that the displacement of all the monitoring points is synchronous. The monitoring period of Points ZG93, ZG118, and XD01 is long, meaning that they are representative and can reflect the entire movement process of the landslide. Therefore, in this study, these three points were taken for detailed analyses (Figure 8). According to the filling scheduling of the reservoir, the monitoring data can be divided into three stages for analysis, as described below.

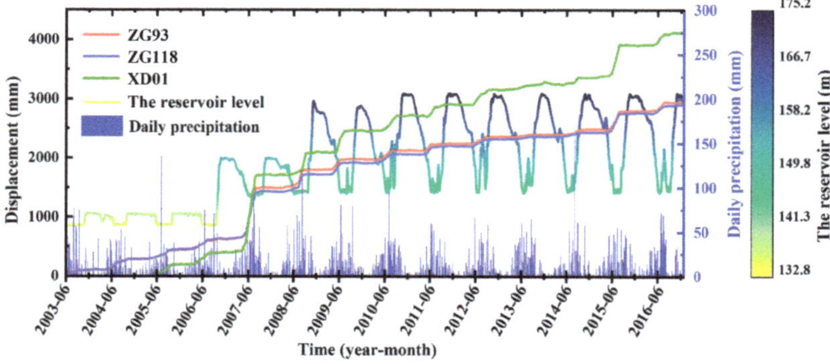

Figure 8. Long term monitoring data of the Baishuihe landslide (displacement, reservoir level, precipitation).

(1) Phase I (from June 2003 to June 2006): The water level of the reservoir started at 135 m in September and reached its highest level of 139 m in October. The maximum displacement of ZG93 and ZG118 was 25.8 mm and 30.6 mm, respectively, during the three impoundment periods. During Phase I, the reservoir basically maintained the highest water level from November to January of the next year. The maximum monthly displacement rates of these three points were below 13 mm/month during this period, which was relatively slow. The water level began to drop in February each year and reached the lowest level (135 m) in July. During this period, the minimum increase of these three points was over 80 mm, and the maximum increase was over 150 mm. Especially in May and June, the rate of increase in landslide displacement was the largest. From the end of July to the beginning of September, the reservoir water level remained at the lowest level, but the landslide displacements first continued to grow rapidly and then basically remained the same. In this stage, the water level of the Yangtze River changed from having the natural water level for many years to the manually adjusted reservoir water level, and the landslide was still in the adaptation period of adjustment of the reservoir's water level. Therefore, we can consider that the deformation of the landslide in this stage was mainly affected by the decline in the reservoir's water level. In particular, the heavy rainfall in July 2005 did not cause an obvious increase in the displacement of the landslide.

(2) Phase II (from July 2006 to June 2008): The water level of the reservoir fluctuated between 145 and 155 m, which dropped from 155 m to 145 m for the first time during April to June 2007. Alternatively, a drastic drop in the water level led to an increase in the hydrodynamic pressure inside the landslide, which caused the displacement of each monitoring point to suddenly increase for the first time, increasing by more than 1000 mm.

(3) Phase III (from July 2008 to December 2016): The water level of the reservoir fluctuated between 145 and 175 m. Before 2015, the annual displacement rate showed a downward trend.

In summary, the fluctuation in the reservoir's water level resulted in the significant extension of the fluctuating range and immersion range of the reservoir's water level, then, in turn, the stress field, seepage field, and rock–soil structure characteristics of the sliding mass changed significantly, which had a significant impact on the evolution process of the Baishuihe landslide. In addition, owing to the different stages of the reservoir's water level operations, there were some differences in the degree of impact on the deformation evolution process of the landslide. Moreover, although the landslide has undergone some adjustment, its shear deformation energy has been released to a certain extent. However, when external effects such as rainfall and the reservoir water level change dramatically again, the landslide will tend to be unstable.

4. Results

4.1. Triggering Factors

Rainfall and periodic fluctuation of reservoir water level are the main inducing factors of landslide deformation [34,35]. The periodic fluctuation of the water level in the Three Gorges causes dynamic osmotic pressure in the slope, resulting in landslide deformation [36]. On the one hand, rainfall can increase the weight of the landslide mass, thus increasing the sliding force of landslides; on the other hand, it can weaken the mechanical strength of the landslide rock and soil mass, resulting in landslide deformation [37,38]. Therefore, rainfall and reservoir water level can be used as trigger factors for landslide deformation [39]. In this research, a total of 10 triggering factors were selected to carry out displacement prediction research, including 5 reservoir level related factors (monthly average water level $\overline{h}$; monthly maximum daily drop of water level $\Delta h_{\max}^{dailydrop}$; monthly maximum daily rise of water level $\Delta h_{\max}^{dailyrise}$; monthly fluctuation of water level Δh^{month}; bimonthly fluctuation of water level Δh^{2month}), 4 rainfall related factors (monthly maximum effective continuous rainfall $q_{continuous}^{effective}$; monthly cumulative rainfall q^{month}; bimonthly cumulative rainfall q^{2month}; monthly maximum daily rainfall $q_{\max}^{day}$), and 1 deformation factor (last monthly velocity of deformation v), as shown in Table 1. In this study, the monitoring data of ZG93 were selected for landslide prediction. In addition, because the monitoring points ZG118 and XD01 have similar deformation characteristics to ZG93, the monitoring data of ZG118 and XD01 were added to increase the sample size and overcome model overfitting errors, as well as to provide a more representative prediction of the overall landslide displacement. The triggering factors are shown in Table 1.

Table 1. Triggering factors used to carry out displacement predictions.

No.	Factors	Category
F1	Monthly average water level ($\overline{h}$) (m)	Reservoir water
F2	Maximum monthly daily drop in water level ($\Delta h_{\max}^{dailydrop}$) (m/day)	Reservoir water
F3	Maximum monthly daily rise in water level ($\Delta h_{\max}^{dailyrise}$) (m/day)	Reservoir water
F4	Monthly fluctuation of the water level (Δh^{month}) (m/month)	Reservoir water
F5	Bimonthly fluctuation of the water level (Δh^{2month}) (m/2 months)	Reservoir water
F6	Maximum monthly effective continuous rainfall ($q_{continuous}^{effective}$) (mm)	Rainfall
F7	Cumulative monthly rainfall (q^{month}) (mm)	Rainfall
F8	Cumulative bimonthly rainfall (q^{2month}) (mm)	Rainfall
F9	Maximum monthly daily rainfall ($q_{\max}^{day}$) (mm)	Rainfall
F10	Monthly velocity (v) (mm/month)	Deformation

4.2. Clustering Results

In the two-step clustering algorithm, the minimum and maximum categories of the triggering factors were set as 2 and 10 respectively. In clustering algorithms, there are two commonly used clustering criteria: the Akaike Information Criterion (AIC) and the Bayesian Information Criterion (BIC). When the number of samples is large, the BIC criterion can effectively avoid the model complexity caused by high model accuracy. Therefore, in this study, the BIC was chosen as the cluster criterion, and the distance measurement method was Euclidean distance. The clustering results of the external triggering factors are shown in Tables 2 and 3. Monthly velocity (v) was clustered into three categories (Low, V1; Medium, V2; High, V3), as shown in Table 4.

Table 2. Clustering results of the reservoir water level factors (ZG93, ZG118, XD01).

No.	Factors	Clustering Results		Count
F1	$\overline{h}$	(135.13~138.95)	High Water Level (F11)	97
		(144.21~158.02)	Medium Water Level (F12)	186
		(160.14~174.74)	Low Water Level (F13)	183
F2	$\Delta h_{\max}^{dailydrop}$	(−0.14~0.58)	Slow Daily Drop (F21)	339
		(0.63~1.87)	Medium Daily Drop (F22)	92
		(1.91~3.69)	Sharp Daily Drop (F23)	35
F3	$\Delta h_{\max}^{dailyrise}$	(−0.43~0.04)	Slow Daily Rise (F31)	129
		(−1.70~−0.49)	Sharp Daily Rise (F32)	337
F4	Δh^{month}	(0~6.18)	Smooth Fluctuation (F41)	349
		(6.59~18.25)	Sharp Fluctuation (F42)	117
F5	Δh^{2month}	(0~6.50)	Non-fluctuation (F51)	250
		(6.68~14.15)	Smooth Fluctuation (F52)	126
		(14.91~28.71)	Sharp Fluctuation (F53)	90

Table 3. Clustering results of the rainfall factors (ZG93, ZG118, XD01).

No.	Factors	Clustering Results		Count
F6	$q_{continuous}^{effective}$	(1.50~30.30)	Light Effective Rainfall (F61)	182
		(31.30~66.00)	Moderate Effective Rainfall (F62)	151
		(67.70~110.50)	Medium Effective Rainfall (F63)	92
		(125.00~239.40)	Heavy Effective Rainfall (F64)	41
F7	q^{month}	(3.10~66.10)	Light Effective Rainfall (F71)	198
		(69.90~163.70)	Moderate Effective Rainfall (F72)	191
		(168.50~291.50)	Medium Effective Rainfall (F73)	60
		(357.50~517.60)	Heavy Effective Rainfall (F74)	17
F8	q^{2month}	(18.40~135.20)	Light Effective Rainfall (F81)	197
		(143.60~362.90)	Moderate Effective Rainfall (F82)	212
		(367.20~726.30)	Heavy Effective Rainfall (F83)	57
F9	$q_{\max}^{day}$	(1.30~25.60)	Light Daily Rainfall (F91)	234
		(26.50~51.30)	Moderate Daily Rainfall (F92)	151

Table 4. Clustering results of the monthly velocity (ZG93, ZG118, XD01).

Monthly Velocity (v) (mm/month)	Clustering Results	Count
(−9.61~21.66)	Low (V1)	358
(22.35~81.89)	Medium (V2)	81
(137.70~313.24)	High (V3)	27

4.3. Association Rules

In the a priori algorithm, the minimum conditional support was set to 0.01 and the minimum rule confidence was set to 100% to ensure that the mining association criteria were absolutely correct. In total, 5447 association rules were generated, most of which were V1 and V2 stages (4247 and 1008, respectively). The main factors controlling V1 deformation of the landslide were smooth fluctuations of the reservoir's water level and light rainfall. The main factors controlling V2 deformation of the landslide were sharp fluctuations of the water level and medium to heavy rainfall. The main factor controlling V3 deformation of the landslide was heavy rainfall. Nevertheless, there may be some time correlation between these nine factors. In general, a drop in reservoir water and heavy rainfall were the main factors causing landslide deformation in the Three Gorges Reservoir area. It can be seen from Figure 7 that the water level of the Three Gorges reservoir has had a period of slow decline (175 m–165 m) from January to April and a rapid decline (165 m–145 m) from April to June since 2008. The heavy rainfall is concentrated from June to September every year. Moreover, this is also a critical period when the landslide produces severe deformation.

The statistical results of the data mining and association rules are shown in Table 5. The total support, average support, and the contribution without support of each triggering factor were counted, and the comprehensive contribution was the mean value of these three contributions. The comprehensive contribution of each factor according to the association rules is shown in Figure 9. Factors with a degree of contribution less than 0.3 were eliminated and were not used as input layers in the prediction model. Therefore, eight triggering factors were taken as the input layer in the V1 and V3 prediction models (F1, F3, F5, F6, F7, F8, F9, and F10), and eight triggering factors were taken as the input layer in the V2 prediction model (F1, F2, F5, F6, F7, F8, F9, and F10).

Table 5. Statistical results of the data mining and association rules.

	Contribution	F1	F2	F3	F4	F5	F6	F7	F8	F9
V1	Association rules	2860	1936	2071	1683	2673	2780	2610	2770	2630
	Total support	4480.98	1867.49	3231.69	1557.06	3723.31	3776.06	3579.76	3800.01	3744.18
	Average support	1.57	0.96	1.56	0.93	1.39	1.36	1.37	1.37	1.42
	Contribution without support	0.67	0.45	0.49	0.40	0.63	0.65	0.61	0.65	0.62
	Comprehensive contribution	0.41	0.23	0.33	0.21	0.36	0.37	0.35	0.37	0.36
V2	Association rules	632	463	308	392	630	654	694	628	725
	Total support	453.26	344.78	195.09	289.57	438.03	467.48	506.75	447.84	478.52
	Average support	0.72	0.74	0.63	0.74	0.69	0.71	0.73	0.71	0.66
	Contribution without support	0.63	0.46	0.31	0.39	0.63	0.65	0.69	0.62	0.72
	Comprehensive contribution	0.36	0.30	0.21	0.27	0.35	0.37	0.39	0.36	0.38
V3	Association rules	130	48	109	0	111	133	126	105	124
	Total support	83.43	29.45	69.32	0	70.54	81.59	80.98	67.26	76.07
	Average support	0.64	0.61	0.64	0	0.64	0.61	0.64	0.64	0.61
	Contribution without support	0.71	0.26	0.60	0	0.61	0.73	0.69	0.58	0.68
	Comprehensive contribution	0.42	0.23	0.37	0	0.38	0.42	0.41	0.37	0.40

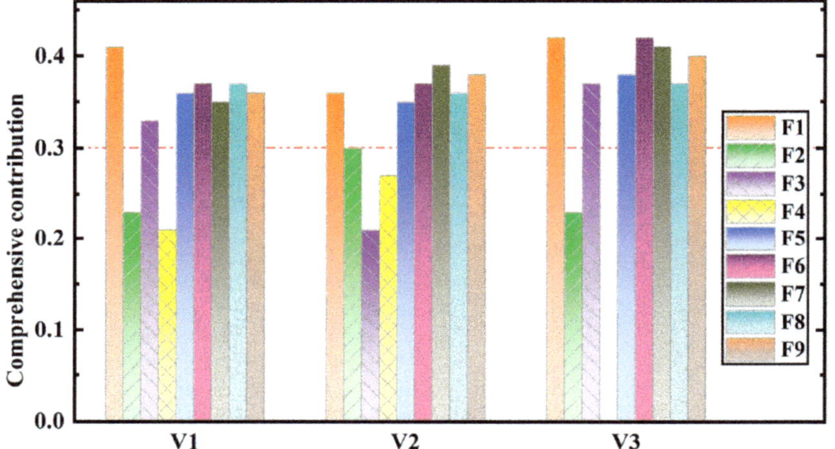

Figure 9. Comprehensive contribution of each factor according to the association rules (V1: Low monthly velocity; V2: Medium monthly velocity; V3: High monthly velocity; F1: Monthly average water level; F2: Maximum monthly daily drop in water level; F3: Maximum monthly daily rise in water level; F4: Monthly fluctuation of the water level; F5: Bimonthly fluctuation of the water level; F6: Maximum monthly effective continuous rainfall; F7: Cumulative monthly rainfall; F8: Cumulative bimonthly rainfall; F9: Maximum monthly daily rainfall).

4.4. Decomposition of Displacement

The non-stationary time series theory indicated that the time series consisted of three parts: the trend term, the periodic term, and the random term. For the landslide displacement, the time series can be divided into three parts: (1) trend displacement, which is controlled by internal factors, such as geological conditions, geomorphology, geological structure, rock and soil properties, etc.; (2) periodic displacement, which is controlled by external factors, such as rainfall, the reservoir's water level, wind load, air temperature, etc.; and (3) random displacement, which is controlled by random factors, such as human activities, engineering construction, vehicle loads, vibration loads, etc., as shown in Figure 10.

$$X(t) = \alpha(t) + \beta(t) + \gamma(t) \tag{11}$$

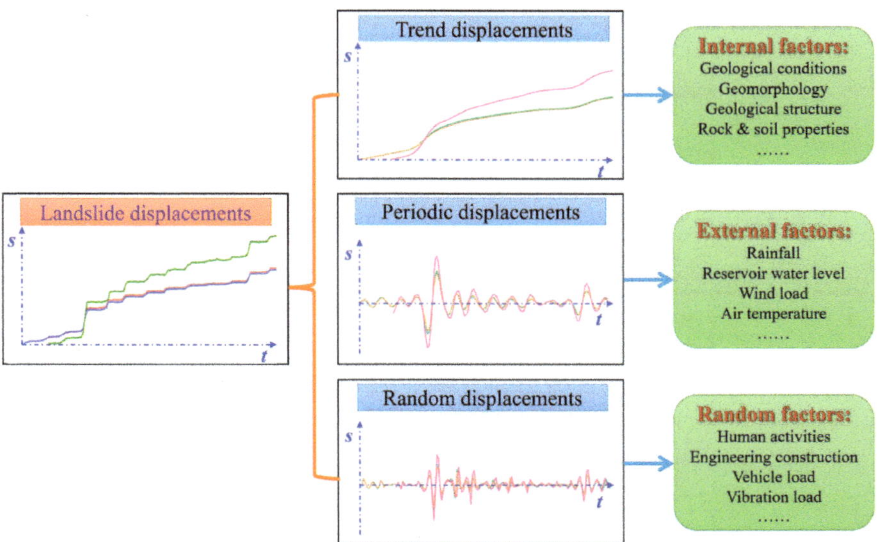

Figure 10. Relationship among trend displacement, periodic displacement, and random displacement.

Here, $X(t)$ denotes the observed value of landslide displacement, and $\alpha(t)$, $\beta(t)$, and $\gamma(t)$ denote the trend, periodic, and random displacements, respectively.

Therefore, K was set to 3 and 2 in the VMD decomposition of the landslide displacements and triggering factors, respectively. The penalty parameter a and the rising step τ ($a = 1.5$ and $\tau = 0.1$) were finally determined through multiple trials as follows. (1) In the displacement decomposition, $a = 1.5$ and $\tau = 0.1$. (2) In the triggering factors decomposing, $a = 700$ and $\tau = 0.5$. The decomposition results are shown in Table 6 and Figure 11.

Table 6. Composition of training and prediction samples.

Samples	Training Samples			Prediction Samples		
Monthly velocity	V1	V2	V3	V1	V2	V3
ZG93	116	30	5	10	2	0
ZG118	119	24	8	10	2	0
XD01	93	22	13	10	1	1
Total samples	328	76	26	30	5	1

4.5. Displacement Prediction

4.5.1. Trend Term Prediction

The displacement of the trend term showed a distinct piecewise function. Therefore, the trend term of ZG93 was divided into three phases: Phase 1 (June 2003~June 2007), Phase 2 (June 2007~June 2014), and Phase 3 (June 2014~December 2016). Multiple fitting results showed that good fitting results can be obtained by using a cubic function and the robust least squares method. The fitting function can be defined as:

$$S = at^3 + bt^2 + ct + d \tag{12}$$

The fitting results and parameters are shown in Figure 12 and Table 7, which indicate that the prediction accuracy's R^2 and the RMSE of the trend term were 99.4% and 4.063.

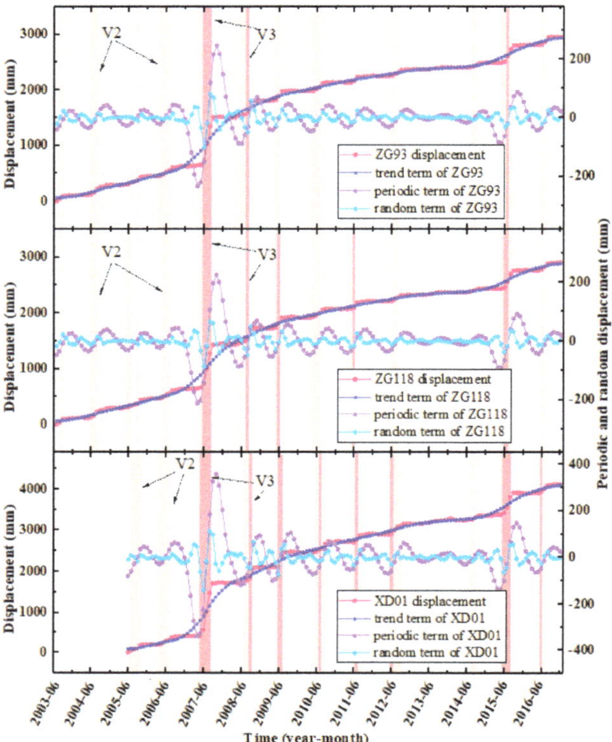

Figure 11. VMD decomposition of displacements at the monitoring points (ZG93, ZG118, XD01).

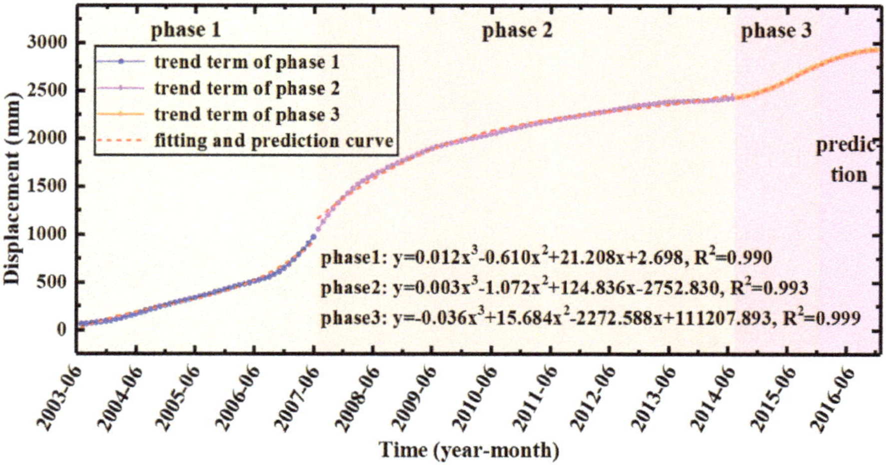

Figure 12. Fitting and prediction curves of the trend term.

Table 7. Parameters of the trend term of displacement based on polynomial fitting.

Phase	a	b	c	d	R^2	MSE	RMSE
Phase 1	0.012	−0.610	21.208	2.698	0.990	518.271	22.766
Phase 2	0.003	−1.072	124.836	−2752.830	0.993	780.995	27.946
Phase 3	−0.036	15.684	−2272.588	111,207.893	0.999	20.876	4.569
All training samples	/	/	/	/	0.994	563.729	23.743
Prediction samples	/	/	/	/	0.991	16.510	4.063

4.5.2. Periodic and Random Term Prediction

The periodic and random displacements were trained and predicted by FOA-BPNN. In general, a model's performance is usually affected by its own structure. Through extensive sensitivity analysis, the most reliable structure can be obtained [40,41]. Therefore, in the process of FOA optimization, 12 different structures with population sizes between 10 and 120 (10 intervals) were tested [42,43]. Each network was executed with 100 repetitions, and the MSE (between the actual and predicted periodic displacements of the landslide) was defined as the objective function used to evaluate the performance error of the model. It is worth noting that each structure was tested five times to evaluate its repeatability. The sensitivity curves are shown in Figure 13, which indicate that the MSE of the model decreases with an increase in the population size. However, because FOA-BPNN integration reduces the error in the training process, the model is less sensitive to population size. The computing time of models with different population sizes is shown in Figure 14. After consideration of the calculation costs and error, the population size of FOA-BPNN model was determined as 10. In total, six FOA-BPNN models were built, including individual periodic prediction models for V1, V2, and V3, and individual random prediction models for V1, V2, and V3, as shown in Figure 15.

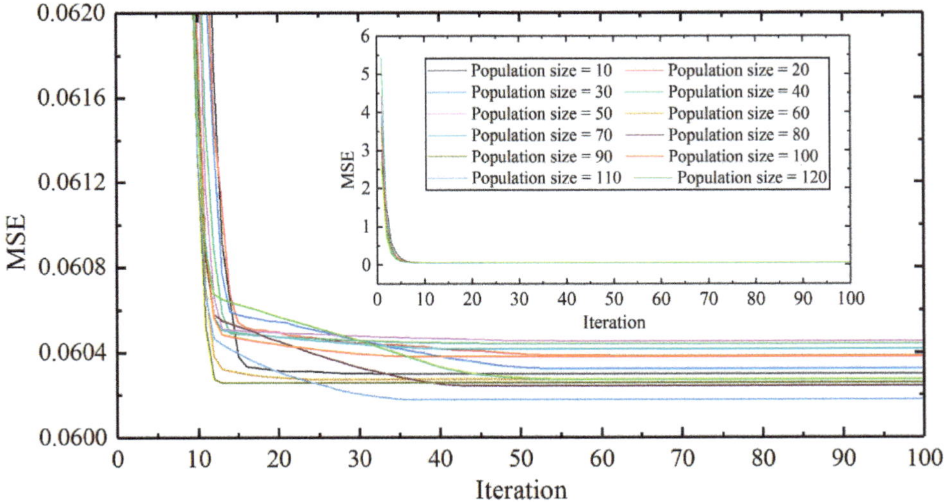

Figure 13. A population-based sensitivity analysis for the FOA-BPNN model.

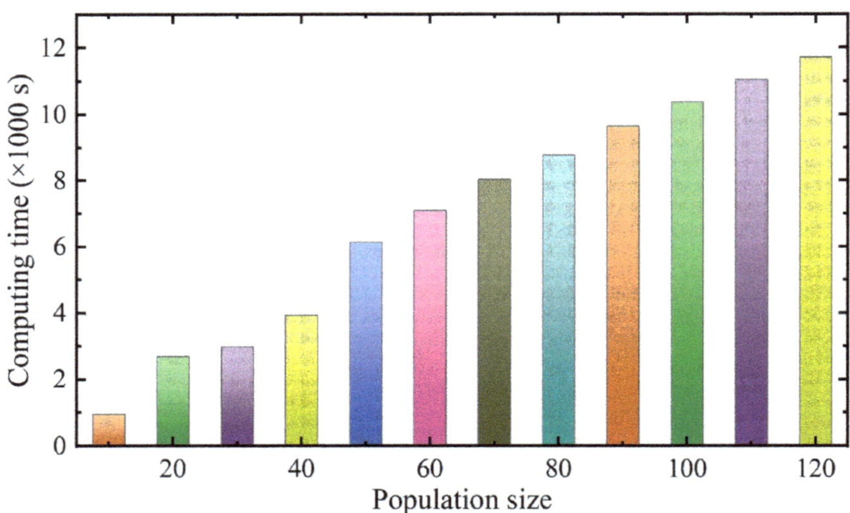

Figure 14. Computing time of different population sizes in MATLAB2019 software.

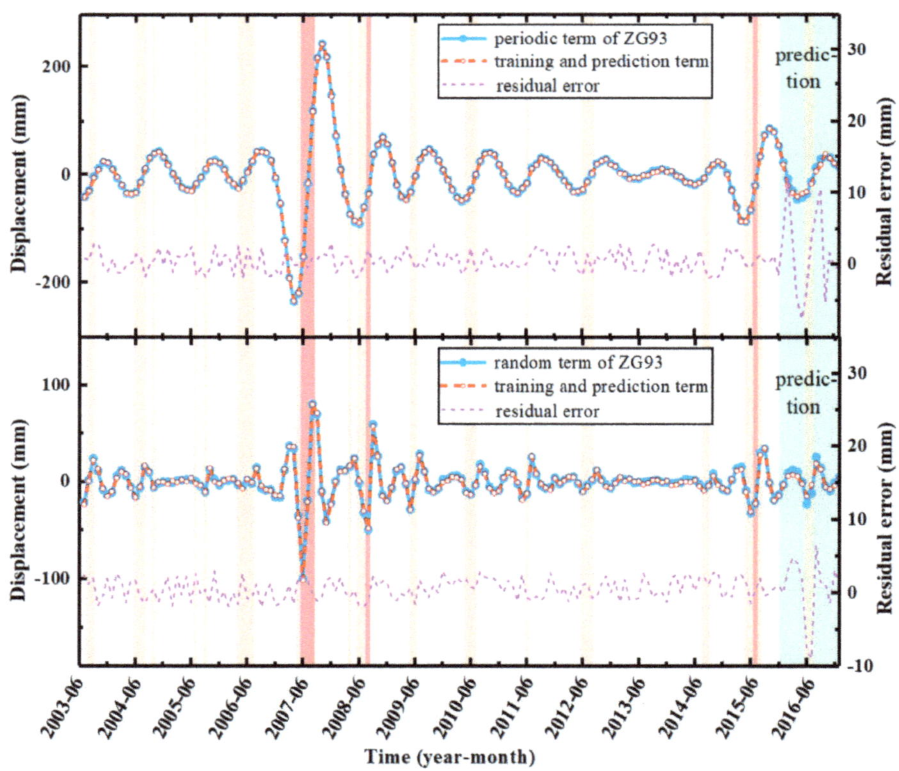

Figure 15. Training and prediction curves of the periodic and random terms.

Figure 15 indicates that the proposed models achieved good prediction results. According to the results of the residual error analysis, in the training process of the model, the residual error of displacement was relatively stable, which also verified the robustness and reliability of the model. For the prediction samples, there were some fluctuations in the residual error. The prediction accuracy of the model will be analyzed in the Discussion section.

4.5.3. Total Displacement

The total displacement prediction results of the landslide can be obtained by superimposing the prediction results of all three types of displacement, as shown in Figure 16, which shows that the prediction model achieved good accuracy for monitoring point ZG93. In June 2007, there was a significant difference between the total displacement training value and the actual value, resulting in the obvious mutation of the residual error. This was because in the three parts of landslide displacement (trend, periodic, random), the trend displacement accounts for more than 85%. In June 2007, it was the boundary between Phase 1 and Phase 2, where there were some differences in the training results of the two polynomial fitting functions, resulting in a large residual error in the total displacement. However, the residual error was relatively stable for the prediction samples.

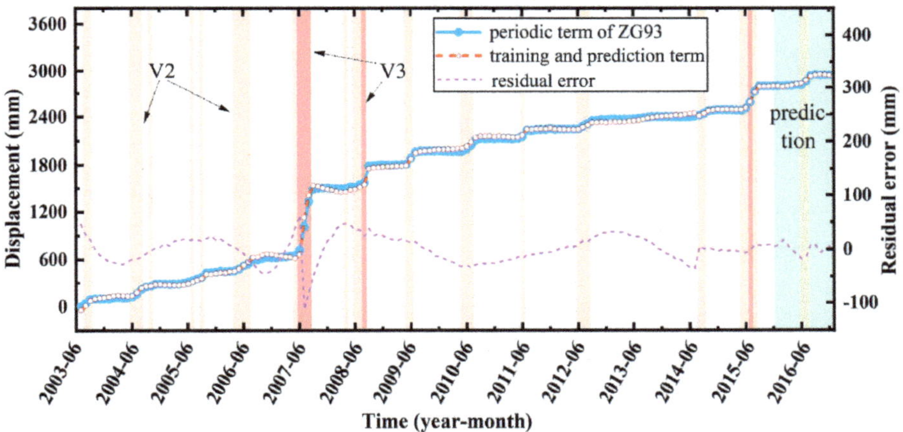

Figure 16. Training and prediction curves of the total displacement.

5. Discussion

As mentioned above, the landslide displacement contains three parts: (1) trend displacement, which is controlled by internal factors; (2) periodic displacement, which is controlled by external factors; and (3) random displacement, which is controlled by random factors. Generally, in predictions of landslide displacement, the selection of triggering factors is based on monitoring data such as rainfall and the reservoir's water level, which are the main factors causing periodic displacement. Therefore, taking these factors as the input of periodic displacement will not only have clear physical significance but will also significantly improve the accuracy of landslide displacement predictions. When the time series analysis method was used to predict the landslide displacement, the displacement trend was relatively easy to predict. Therefore, choosing the appropriate periodic displacement prediction model is the key to improving the effect of landslide displacement predictions. Moreover, landslide prediction models have experienced rapid development in the past 50 years, and various machine learning models have been widely used for predicting landslide displacements. However, each algorithm has its limitations. For instance, SVM has low computational complexity but it is sensitive to the choice of parameters and kernel function. The decision tree model does not need any prior assumptions on the data, but the required sample size is relatively large, and its ability to deal with missing values is quite limited. ELM uses the principle of least squares and a pseudo-inverse matrix to solve the problem, which is only suitable for single-hidden-layer neural networks. BPNN has strong self-learning, self-adaptive ability, and good generalization ability but it is prone to slow convergence. In this study, based on the VMD and data mining results, the FOA-BPNN was used to predict the periodic and random terms of monitoring point ZG93′s displacement. The BPNN, SVM, and ELM algorithms were chosen as the comparison models (Models 2–4). The performance of various displacement prediction models of the Baishuihe landslide are shown in Table 8 and Figure 17. The prediction accuracy of the FOA-BPNN model was the highest. The R^2 reached 0.977 and its RMSE was only 10.041. In contrast, the proposed model could improve the accuracy of landslide displacement predictions.

Table 8. Performance of various displacement prediction models of the Baishuihe landslide.

Model	Algorithm's Combination	Prediction Term		
		R^2	MSE	RMSE
Model 1	VMD + FOA-BPNN	0.977	100.828	10.041
Model 2	VMD + BPNN	0.923	340.481	18.452
Model 3	VMD + SVM	0.944	282.566	16.81
Model 4	VMD + ELM	0.877	940.462	30.667

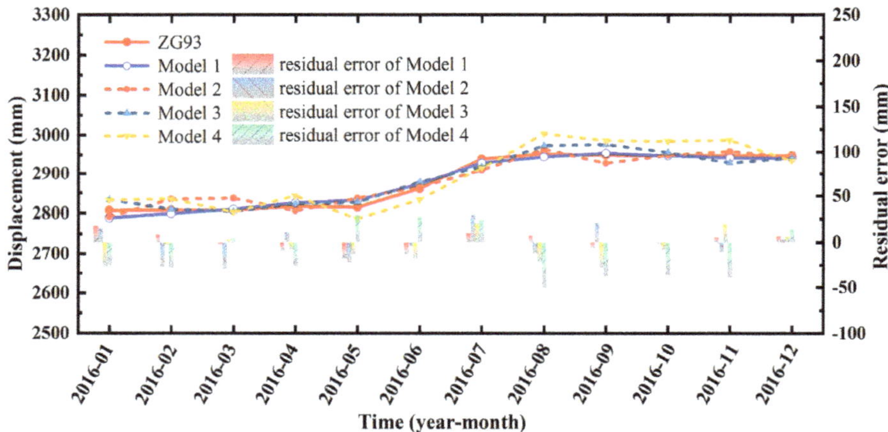

Figure 17. Prediction curves of the total displacement.

In this study, ZG93's monitoring data were selected for predicting displacement, and the monitoring data of points ZG118 and XD01 were added to increase the sample size and overcome the model's overfitting error, and to provide a better representative prediction of the overall landslide displacement. The accuracy of various models in terms of predicting ZG93's displacement has been discussed. The monitoring points ZG118 and XD01 in 2016 were used for the model validation. The measured and predicted displacements of ZG93, ZG118, XD01 are shown in Figure 18. The R^2 values between the measured and predicted displacements of ZG118 and XD01 were 0.977 and 0.978, respectively. The RMSE of these two monitoring points was 12.40 and 16.04, respectively. In the previous study [10], cumulative displacement was divided into trend term and periodic term by time series model and moving average method. A cubic polynomial model was proposed to predict the trend term of displacement. Then, multiple algorithms were used to determine the optimal support vector regression (SVR) model and train and predict the periodic term. In this paper, data mining technology is used to screen the trigger factors of periodic items, and the more advanced FOA optimization algorithm is used to optimize the parameters of the machine learning model. Furthermore, this paper uses a VMD model to divide the landslide displacement data, which makes great progress compared with the moving average model, and will be more conducive to the integration and automation of the landslide prediction model. Therefore, the prediction accuracy obtained in this paper

(R^2 = 0.977 and 0.978) is significantly higher than that of previous studies (R^2 = 0.963 and 0.951). In general, the model proposed in this study has achieved good results in terms of predicting the displacement of different monitoring points of the landslide, which has high practicability and application value in the study of landslide displacement predictions. However, it is worth noting that due to the small amount of displacement data in the V3 state of the monitoring point (Table 6), the prediction results of XD01 have obvious errors for July 2016. Therefore, in order to obtain satisfactory prediction results, the monitoring data of various states should be supplemented as much as possible.

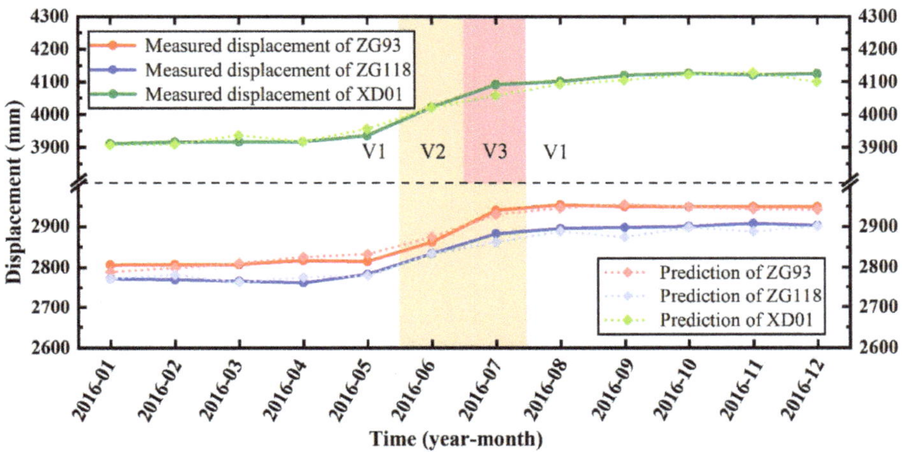

Figure 18. Measured and predicted displacements of ZG93, ZG118, and XD01.

6. Conclusions

In this paper, the Baishuihe landslide in the Three Gorges Reservoir area was taken as an example. Data mining and deep learning were used for displacement prediction. The following conclusions can be reached:

(1) Using VMD to decompose the displacement of Baishuihe landslide can correspond to the triggering factors, which had clear physical significance.

(2) The association rules showed that the main factors controlling the V2 and V3 deformation of the landslide were the sharp fluctuation of reservoir water level and medium–heavy rainfall.

(3) R^2 between the measured and prediction displacements of ZG118 and XD01 were 0.977 and 0.978. RMSE of these two monitoring points were 12.40 and 16.04, respectively.

(4) An integrated approach for landslide displacement prediction including data mining and deep learning was proposed, which could guide the managers of geological disasters to improve the prediction accuracy, so as to reduce the losses caused by landslides.

Author Contributions: Writing—original draft, F.M.; data curation, X.X.; writing—review and editing, Y.W.; investigation, F.Z. All authors have read and agreed to the published version of the manuscript.

Funding: This research was supported by the National Natural Science Foundation of China (42007267, 41977244), Science and Technology Project of Hubei Provincial Department of Natural Resources (ZRZY2020KJ12), and the National Key R&D Program of China (2017YFC-1501301).

Institutional Review Board Statement: Not applicable.

Informed Consent Statement: Not applicable.

Data Availability Statement: Not applicable.

Conflicts of Interest: The authors declare no conflict of interest.

References

1. Hong, H.; Pourghasemi, H.; Pourtaghi, Z. Landslide susceptibility assessment in Lianhua County (China): A comparison between a random forest data mining technique and bivariate and multivariate statistical models. *Geomorphology* **2016**, *259*, 105–118. [CrossRef]
2. Juang, C.H.; Dijkstra, T.; Wasowski, J.; Meng, X. Loess geohazards research in China: Advances and challenges for mega engineering projects. *Eng. Geol.* **2019**, *251*, 1–10. [CrossRef]
3. Wu, Y.; Miao, F.; Li, L.; Xie, Y.; Chang, B. Time-varying reliability analysis of Huangtupo Riverside No. 2 landslide in the Three Gorges Reservoir based on water-soil coupling. *Eng. Geol.* **2017**, *226*, 267–276. [CrossRef]
4. Tang, H.; Wasowski, J.; Juang, C.H. Geohazards in the three Gorges Reservoir Area, China–Lessons learned from decades of research. *Eng. Geol.* **2019**, *261*, 105267. [CrossRef]
5. Miao, F.; Wu, Y.; Li, L.; Tang, H.; Li, Y. Centrifuge model test on the retrogressive landslide subjected to reservoir water level fluctuation. *Eng. Geol.* **2018**, *245*, 169–179. [CrossRef]
6. Li, H.; Xu, Q.; He, Y.; Fan, X.; Li, S. Modeling and predicting reservoir landslide displacement with deep belief network and EWMA control charts: A case study in Three Gorges Reservoir. *Landslides* **2020**, *17*, 693–707. [CrossRef]
7. Intrieri, E.; Carlà, T.; Gigli, G. Forecasting the time of failure of landslides at slope-scale: A literature review. *Earth-Sci. Rev.* **2019**, *193*, 333–349. [CrossRef]
8. Lenti, L.; Martino, S. The interaction of seismic waves with step-like slopes and its influence on landslide movements. *Eng. Geol.* **2012**, *126*, 19–36. [CrossRef]
9. Yang, B.; Yin, K.; Lacasse, S.; Liu, Z. Time series analysis and long short-term memory neural network to predict landslide displacement. *Landslides* **2019**, *16*, 677–694. [CrossRef]
10. Miao, F.; Wu, Y.; Xie, Y.; Li, Y. Prediction of landslide displacement with step-like behavior based on multialgorithm optimization and a support vector regression model. *Landslides* **2018**, *15*, 475–488. [CrossRef]
11. Huang, F.; Huang, J.; Jiang, S.; Zhou, C. Landslide displacement prediction based on multivariate chaotic model and extreme learning machine. *Eng. Geol.* **2017**, *218*, 173–186. [CrossRef]
12. Shihabudheen, K.V.; Pillai, G.N.; Peethambaran, B. Prediction of landslide displacement with controlling factors using extreme learning adaptive neuro-fuzzy inference system (ELANFIS). *Appl. Soft Comput.* **2017**, *61*, 892–904.
13. Lu, X.; Miao, F.; Xie, X.; Li, D.; Xie, Y. A new method for displacement prediction of "step-like" landslides based on VMD-FOA-SVR model. *Environ. Earth Sci.* **2021**, *80*, 542. [CrossRef]
14. Ska, B.; Dbk, A.; Tsa, C. Investigating the potential of a global precipitation forecast to inform landslide prediction. *Weather Clim. Extrem.* **2021**, *33*, 100364.
15. Deng, L.; Smith, A.; Dixon, N.; Yuan, H. Machine learning prediction of landslide deformation behaviour using acoustic emission and rainfall measurements. *Eng. Geol.* **2021**, *293*, 106315. [CrossRef]
16. Goetz, J.N.; Brenning, A.; Petschko, H.; Leopold, P. Evaluating machine learning and statistical prediction techniques for landslide susceptibility modeling. *Comput. Geosci.* **2015**, *81*, 1–11. [CrossRef]
17. Chen, H.; Zeng, Z. Deformation prediction of landslide based on improved back-propagation neural network. *Cogn. Comput.* **2013**, *5*, 56–62. [CrossRef]
18. Zhang, Y.; Tang, J.; Liao, R.; Zhang, M.-F.; Zhang, Y.; Wang, X.-M.; Su, Z.-Y. Application of an enhanced bp neural network model with water cycle algorithm on landslide prediction. *Stoch. Environ. Res. Risk Assess.* **2021**, *35*, 1273–1291. [CrossRef]
19. Liu, Y.; Xu, C.; Huang, B.; Ren, X.; Liu, C.; Hu, B.; Chen, Z. Landslide displacement prediction based on multi-source data fusion and sensitivity states. *Eng. Geol.* **2020**, *271*, 105608. [CrossRef]
20. Zhou, C.; Yin, K.; Cao, Y.; Intrieri, E.; Ahmed, B.; Catani, F. Displacement prediction of step-like landslide by applying a novel kernel extreme learning machine method. *Landslides* **2018**, *15*, 2211–2225. [CrossRef]
21. Li, L.; Wu, Y.; Miao, F.; Xue, Y.; Huang, Y. A hybrid interval displacement forecasting model for reservoir colluvial landslides with step-like deformation characteristics considering dynamic switching of deformation states. *Stoch. Environ. Res. Risk Assess.* **2021**, *35*, 1089–1112. [CrossRef]
22. Xu, S.; Niu, R. Displacement prediction of Baijiabao landslide based on empirical mode decomposition and long short-term memory neural network in Three Gorges area, China. *Comput. Geosci.* **2018**, *111*, 87–96. [CrossRef]
23. Zhou, C.; Yin, K.; Cao, Y.; Ahmed, B. Application of time series analysis and PSO–SVM model in predicting the Bazimen landslide in the Three Gorges Reservoir, China. *Eng. Geol.* **2016**, *204*, 108–120. [CrossRef]
24. Li, X.Z.; Kong, J.M. Application of GA-SVM method with parameter optimization for landslide development prediction. *Nat. Hazards Earth Syst. Sci.* **2014**, *14*, 525. [CrossRef]
25. Wang, J.; Di, Y.; Rui, X. Research and application of machine learning method based on swarm intelligence optimization. *J. Comput. Methods Sci. Eng.* **2019**, *19*, 179–187. [CrossRef]
26. Guo, Z.; Chen, L.; Gui, L.; Du, J.; Yin, K.; Do, H.M. Landslide displacement prediction based on variational mode decomposition and WA-GWO-BP model. *Landslides* **2020**, *17*, 567–583. [CrossRef]
27. Hong, H.; Chen, W.; Xu, C.; Youssef, A.M.; Pradhan, B.; Bui, D.T. Rainfall-induced landslide susceptibility assessment at the Chongren area (China) using frequency ratio, certainty factor, and index of entropy. *Geocarto Int.* **2017**, *32*, 139–154. [CrossRef]
28. Wu, X.; Zhan, F.; Zhang, K.; Deng, Q. Application of a two-step cluster analysis and the apriori algorithm to classify the deformation states of two typical colluvial landslides in the three gorges, china. *Environ. Earth Sci.* **2016**, *75*, 146. [CrossRef]

29. Agrawal, R.; Srikant, R. Fast algorithms for mining association rules. *Int. Conf. Very Large Data Bases VLDB* **1994**, *1215*, 487–499.
30. Dragomiretskiy, K.; Zosso, D. Variational mode decomposition. *IEEE Trans. Signal Processing* **2013**, *62*, 531–544. [CrossRef]
31. McClelland, J.; Rumelhart, D.; PDP Research Group. Parallel distributed processing. *Explor. Microstruct. Cogn.* **1986**, *2*, 216–271.
32. Pan, W.T. A new fruit fly optimization algorithm: Taking the financial distress model as an example. *Knowl. Based Syst.* **2012**, *26*, 69–74. [CrossRef]
33. Miao, F.; Wu, Y.; Li, L.; Liao, K.; Xue, Y. Triggering factors and threshold analysis of baishuihe landslide based on the data mining methods. *Nat. Hazards* **2021**, *105*, 2677–2696. [CrossRef]
34. Song, K.; Wang, F.; Yi, Q.; Lu, S. Landslide deformation behavior influenced by water level fluctuations of the Three Gorges Reservoir (China). *Eng. Geol.* **2018**, *247*, 58–68. [CrossRef]
35. Huang, D.; Gu, D.M.; Song, Y.X.; Cen, D.F.; Zeng, B. Towards a complete understanding of the triggering mechanism of a large reactivated landslide in the Three Gorges Reservoir. *Eng. Geol.* **2018**, *238*, 36–51. [CrossRef]
36. Wang, G.; Sassa, K. Factors affecting rainfall-induced flowslides in laboratory flume tests. *Geotechnique* **2001**, *51*, 587–599. [CrossRef]
37. Wang, J.; Xiao, L.; Zhang, J.; Zhu, Y. Deformation characteristics and failure mechanisms of a rainfall-induced complex landslide in Wanzhou County, Three Gorges Reservoir, China. *Landslides* **2020**, *17*, 419–431. [CrossRef]
38. Wu, L.Z.; Zhu, S.R.; Peng, J. Application of the Chebyshev spectral method to the simulation of groundwater flow and rainfall-induced landslides. *Appl. Math. Model.* **2020**, *80*, 408–425. [CrossRef]
39. Xiong, X.; Shi, Z.; Xiong, Y.; Peng, M.; Ma, X.; Zhang, F. Unsaturated slope stability around the Three Gorges Reservoir under various combinations of rainfall and water level fluctuation. *Eng. Geol.* **2019**, *261*, 105231. [CrossRef]
40. Mehrabi, M.; Pradhan, B.; Moayedi, H.; Alamri, A. Optimizing an adaptive neuro-fuzzy inference system for spatial prediction of landslide susceptibility using four state-of-the-art metaheuristic techniques. *Sensors* **2020**, *20*, 1723. [CrossRef]
41. Moayedi, H.; Mehrabi, M.; Bui, D.T.; Pradhan, B.; Foong, L.K. Fuzzy-metaheuristic ensembles for spatial assessment of forest fire susceptibility. *J. Environ. Manag.* **2020**, *260*, 109867. [CrossRef] [PubMed]
42. Moayedi, H.; Mehrabi, M.; Kalantar, B.; Mu'Azu, M.A.; Rashid, A.S.A.; Foong, L.K.; Nguyen, H. Novel hybrids of adaptive neuro-fuzzy inference system (ANFIS) with several metaheuristic algorithms for spatial susceptibility assessment of seismic-induced landslide. Geomatics. *Nat. Hazards Risk* **2019**, *10*, 1879–1911. [CrossRef]
43. Moayedi, H.; Mehrabi, M.; Mosallanezhad, M.; Rashid, A.S.A.; Pradhan, B. Modification of landslide susceptibility mapping using optimized PSO-ANN technique. *Eng. Comput.* **2019**, *35*, 967–984. [CrossRef]

Article

Slope with Predetermined Shear Plane Stability Predictions under Cyclic Loading with Innovative Time Series Analysis by Mechanical Learning Approach

Tingyao Wu [1], Hongan Yu [2], Nan Jiang [1,*], Chuanbo Zhou [1] and Xuedong Luo [1]

[1] Faculty of Engineering, China University of Geosciences (Wuhan), Wuhan 430074, China; wutingyao@cug.edu.cn (T.W.); cbzhou@cug.edu.cn (C.Z.); cugluoxd@foxmail.com (X.L.)
[2] CCCC Second Highway Consultants Co., Ltd., Wuhan 430056, China; hh5-106yha@163.com
* Correspondence: happyjohn@foxmail.com

Abstract: We propose a mechanical learning method that can be used to predict stability coefficients for slopes where slopes with predetermined shear planes are subjected to cyclic seismic loads under undrained conditions. Firstly, shear tests with cyclic loading of different parameters were simulated on designated slip zone soil specimens, in which the strain softening process leading to landslide occurrence was closely observed. At the same time, based on the limit equilibrium analysis of the Sarma method, the variation of slope stability coefficients under different cyclic loads was investigated. Finally, a Box–Jenkins' modeling approach is used to predict the data from the time series of slope stability coefficients using a mechanical learning approach. The simulation results show that (1) reduction in coordination number can be an accurate indicator of the level of strain softening and evolutionary processes; (2) the gradual reduction of shear stress facilitates the soil strain softening process, while different cyclic loading stress amplitudes will result in rapid penetration or non-penetration of the fracture zone by means of particulate flow. Although the confining pressure of the slip zone soil can inhibit the increase of fractures, it has a limited inhibitory effect on strain softening; (3) based on field observations of the slope stability factor and stress field, two possible landslide triggering mechanisms are described. (4) Mechanical learning of time series can accurately predict the changing pattern of stability coefficients of slopes without loading. This study establishes a potential bridge between the geological investigation of landslides and the theoretical background of landslide stability coefficient prediction.

Keywords: slip zone soils; cyclic loading; strain softening; fracture zone; landslide triggering mechanisms; mechanical learning

Citation: Wu, T.; Yu, H.; Jiang, N.; Zhou, C.; Luo, X. Slope with Predetermined Shear Plane Stability Predictions under Cyclic Loading with Innovative Time Series Analysis by Mechanical Learning Approach. *Sensors* **2022**, *22*, 2647. https://doi.org/10.3390/s22072647

Academic Editor: Francesca Cigna

Received: 28 February 2022
Accepted: 28 March 2022
Published: 30 March 2022

1. Introduction

Landslides are geological phenomena that can occur on land, but also under the seabed due to earthquakes, tsunamis, etc. At the same time, they include various types of movements, such as slope failures, rock falls, or mudslides [1]. Landslide were also defined as a mass of rock, debris, or earth, moving or sliding down a slope [2]. Landslides are one of the most frequent geological hazards. It has long been believed that one of the main triggers of landslides is earthquakes, which can induce large-scale and catastrophic landslides with the loss of human lives or property damage [3,4]. According to statistics for many countries, landslides cause serious social and economic impacts globally, for example, between 2007 and 2015, more than 7200 landslides were recorded worldwide, causing more than 26,000 deaths and costing more than $1.8 billion in damages [5]. Meanwhile, Japan is known to contain a frequency of catastrophes, such as earthquakes, volcanic eruptions, landslides, and typhoons [6–11]. According to the Japan Meteorological Agency (JMA), 230 aftershocks occurred between 6 September and 11 September 2018, killing over 44 people and injuring over 660 [12]; this also had a significant impact on Hokkaido's infrastructure, causing a combined economic loss of US$2 billion in damage to transportation and utilities [13]. In addition, the transboundary Koshi River basin, located in the central Himalayas, is shared by China, Nepal and India. The triggering mechanism and landslide number were counted based on remote sensing findings, providing information that 5858 rainfall-triggered landslides occurred in the study area between 1992 and 2015, and an additional 14,127 seismic landslides were mapped after the 2015 Gorkha earthquake [14]. In China, landslides are the second most frequent natural cause of damage to man-made structures after earthquakes, so it is foreseeable that the potential increase in the number of extreme weather events, combined with the concentration of population and infrastructure in mountainous areas, will lead to an increase in landslide-related casualties in the future [15–17]. A better understanding of landslide triggering mechanisms and monitoring the early movement of soil mass, along with effective evacuation strategies as early as possible, are essential for landslide mitigation [18].

Landslides are related to different factors, such as topography, geology, tectonic history, weathering erosion history and land use. However, landslides are usually considered to be triggered by only one factor [19]. A trigger is considered to be an external stimulus, such as a strong rainfall event, an earthquake of different magnitudes, a volcanic eruption, a storm, or rapid flow in the form of erosion leading to a rapid increase in stress or strain on the landslide and a decrease in the slip zone soil material [19–21]. Based on classical elastoplastic mechanics theory, the strain softening process is reduced to a series of brittle plasticity processes, and thus the solution for the strain softening problem is reduced to the solution of a series of brittle plasticity problems, also, the strain softening behavior is the result of the soil reaching the peak stress point and then the stress decreasing as the strain continues to increase, i.e., the strength of the soil decreases with increasing strain, as a result of the spatial rearrangement of the soil particles and the forces between the phases and particles in the soil [22]. Moreover, higher cyclic stress ratios accelerate the softening behavior of soils, and different super consolidation ratios have a great influence on the softening of soils [23–25], and changes in the principal stress direction can cause structural remodeling of clays, which leads to a reduction in clay strength. Therefore, it is essential to investigate the strain softening characteristics of soil under cyclic loading. Examining the rupture mechanism is necessary if the stress state of the soil wants to be understood, especially the characteristics of the particle arrangement inside the soil. However, it is not practical to directly measure the soil distribution characteristics, especially the soil particle arrangement. Considering the close relationship between stress and strain, the majority of laboratory experiments have specialized in the characterization of displacements. Mounting strain gauges and displacement gauges on the surface of the device loaded with soil material are the most straightforward and simplified methods for measuring stress and deformation in laboratory tests. There are also some parts that use non-touch optical technology, such as electronic photography and other related technologies [26,27]. However, the placement of CT is somewhat limited due to the three-dimensional stress state and difficult exposure of the slip zone soils [28,29]. The main point to be made is that stress-strain relationships in soils are often full of randomness and strongly depend on the stress state and structural characteristics of the various materials. Although the empirically determined principal structure equations can be easily derived from the stress-strain values of soils, they are sometimes questionable.

Meanwhile, in the study of landslide mechanisms, typically landslide trigger and formation mechanisms are studied by various methods, such as traditional field surveys, satellite remote sensing, and three-dimensional imaging by drones. Numerical simulations and model experiments have also been used, but the studies have focused on macroscopic and microscopic parameters for the analysis. As in field testing, regarding the characterization of landslide studies, size and velocity, depth, impact pressure, or displacement all play an important role, and different types of mass movements are different. Volume may be a more important landslide feature than size, but this is difficult to measure because it requires specific geophysical or geotechnical methods. In order to solve the above-mentioned problems, researchers generally also go through numerical simulation software to study the triggering mechanism of landslides, for example, there are now two main methods for numerical simulation of landslides: continuum (finite element) and discontinuity (discrete unit). However, most previous studies have focused on the analysis of slope stability or dynamic response characteristics. To our knowledge, no attempt has been made to use PFC3D for inverse analysis of the microscopic mechanisms of landslides. Although earthquakes are the most important landslide causative factor, the relationship between cyclic microseismic loading and landslides has rarely been analyzed in detail, and no specific vibration control thresholds have been identified for specific shallow landslides [30,31]. At the same time, the precise onset of landslide damage is often unknown. In addition, the most destructive landslides are usually those caused by damage associated with deep-seated landslides. For such landslides, the relationship between cyclic micro seismic loading parameters and landslide occurrence is very complex [32,33].

With regard to the safety protection of slopes, the first thing to do is to clarify the safety state of the slope, but most of the existing research tools are used to obtain the stability state of the slope through the traditional quantitative solution approach, which has certain limitations. The safety of slopes contains many parameters and randomly variable laws, which leads to a relatively complex calculation process, and with the rapid rise of machine learning technology, a new way of thinking is proposed for the study of slope safety and stability. Decision trees [34], random forests, support vector machines [35–37] and plain Bayesian [38] algorithms have been widely used in slope research. By collecting and analyzing 250 slope data, 31 variables were identified and their functional relationships were explained with the help of Principal Component Analysis (PCA), thus constructing an algorithmic model to assess the stability of slopes with good generalization to the test data [39]. The slopes of the Klang Valley were studied with the help of a back-propagation neural network and calculated the weights of 11 relevant influencing factors, such as slope rate and slope height [40]. Genetic algorithms have been utilized to study specific problems in geotechnical slopes [41]. Multiple logistic regressions were carried out, while various methods, such as random forests, K-nearest neighbors' algorithms and decision trees were also used to explore the steady state of the slopes [42].

However, various machine learning tools have emerged, mainly artificial neural networks and multiple regression analysis as the two classical models of machine learning, which have been widely used in solving various prediction problems. However, the weight analysis of the influence factors of these two models is not comprehensive enough, lacking a comprehensive comparison between the weight analysis based on correlation tests and the weight analysis based on intelligent models, while not considering the randomness and periodicity of time series data, resulting in large errors in forecasting.

Therefore, based on the work of Wong et al. (1998) and Zhang et al. (2021), the focus of this study is on the quantification of damage extent and estimation of shear stress in slip zone soils. Then, the effect of different parameters on the shear strain-softening behavior of the slip zone soils can be investigated, including the cyclic loading parameters (cyclic loading stress amplitude and the number of cyclic loading) and the confining pressure applied to the numerical soil model. The relationship between shear stress and soil fine-scale damage variables will also be discussed in detail in the paper. Meanwhile, comparing the results of discrete element numerical simulations with the changes in slope stability coefficients, the influence of simulation results on landslide formation will be interpreted. Finally, in combination with the Box–Jenkins modeling approach, the stability of slopes under different cyclic loads are systematically and comprehensively predicted from time series data of slope stability coefficients using a mechanical learning approach.

2. Numerical Methodology

2.1. Conceptual Model for Cyclic Shear Failure Behavior in Slip Zone Soils

The simulated situation can be interpreted as an advancing layered rock landslide, where the deformation is progressive from the trailing edge of the landslide to the front of the level flattening type cascading rock landslide, meanwhile, a detailed scheme for the evolutionary model of advancing layered rock landslides is presented in the specific work [43,44], as shown in Figure 1, where Figure 1 is a simplified model of a conceptual mechanics model that is designed to give the reader a more intuitive understanding of the landslide triggering mechanism. The gray grid in Figure 1 is the slope surface and bedrock, and the red part is called the anchored section of the landslide, which is the slip zone soil not affected by cyclic loading, but as the landslide evolution stage develops, the red part gradually becomes white, which means that the slip zone soil is affected by cyclic loading and its mechanical strength gradually weakens, and the unaffected part of the slip zone soil gradually becomes less and the safety factor of the landslide gradually decreases, that is, progressively enlarging the weakened upper zone while reducing the size of the lower locking block. The causes of landslides are mainly due to the strain softening rate of the slip zone soil at the trailing edge of the landslide under the cumulative effect of the cyclic loading. Moreover, under the action of long-term tension stress, deep and large tension fractures are produced at the trailing edge of the landslide. The mechanical properties of the trailing edge of the landslide are weaker than that of the leading edge of the landslide, and there is a phenomenon of gradually spreading development from the trailing edge to the leading edge, which means that the trailing edge of the landslide is destroyed first and the leading edge of the landslide is destroyed later. The process of landslide being triggered by cyclic loading has the characteristics of the trailing edge of landslide extruding to the leading edge of the landslide and locking of the leading edge. The evolutionary development of the advancing landslide can be divided into the following stages: (i) tension fractures are formed under the cumulative effect of the cyclic loading, and the fractures develop from the slope to the depth until they cut the underlying soft interlayer, and the tension fractures develop at the trailing edge of the landslide (Figure 1a); (ii) The mechanical structure of the soil in the slip zone at the trailing edge of the landslide is gradually destroyed, and the mechanical strength of the slip zone soil gradually decreases. As the landslide evolution advances, the weakening zone of the slip zone soil also gradually expands toward the leading edge (Figure 1b); (iii) with the expansion of the weakening zone, the length of the landslide locking section decreases accordingly. When the weakening zone expands to a certain length and the landslide resistance is insufficient to resist the sliding force of the landslide, the locking section is sheared out and the landslide is triggered along the penetrating slip surface (Figure 1c).

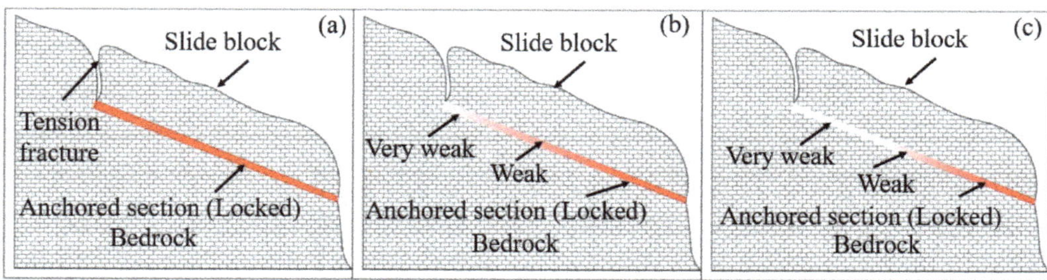

Figure 1. Evolutionary model of advancing layered rock landslide: (**a**) undamaged stage; (**b**) initial stage of damage; (**c**) end stage of damage.

Based on the assumption of cyclic shear failure behavior, the conceptual model has to be able to (i) allow the rupture of interparticle links and (ii) characterize the microstructural features of the soil. For the first aspect, the discrete element method (DEM) was used considering its unique ability to characterize microfractures. The most commonly used DEM model for simulating soil strain softening is the linear Parallel Bond Model (PBM), the linear PBM includes two classical interface behaviors: a linear-elastic model interface containing only forces and a viscoelastic model interface containing both forces and moments. In the first interface, rotation is not applicable and only one direction of motion can be performed according to the forces, while the second model interface can accommodate the rotational action of the forces and also the motion under the forces, while the second model interface is automatically transformed into the first model interface when the forces between the particles exceed the defined yield value of the connection bonds. Meanwhile, the basic object studied by PFC3D is the contact between particles and particles, which can directly simulate the physical problems of motion and interaction between particles. The large particles of arbitrary shapes can be created by connecting two or more small particles, and the combined particles made by the connection can be studied as independent particle bodies. It avoids the study of mechanical properties of materials by obtaining an intrinsic model from traditional empirical data while studying the mechanical behavior of bulk media from a fine-scale perspective, and the method overcomes the macroscopic continuity assumption of traditional mechanical models of continuous media [45]. The engineering properties of soils are simulated numerically at the fine view level and the macroscopic mechanical behavior is analyzed by using the study of fine view parameters.

For this purpose, the slip zone soil fabrics are prepared and then covered with the base PBM, the grain boundary properties are simulated by modifying the intra-grain contact (see Figure 2c), and the length of the discrete microfractures are described by the midpoint of the line segment connected by two centers of mass and the average of their radii, respectively. By applying different types of cyclic loading to the numerical soil model, the accumulation of micro-fractures in the soil gradually forms macro-fractures, which eventually lead to shear failure damage of the slip zone soil due to soil strain softening. Therefore, it is promising to use this PBM to study the interaction and connection of microfractures under the action of cyclic loading. In order to consider the mechanical properties of shear failure of slip zone soil under cyclic loading, a detailed scheme for numerical modeling of slip zone soils is proposed. The concept of observing soil strain softening behavior tests and numerical modeling of slip zone soils at a laboratory scale is shown in Figure 2. The size of the discrete element model of slip zone soil is 150 mm × 150 mm × 150 mm (X∗Y∗Z), which means it is a tiny unit in the slip zone soil. The purpose of this selection is to analyze the variation of shear strength of slip zone soil particles under different dynamic loads from a fine viewpoint. Table 1 lists the calibrated microparameters of the slip zone soils. Slip zone soil was collected from an open pit of mine in Tieshan District, Huangshi City, Hubei Province, China.

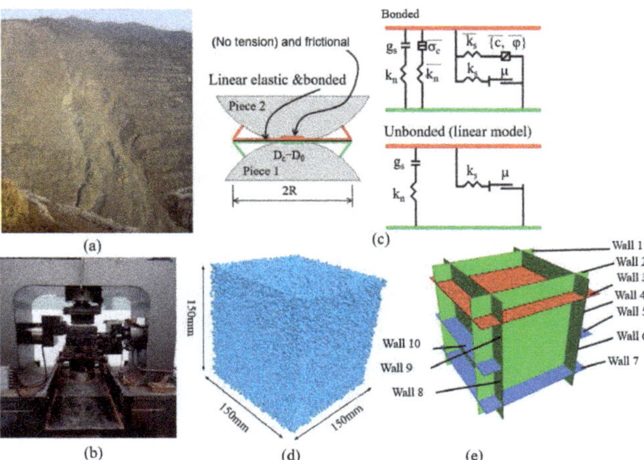

Figure 2. Concept of observing strain softening behavior tests and numerical modeling of slip zone soils at laboratory scale: (**a**) the slope with slip zone soil; (**b**) laboratory direct shear test; (**c**) linear Parallel Bond Model (Table 1 shows the meaning of the symbols of the parameters in the model); (**d**) numerical model of slip zone soil; (**e**) numerical model of walls for applying servo-control stress.

Table 1. Calibrated microscopic parameters of PFC3D particles.

Microparameter of Slip Zone Soil	Unit	Value
Minimum particle radius, R_{min}	mm	1
Maximum particle radius, R_{max}	mm	2
Density, ρ	g/cm^3	2.36
Particle-particle contact modulus, Ec	GPa	0.03
Friction coefficient, μ	-	1
Particle normal stiffness to shear stiffness, kn/ks	-	1.3
Parallel bond normal to shear stiffness ratio, $\overline{kn}/\overline{ks}$	-	1.0
Parallel bond friction angle, $\overline{\varphi}$	deg	25
Parallel bond connection modulus, $\overline{E}$	GPa	0.01
Parallel bond tensile strength $\overline{\sigma_c}$	MPa	16.5
Parallel bond cohesion $\overline{C}$	MPa	16.5

2.2. Measurement Indicators- Coordination Number

The numerical simulation software PFC3D contains a set of more effective statistical methods to record the changes in variables between particles within the numerical model throughout the numerical simulation process, such as the forces and deformations, as well as the development of fractures [46]. One advantage of DEM simulations is the ability to obtain information that cannot be acquired from continuum-based techniques or physical experiments such as fabric analysis. In this regard, fabric tensor and coordination number provide a global description of contact orientations and packing stability. This criterion was used to assess the onset of instability (liquefaction) as loading progressed. Assuming that there are N particles in the measurement area, the coordinate number C_n is defined as the average number of active contacts per body, and is computed as [47]:

$$C_n = \sum_N n_c \Big/ N \quad (1)$$

where n_c is the number of contacts per particle. Once the contact force between the particles exceeds the limit value defined before the numerical calculation, then the contact between the particles will break due to the excessive force, followed by microscopic fractures being generated, while the number of contacts between the particles will be further reduced and the number of defined coordinates will gradually decrease. The decrease of coordination number implies breakage of particle contacts, which is called a fracture.

2.3. Verification of the Accuracy of the Numerical Model and Details of the Applied Cyclic Loading

The numerical simulation test process for slip zone soil materials consists of three parts: (i) stress initialization of the numerical model, (ii) servo control of the numerical model in predetermined confining pressure, and (iii) implementation of cyclic loading.

2.3.1. Stress Initialization of the Numerical Model

The specimen consists of several particles, and after generating the particles, the particles and the wall are attributed with their own mechanical properties, and simultaneously with the connection properties between the particles and between the wall and the particles. Then, a certain porosity is attributed between the particles, and the particles are allowed to adjust freely to reach an equilibrium state without confining stress, i.e., stress initialization of the numerical model.

2.3.2. Servo Control of the Numerical Model in Predetermined Confining Pressure

Following the generation of the sample, a servomechanism was applied iteratively to isotropically consolidate the specimen to the desired confining stress. The servomechanism uses the feedback of the stresses on the walls to determine if the isotropic stress is more or less than the desired value and adjusts the wall positions accordingly (Itasca, 2014).

2.3.3. Implementation of Cyclic Loading

Seismic waves generated by blasting are random waves, and it is better simulated if the original waveform is fed into the numerical calculation model for blasting vibration. However, because of the difficulty of frequency conversion and amplitude variation during the numerical test, all sinusoidal waveforms were used for the input vibration waveforms in the numerical test. In addition, in order to better compare the effects of amplitude and number of cycle loads on slope stability, simple waves were used for all studies in the full paper. At the same time, sinusoidal waves have most of the properties that seismic waves have and also have the advantage of simplifying calculations [48–50]. Therefore, after the consolidation phase, the sample was subjected to a strain-controlled sinusoidal cyclic loading pattern. During loading, the volume of the sample was conserved to simulate undrained conditions. The implementation of cyclic loading applied to the numerical model is the key to this simulation. Cyclic loading in laboratory tests is mainly controlled by inputting vibration waves to the test apparatus, while in PFC3D, it is mainly through the definition of wall velocity to achieve cyclic shear loading, which is essentially a displacement-controlled loading method. In the PFC3D simulation, the stress values between the loading wall and the particles are monitored to obtain the shear stress in the soil during cyclic loading. Based on the vibration parameters in the numerical model, a direct shear test is conducted mainly by the lower shear box, and a sine wave is an input to adjust the motion direction of the wall in real time, to achieve cyclic loading of the slip zone soil material. That is, the same shear rate is specified for walls 1#, 2#, 3#, 4#,5#, 9# and 10# in Figure 2. Figure 2 shows a schematic diagram of the cyclic loading process of the numerical model. During the numerical simulation, the PFC3D program mainly includes the measurement of the numerical model shear stress and the application of the wall velocity during cyclic loading. As shown in Figure 2 for the shear box, the force on the lower shear box is the sum of the contact force on wall 8 and the contact force on wall 6, while the force on the upper shear box is the sum of the contact force on wall 9 and the contact force on wall 4, and the number of cycle loading include 750, 1500, 2250, 3000, 3750, 4500, 5250, 6000, 7500, 9000, 11,500, 12,000, and cyclic loading stress amplitude include 0.5 cm/s, 0.9 cm/s, 1.5 cm/s, 2.2 cm/s, where cyclic loading stress amplitude is equivalent to the peak ground velocity (PGV) of a natural earthquake.

The applied maximum shear strain (γ_{max}) follows the periodic (sinusoidal) pattern shown in the left part of Figure 3. By controlling strain, the displacement can be controlled, and then the velocity amplitude (cyclic loading stress amplitude) can be obtained by calculation. From the sine wave in Figure 3, it can be seen that when the maximum strain amplitude reaches the maximum shear strain (γ_{max}), then the value remains constant until the cyclic loading stops. Cyclic loading is applied gradually and the strain amplitude can be varied to cover a wide range of strain amplitudes. The use of numerical simulations makes it easy to repeat the test on the same specimen while changing one parameter of the sine wave compared to laboratory tests. Once the specimen (at a specific porosity) is generated and isotopically consolidated, the exact same specimen can be tested numerous times and subjected to various signals without changing the initial conditions.

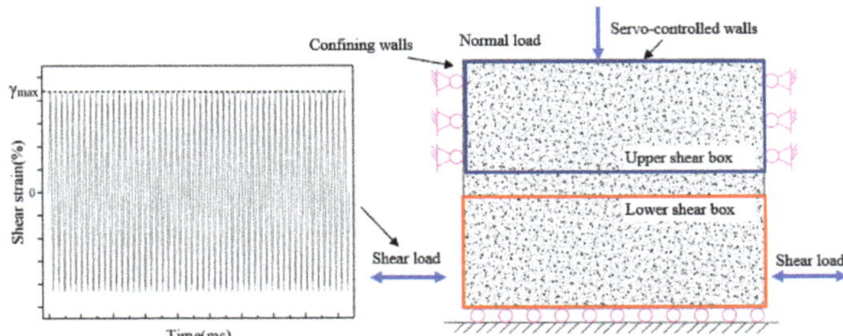

Figure 3. Schematic diagram of the cyclic loading process of the numerical model.

2.4. Comparison of Numerical Simulation and Laboratory Tests

The contact and interaction between numerous particles in a PFC model often exhibits macroscopic mechanical behavior. In order to obtain a suitable set of microscopic parameters matching the particles, the calibration of the numerical model is usually tested by comparing the parameters of the macroscopic mechanical behavior of the numerical simulation with the results obtained in the laboratory [46]. The mechanical parameters involved in PFC3D are fine parameters characterizing the properties of the particles, which have a random nature and a complex relationship with the macroscopic mechanical properties, and the calibrated parameters are generally considered reasonable when the obtained macroscopic mechanical properties are consistent with the actual test results through basic mechanical tests [51]. Therefore, based on the macro-mechanical parameters of slip zone soil, the shear stress-strain curve of the slip zone soil is simulated to find and determine the fine-scale parameters of the shear strength of the numerical model corresponding to the appropriate curve. Combined with the recommendations of the work of Hofmann et al. (2015) on the calibration process of the numerical model [52], the numerically simulated mechanical curves that match well with the laboratory tests are obtained, as shown in Figure 4. Table 1 lists the microparameters related to the slip zone soils.

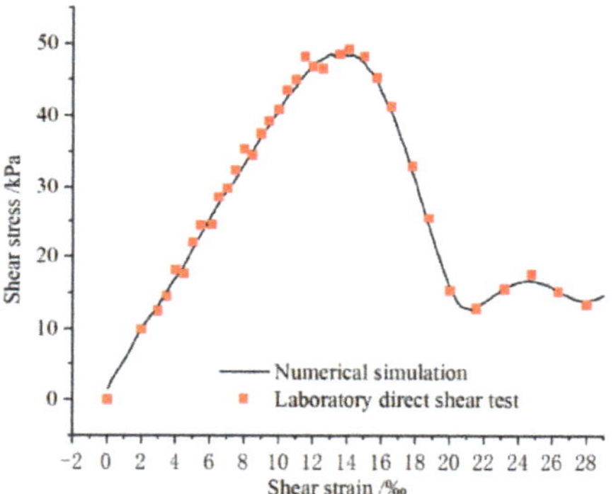

Figure 4. Comparison of straight shear stress-strain curves from laboratory direct shear test and numerical simulation.

3. Analysis for Strain Softening Characteristics of Slip Zone Soil

3.1. Coordination Number Analysis for Strain Softening Characteristics

The coordination number varies widely under different working conditions, and in order to distinguish the degree of variation of the coordination number, the percentage of coordination number is defined as P_c.

$$P_c = (C_n - C_{ni})/C_{ni} * 100 \tag{2}$$

where C_n and C_{ni} are the current and initial coordination numbers in the simulation model, respectively. In the different conditions, the values of P_c of the numerical model are plotted versus the cyclic loading times in Figure 5. In this study, we consider a particle contact breakage rate of 5% (i.e., $P_c = -5\%$) as the maximum limit that can be achieved by particle motion. When the P_c reduction is greater than 5%, significant breakage is assumed. After close observation of Figure 5, when the cyclic loading stress amplitude is 0.5 cm/s, the initial change of P_c does not exceed 10%, which is because this is a small loading rate, which only makes a part of the particles in the numerical soil model move, and even with the increase of cyclic loading, the change of P_c is not significant. Meanwhile, the rate of P_c reduction gradually increased when the cyclic loading stress amplitude increased from 0.9 cm/s to 2.2 cm/s.

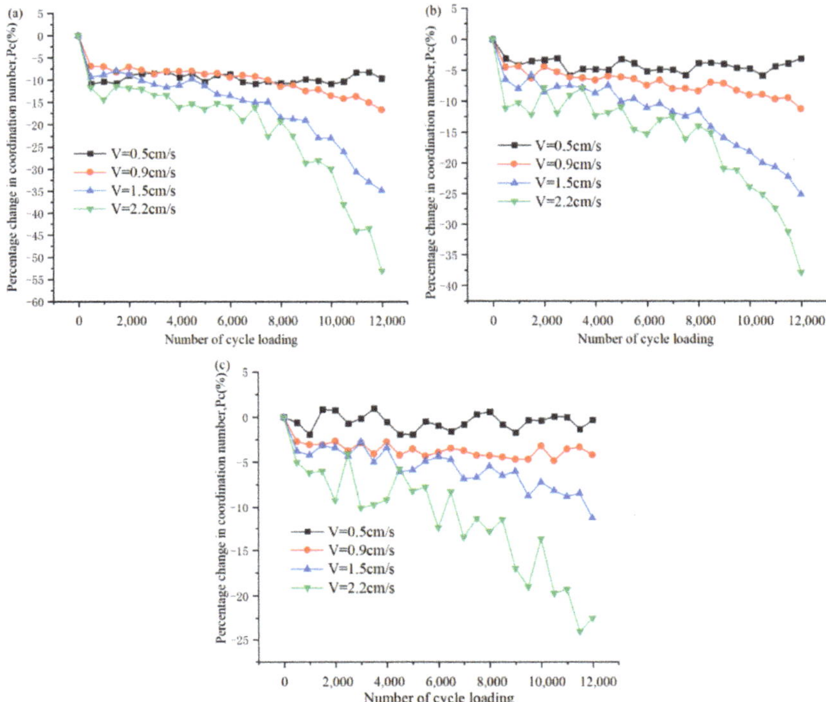

Figure 5. Change of coordination number characterizing the softening process of the slip zone soil: (**a**) 50 kPa; (**b**) 100 kPa; (**c**) 200 kPa.

In order to highlight the level of strain softening of the slip zone soil material under cyclic loading, when the number of cyclic loading is 12,000, the 100% stacking bar graph in Figure 6 shows the final P_c estimates for the slip zone soil material under different operating conditions, and their values are indicated separately in the corresponding segments. For example, when the slip zone soil material is subjected to a predetermined confining pressure of 50 kPa, at cyclic loading stress amplitudes of 1.5 cm/s and 2.2 cm/s, it can be seen that the strain softening of the slip zone soil corresponds to P_c values of −34.8% and −53.0% in Figure 6. On the other hand, it is obvious from Figure 6 that the cyclic loading stress amplitude is the main factor controlling the number of fractures generated. When observing the effect of different loading stress amplitudes on the P_c of fractures, i.e., the P_c values at different loading stress amplitudes in Figure 5a–c, it is not difficult to find that the change in the fracture is greatest for loading stress amplitude equal to 2.2 cm/s compared with other cyclic loading stresses, which shows that the soil strain softening effect becomes more pronounced with the increase in the number of cyclic loading at larger cyclic loading stresses. This indicates that the cyclic loading stress amplitude plays a deterministic role in the occurrence of fractures (level of strain softening). On the other hand, as the predetermined confining pressure of the slip zone soil increases, i.e., when the confining pressure of soil increases from 50 kPa to 200 kPa (as observed in Figure 5a,c, the number of cyclic loading is 12,000), it is easy to find that when the loading stress amplitude is 0.9 cm/s, the P_c values are −16.6% and −4.12% respectively, and the above discussion reveals that when the confining pressure is reduced from 200 kPa to 50 kPa (the relationship between the confining pressures values is four times), fractures are four times more easily generated in the 50 kPa model than in the 200 kPa model. However, when observing Figure 6, when the cycle loading stress amplitude increase from 0.5 cm/s to 2.2 cm/s (the relationship between the cycle loading stress amplitude values is almost four times), the multiplication of the increase in P_c values is 5.6, 12.5, and 84.9 for the confinement of 50 kPa, 100 kPa, and 200 kPa, respectively. Therefore, we are able to reach the interesting conclusion that although the confining pressure of the slip zone soil can inhibit the increase of fractures, it has a limited inhibitory effect on strain softening, and it is the cyclic loading stress amplitude that is the most key factor in strain softening process of soil.

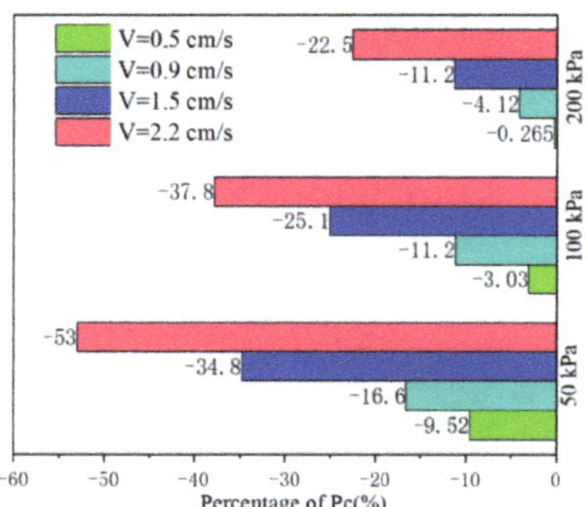

Figure 6. Comparison of the percentage reduction of the coordination number (the number of cyclic loading is 12,000).

3.2. Stress Analysis for Strain Softening Characteristics of Slip Zone Soil

The above discussion reveals that the variation of C_n is an indicative parameter of slip zone soil under the action of cyclic loading and predetermined confining pressure, while the essential cause of the strain softening characteristics of slip zone soil is the variation of shear stress. The variation trend of shear stress of slip zone soil obtained by using the program servo control in numerical simulation, the change law of shear stress of slip zone soil under different predetermined confining pressure and the different number of cyclic loading are discussed, as shown in Figure 7, the change of fractures in the numerical model of slip zone soil is shown in Figure 8, in which red fractures represent tensile fractures and green fractures represent shear fractures, also the information of the numerical model is shown in Figure 2.

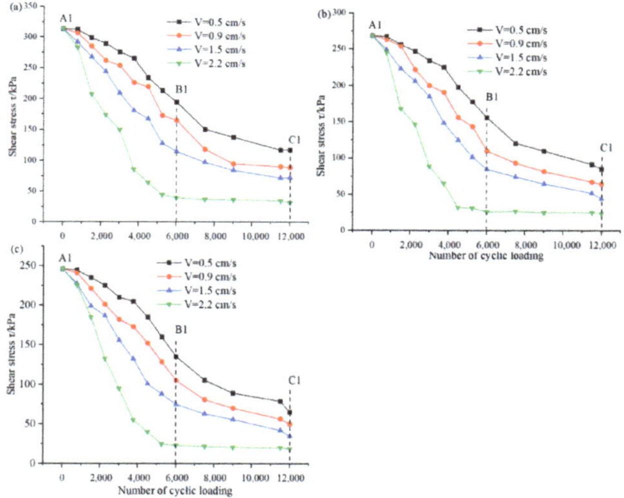

Figure 7. Evolutions of shear stress of slip zone soil under different cycle loading in (**a**) confining pressure = 200 kPa, (**b**) confining pressure = 100 kPa, (**c**) confining pressure = 50 kPa.

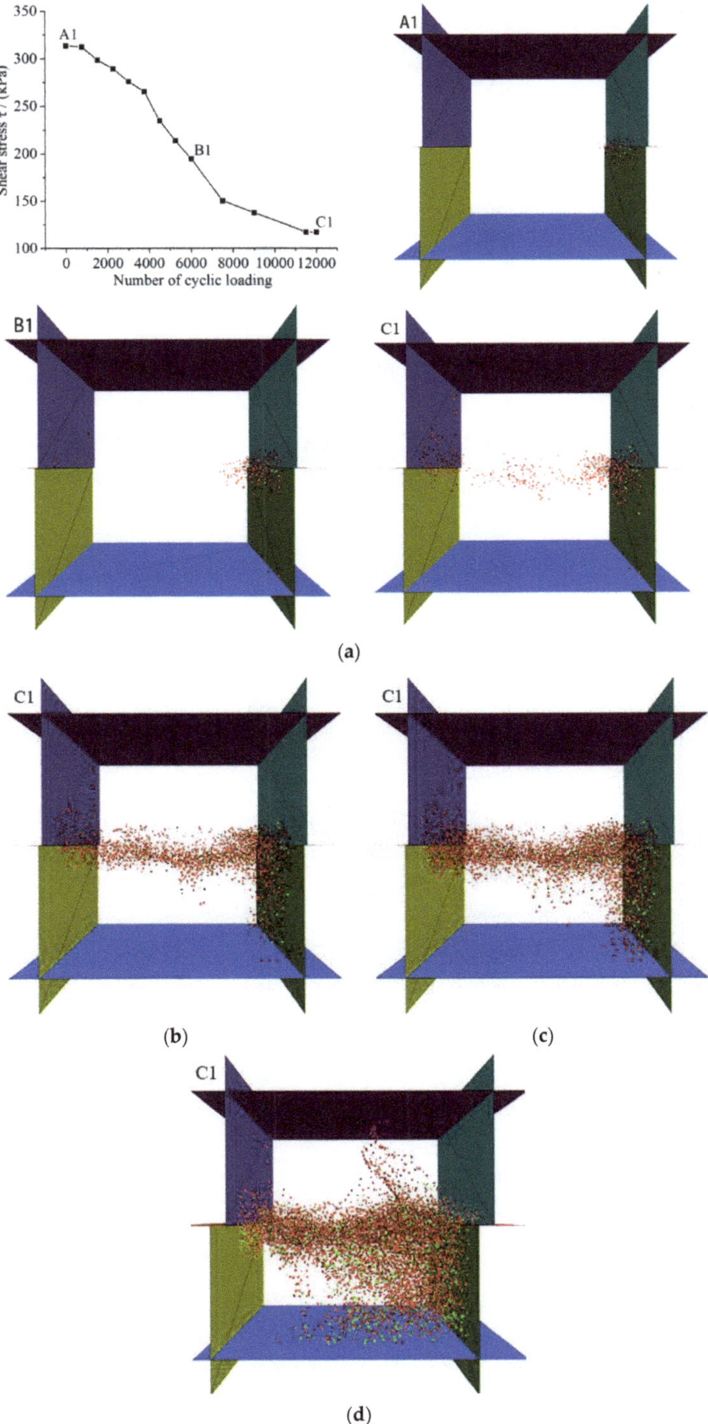

Figure 8. *Cont.*

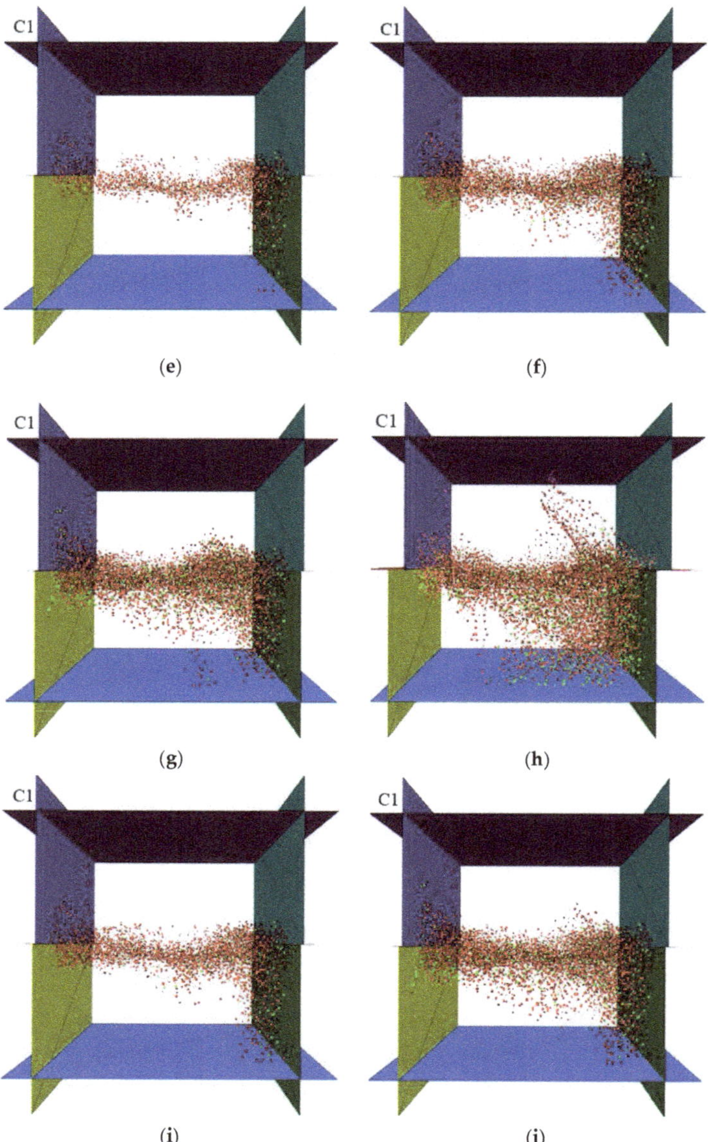

Figure 8. *Cont.*

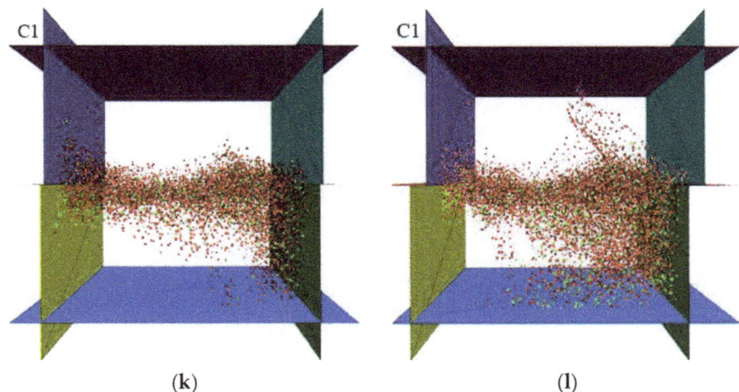

Figure 8. Change of fractures in the numerical model of slip zone soil under different cycle loading in (**a**–**d**) confining pressure = 200 kPa, (**e**–**h**) confining pressure = 100 kPa, and (**i**–**l**) confining pressure = 50 kPa. (**a**) V = 0.5 cm/s, confining pressure = 200 kPa. (**b**) V = 0.9 cm/s, confining pressure = 200 kpa. (**c**) V = 1.5 cm/s, confining pressure = 200 kPa. (**d**) V = 2.2 cm/s, confining pressure = 200 kPa. (**e**) V = 0.5 cm/s, confining pressure = 100 kPa. (**f**) V = 0.9 cm/s, confining pressure = 100 Kpa. (**g**) V = 1.5 cm/s, confining pressure = 100 kPa. (**h**) V = 2.2 cm/s, confining pressure = 100 kPa. (**i**) V = 0.5 cm/s, confining pressure = 50 kPa. (**j**) V = 0.9 cm/s, confining pressure = 50 kPa. (**k**) V = 1.5 cm/s, confining pressure = 50 kPa. (**l**) V = 2.2 cm/s, confining pressure = 50 kPa.

From a close view of Figures 7 and 8, it can be seen that the peak shear strength of the slip zone soil material increases with an increase in the predetermined confining pressure, while the cyclic loading stress and the number of cyclic loading also have a positive effect on the strain softening of the slip zone soil. When the cyclic loading stress amplitude is low (less than or equal to 1.5 cm/s), the shear strength decay of the slip zone soil is dominated by the cumulative weakening effect. That is, with the operation of the cyclic loading, the strain softening of the slip zone soil gradually develops and the shear stress gradually decreases, while at a higher cyclic loading amplitude (equal to 2.2 cm/s), the mechanical parameter weakening of the slip zone soil shows the inertial damage characterized by the action of cyclic loading. In this form, there are only two stages of weakening of the slip zone soil mechanical parameters. In the first stage, there is a tendency for rapid decay with the operation of cyclic loading, which means that the slip zone soil is immediately damaged by the loading in this stage. In the second stage, the strength parameters of the slip zone soil have reached the damage limit range, and basically do not change significantly with the increase of the number of cyclic loading. Through the A1 in Figure 8a, it is not difficult to find that fractures appear first at the intersection of the upper and lower shear boxes, which is due to the joint motion of the left upper retaining wall and the right upper retaining wall, which causes the wall to drive the particles attached to the retaining walls on both sides to carry out horizontal motion, thus indirectly transferring them to the center. So based on the diagram of fractures of A1 in Figure 8a, it is easy to know that the first location of the fractures is almost at the junction of the upper and lower shear boxes. When the number of cyclic loading is increased to 6000, the number of fractures appearing at the junction of the shear box further increases, while the location where the fractures exist expands from the junction to the middle of the numerical model, which is caused by the rearrangement and movement of the particles in the middle of the numerical model. Moreover, with the action of cyclic loading, the interaction between the numerical model particles makes more and more particles subject to cyclic loading, and the chain reaction caused by their mutual motion increases geometrically. Therefore, when the number of cyclic loading is 12,000, there is an abrupt increase in the number of fractures in the numerical model of the soil in the slip zone, and a penetrating fracture zone appears inside the model as can be seen by C1 in Figure 8a. The analysis and extension for 200 kPa is applied to 100 kPa and 50 kPa.

There is no doubt that the smaller the predetermined confining pressure, the smaller the restraint on the rate of fractures growth, and thus the more pronounced the increase of fractures in the slip zone soil, i.e., the greater the degree of slip zone soil strain softening.

On the other hand, the increase of the cyclic loading stress amplitude also contributes to the penetration of the slip zone soil fractures. From Figure 8a–l, it is not surprising to observe that the greater the cyclic loading stress amplitude, the more obvious the failure damage behavior of the slip zone soil, the more fractures increase, so the larger the macroscopic fracture zone range of the slip zone soil is under the cyclic loading. Interestingly, when the loading stress amplitude is equal to 2.2 cm/s, although the shear stress in the slip zone soil has stabilized after reaching a certain value under different limiting pressures, the number of fractures still increases with the number of cyclic loading. This is because as the number of cyclic loading increases, although the development of fracture is gradually restrained, the cyclic loading causes more particles close to the wall to form a directional arrangement, and the next cyclic loading to be transmitted to the middle of the numerical model needs to pass through particles that have already undergone shear damage behavior.

Moreover, the residual shear strength of this part of the particles is also the constant friction coefficient between the particles that will continuously consume the energy of the cyclic loading, thus leading to the increase of fractures in the slip zone soil model becomes more and more difficult as the number of cyclic loading increases. Therefore, according to the measurement mechanism of shear stress mentioned above, when the macroscopic fracture zone in the middle part of the slip zone soil model has been formed, the connection between particles far from the interface part becomes more vulnerable to disruption than at the beginning of cyclic loading, so once the larger and dense fracture zone is penetrated, the slip zone soil shear stress values are not significantly correlated with the growth of fractures anymore. That is, landslide occurrence is particularly potent in practical engineering when significant deformation of the slip zone soil occurs due to particle rotation and sliding, followed by a penetrating fracture zone. In general, the development of soil stresses and fractures in the slip zone under different confining pressures and different cyclic loading exhibited a comparable trend of evolution. It is shown that the cyclic loading stress field varies similarly for slip zone soils, and the shear stress decreases monotonically under the cyclic loading. At the same time, penetration of the fracture zone is more likely to occur under higher cycle loading stress and lower confining pressures, as in (Figure 8g,h,k,l). Moreover, after the formation of the macroscopic fracture zone of the slip zone soil, the damage to the strain softening of the slip zone soil by cyclic loading will gradually develop towards the internal changes of the soil. As for A3 in Figure 8, the increase in the number of cyclic loading and the increase in the cyclic stress amplitude will accelerate the strain softening rate of the soil.

4. Impacts of Landslide Formation

4.1. Mechanical Model of Landslide Mechanism

It is well accepted that the trigger mechanism of landslides is reflected by the change in the stability coefficient of the slope. The stability coefficient of the slope is calculated according to the limit equilibrium analysis of the Sarma method, where the change of shear strength is obtained in the previous Section 3.2. Meanwhile, Figure 9 shows the force applied to any block i in the landslide body, which was used for studying the theoretical principle of the Sarma method, the horizontal seismic inertial force is applied to the block, the transient stability factor on the slip zone soil is equal to one.

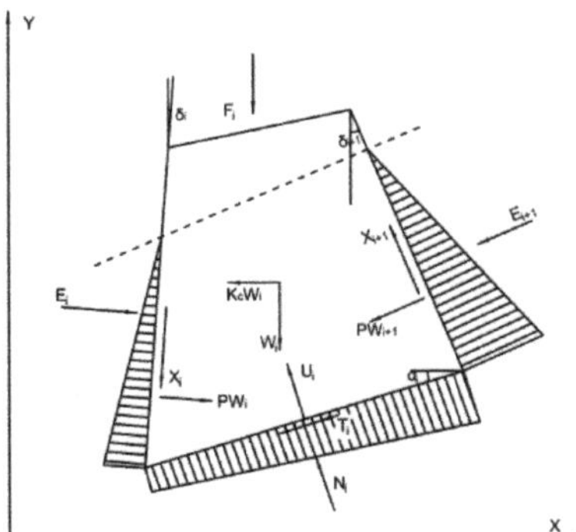

Figure 9. Schematic diagram of the force applied to any block i in the landslide body.

According to the principle of static balance, the following Equations (2)–(4) are obtained.

$$\sum X = 0, \sum Y = 0 \tag{3}$$

$$T_i \cos\alpha_i - N_i \sin\alpha_i - K_c W_i - FX_i - X_{i+1}\sin\delta_{i+1} + X_i \sin\delta_i - E_{i+1}\cos\delta_{i+1} + E_i \cos\delta_i = 0 \tag{4}$$

$$T_i \sin\alpha_i - N_i \cos\alpha_i - W_i + FY_i - F_i + X_{i+1}\cos\delta_{i+1} - X_i \cos\delta_i - E_{i+1}\sin\delta_{i+1} + E_i \sin\delta_i = 0 \tag{5}$$

where $K_c W_i$ is the horizontal seismic inertial force acting on the block. T_i and N_i are, respectively, the shear force and normal force acting on the bottom surface of the block i; W_i is the weight of block i; X_i and X_{i+1} are, respectively, acting on the shear force of the block i side and the $i + 1$ th side; is E_i the normal forces acting on the block i side, and E_{i+1} is the normal forces acting on the block $i + 1$ side; F_i is the external load acting on the top of the slope; the angle between the block i side and the vertical direction is δ_i, the angle between the block $i + 1$ side and the vertical direction is $\delta_i + 1$ the angle between the sliding surface of the block i and the horizontal direction is α_i;

According to the Mohr-Coulomb criterion, Equations (5)–(7) can be obtained:

$$T_i = (N_i - U_i)\tan\phi_{Bi} + c_{Bi} b_i \sec\alpha_i \tag{6}$$

where B_i is the width acting on the bottom surface of block i; U_i is the water pressure acting on the bottom surface of block i, and φ_{Bi}, C_{Bi} is the shear strength parameter of the bottom surface of block i.

$$X_i = (E_i - PW_i)\tan\phi_{Si} + c_{Si} d_i \tag{7}$$

$$X_{i+1} = (E_{i+1} - PW_{i+1})\tan\phi_{Si+1} + c_{Si+1} d_{i+1} \tag{8}$$

where d_i is the length of the block i side, d_{i+1} is the length of the block $i + 1$ side; PW_i is the water pressure acting on the side of the block i, and PW_{i+1} is the water pressure acting on the side of the block $i + 1$, φ_{Si}, C_{Si} are the shear strength parameter of the block i side. Different horizontal seismic inertial forces are calculated through different stability coefficients. When the horizontal seismic inertial force acting on the block is equal to 0, the stability coefficient of the slope is the slope stability coefficient in the natural state.

4.2. Calculation of Stability Coefficient

The section of the slope with slip zone soil is selected as the research section, as shown in Figure 1, in which it is assumed that the slope is a laminated structure, the thickness of the slip zone soil is relatively uniform, and the shear strength parameters of the soil material in the slip zone at different places are the same. In order to reasonably analyze the relationship between the shear strength of the slip zone soil and slope stability, the change of slope stability is discussed and analyzed under different numbers and stress amplitude of cyclic loading. Figure 10 shows a schematic diagram of a section of the slope with the slip zone soil. The sliding section examined includes the slip zone soil, the unloading boundary line and the bedrock. The potential landslide body is divided into strips, and the potential landslide body is divided into eight vertical strips. The cohesion of the potential landslide body and bedrock is 466 KPa and the internal friction angle is 29°, meanwhile, the cohesion between strips from bar 1 to bar 8 is 400 KPa and the internal friction angle is 25°. To simplify the calculation process, the strength parameters of bars 1–8 adopt the uniformity value, the strength parameters of the slip zone soil are the data obtained from the above Section 3.2.

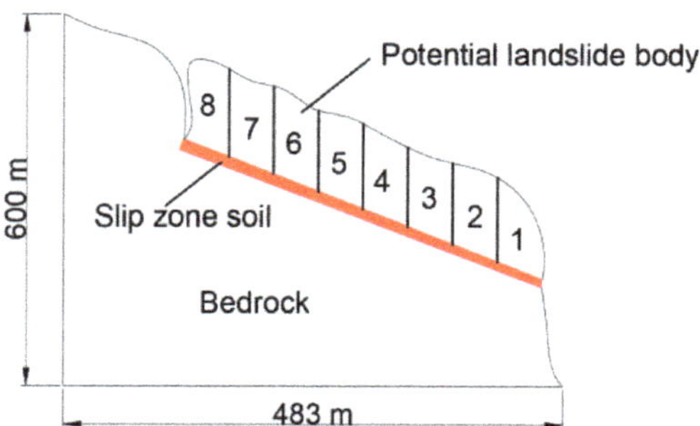

Figure 10. Schematic diagram of section of slope with the slip zone soil.

According to the results obtained by analyzing the slope stability, the change of slope safety factor is statistically analyzed under different loading numbers and different loading stress amplitude, whose change is shown in Figure 11 and Table 2.

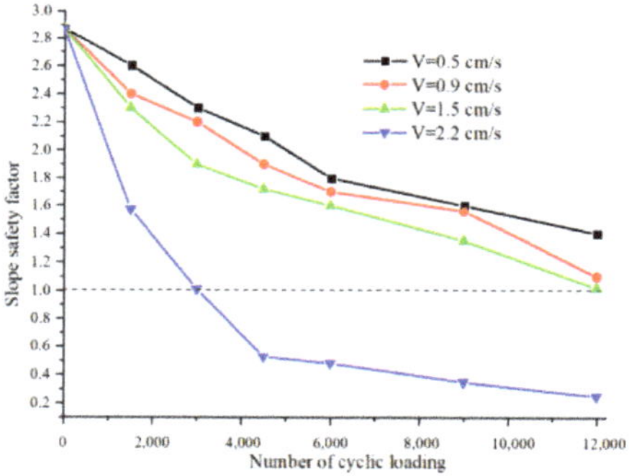

Figure 11. Change of slope safety factor.

Table 2. Change of slope safety factor under different loading number and different loading stress amplitude.

Number of Case	Loading Stress Amplitude/(cm/s)	Loading Number	Slope Safety Factor	Number of Case	Loading Stress Amplitude/(cm/s)	Loading Number	Slope Safety Factor
0	0.5	0	2.87	13	1.5	1500	2.3
1	0.5	1500	2.6	14	1.5	3000	1.9
2	0.5	3000	2.3	15	1.5	4500	1.72
3	0.5	4500	2.1	16	1.5	6000	1.6
4	0.5	6000	1.8	17	1.5	9000	1.35
5	0.5	9000	1.6	18	1.5	12,000	1.02
6	0.5	12,000	1.4	19	2.2	1500	1.57
7	0.9	1500	2.4	20	2.2	3000	1.01
8	0.9	3000	2.2	21	2.2	4500	0.52
9	0.9	4500	1.9	22	2.2	6000	0.48
10	0.9	6000	1.7	23	2.2	9000	0.35
11	0.9	9000	1.56	24	2.2	12,000	0.25
12	0.9	12,000	1.1				

The stability coefficient of the slope decreases with the increase of stress amplitude and the number of cyclic loading, as can be seen in Figure 11. The calculation in Section 3.2

shows that the strain of the slip zone soil gradually softens under the cyclic loading. The change of shear stress is not only caused by the confining pressure of soil but also the cyclic loading stress plays a key role in the expansion of fractures. However, sometimes the gradual decrease of shear stress is not necessarily all the effect of cyclic loadings, such as the continuous erosion of groundwater and the gradual increase of pore water pressure, which will lead to the gradual decrease of shear stress in the slip zone soil. Meanwhile, by a close analysis of Figure 11, the continuous development and extension of fractures lead to the continuous reduction of the landslide anchorage section, and the stability coefficient of the slope is decreasing, which is one of the key driving factors for landslide occurrence.

Therefore, two possible landslide triggering mechanisms are described. For mode 1, by analyzing A1 and A2 in Figure 8a–c,e–g,i–k, the shear stress gradually decreases with the increase in the number of cyclic loading, and the fracture is also gradually expanded into a penetrating fracture zone, that is, the strain softening behavior occurs slowly with cyclic loading. Additionally, the same analysis applies to the stability coefficient of the slope when cyclic loading stress amplitude is less than or equal to 1.5 cm/s in Figure 11. On the other hand, model 2 can be easily obtained by analyzing A3 in Figure 8d,h,l, it can be seen that when the shear stress no longer increases with the number of cyclic loading, the number of fractures still continues to increase, indicating that the landslide has been triggered. Mode 2 shows that when the cyclic loading stress is larger, the shear stress in the slip zone soil tends to decay rapidly as the number of cyclic load loading increases. Therefore, this phenomenon infers that if the amplitude and number of cycle loading stress are large enough, the strain rate of the slip zone soil will gradually accelerate and a large shear fractures zone is likely to occur, followed by a landslide occurrence, which means that the strain softening behavior of slip zone soils is the result of inertial forces, which expand at a very fast development rate, which is also verified by the data in Figure 11 that is equal to 2.2 cm/s.

4.3. Application of Mechanical Learning-Based Time Series Analysis to Slope Stability Prediction

Time series analysis is a highly applied branch of probability statistics, with mathematical tools and theories used in many fields, such as finance and economics, meteorology and hydrology, signal processing and mechanical vibration [53,54]. Although the analysis of long-term trends and cyclical fluctuations control the basic style of time series movements, it is after all not the whole picture of time series movements, and it is more reasonable and superior to use the theory of stochastic processes and statistical theory to examine the time series of long-term trends, seasonal variations and other factors that act together. The analysis of time series is based on the theory of stochastic processes and statistical theory, leading to the stochastic analysis of time series. Stochastic time series analysis enables, on the one hand, the creation of mathematical models that more accurately reflect the dynamic dependencies contained in the series and thereby forecast the future of the system and, on the other hand, statistical methods that more accurately reveal the dynamic structure and laws of the system.

Stochastic analysis of time series usually utilizes the Box–Jenkins modeling approach. The steps for modeling using the Box–Jenkins method are as follows.

(1) Calculate the correlation coefficient and the bias correlation coefficient for a sample of the observed series.

The easiest way to determine whether the data is smoothed or not is to use the image and other analyses using the sample autocorrelation function and sample partial autocorrelation function. When we have a sample series x_t, we can calculate the covariance of the samples as shown in Equations (8)–(12).

$$\overline{\mu} = \frac{\sum_{t=1}^{n} x_t}{n} \tag{9}$$

$$s^2 = \frac{\sum_{t=1}^{n} (x_t - \overline{\mu})(x_{(t)} - \overline{\mu})}{n-1} \tag{10}$$

$$\gamma_k = 1/n\sum_{t=k+1}^{n} (x_t - \overline{\mu})(x_{(t-k)} - \overline{\mu}) \tag{11}$$

$$\rho_k = \frac{\gamma_k}{s^2} \tag{12}$$

where n is the number of sample series,$\overline{\mu}$ is average, s^2 is variance, γ_k is covariance, ρ_k is sample autocorrelation function.

$$X_t = \Phi_{k1}x_{t-1} + \Phi_{k2}x_{t-2} + \ldots + \Phi_{kk}x_{t-k} + u_t \tag{13}$$

Each regression coefficient in the autocorrelation function represents the autocorrelation coefficient between x_t and x_{t-k} after excluding the effect of its intermediate variables $x_{t-1}, x_{t-2},\ldots, x_{t-k+1}$, i.e., the partial autocorrelation function.

Each of the regression coefficients in the autocorrelation function (Φ_{kk}) represents the autocorrelation coefficient between x_t and x_{t-k} after excluding the effect of its intermediate variables $x_{t-1}, x_{t-2},\ldots, x_{t-k+1}$, i.e., Φ_{kk} is the partial autocorrelation function.

(2) Pattern recognition: check whether the sequence is a smooth non-white noise sequence. If the series is a white noise series, the modeling is finished; if the series is a non-stationary series, the modeling method of non-stationary time series is used to build an autoregressive model (ARIMA model) or auto-regressive sliding average model (MA model); if the series is a stationary series, an auto-regressive sliding average model (ARMA model) is built.

(3) Initial order and parameter estimation: after the model is identified, the highest order of the model to which it belongs is framed; then the model is fitted and tested from low to high order within the identified type.

There are a number of criteria that can be used to model {εt}. The criteria are mainly based on the following function (13) shown below.

$$\delta(p') = n\log\left(\widehat{\sigma_{p'}^2}\right) + p'g(n) \tag{14}$$

{εt} is the residual of the data x_t, $\delta(p')$ is a function of constant order, p' is a constant order, and $\left(\widehat{\sigma_{p'}}\right)$ is an estimate of the variance of the residual obtained when taking order p', which is the Bayesian information criterion (BIC) criterion when $g(n)$ = log n, if the value of $\delta(p')$ is smaller, then the prediction model is better.

(4) Goodness-of-fit test: different models are compared using the fixed-order method to determine the most suitable model.

(5) Fit test: the selected model is tested for fit and parameters to further determine the most appropriate model from the selected model.

(6) Prediction: using the prediction model developed, the data is then predicted.

The order in the model identification process was first set to 1, and then the ARIMA (1,1,1)(1,1,1) model was established, and the display model fit measure was selected through Equations (1)–(6), which ultimately allowed the parameter estimate value and standard BIC values to be obtained, as shown in Tables 3 and 4, and the most appropriate model was selected to determine the most appropriate model by obtaining estimates of the model parameters, and the probabilities of the statistics of the independent variables.

Table 3. Statistics of the parameter estimate value and probability of the data statistic for ARIMA (1,1,1)(1,1,1) model.

Type of Model			Parameter Estimate Value	Probability of the Data Statistic
	Constant		0.227	0.616
AR		Lag1	0.019	0.762
	Difference		1	
MA		Lag1	0.714	0.000

Table 4. Statistics of the Stationary R-squared and normalized BIC value.

Type of Model	Stationary R-Squared	Normalized BIC	Sratistics	Probability of t Data Statistic
AR	0.425	10.16	2.526	0.748
MA	0.225	12.16	3.526	0.648

As can be seen from Table 3, the parameter estimate for AR(1) is 0.019 and the probability of the data statistic is 0.762, which means that the original hypothesis, that AR(1) is zero, is accepted. The parameter estimate for MA(1) is 0.714 and the probability of the data statistic is 0, which rejects the original hypothesis that MA(1) is zero. So, the model is not optimal and the analysis of the data is not very appropriate. Therefore, the AR model was selected for another adjustment to obtain the most appropriate model. After the analysis in Tables 3 and 4 above, it is clear that the data is a non-stationary time series, so the goodness of fit of the ARIMA(1,1,0)(0,1,1) model is established, including all values of the goodness of fit adjusted for R-Square, normalized BIC, etc. The standard BIC value of 8.160 can be seen in Table 5, which is somewhat smaller than the standard BIC value of the ARIMA(1,1,1)(1,1,1) model. At the same time, the probability of the data statistic is 0.848.

Table 5. Statistics of the Stationary R-squared and normalized BIC value for ARIMA(1,1,0)(0,1,1) model.

Stationary R-Squared	Normalized BIC	Sratistics	Probability of the Data Statistic
0.747	8.16	9.526	0.848

Therefore, this model is a suitable model, so the data in Table 2 are modeled and predicted according to this validated model, as shown in Figure 12. It is easy to see through Figure 12 that the stability coefficients of slopes under different load cases are well predicted, while the difference error between the two is very small. This model can be used for the prediction of stability coefficients of slopes, and also provides an accurate and efficient mechanical learning method for stability analysis of slopes under cyclic loading.

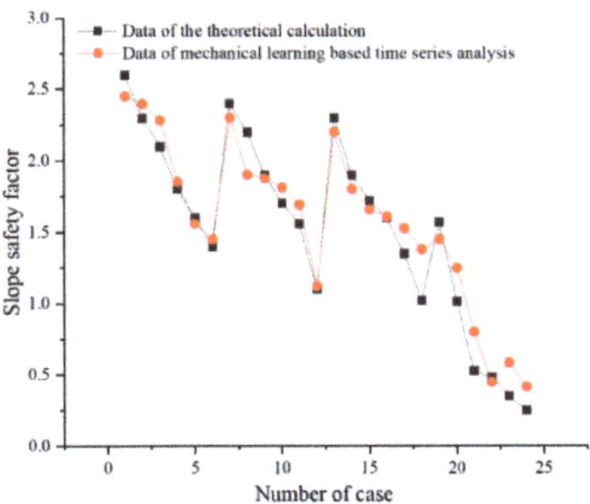

Figure 12. Mechanical learning-based time series analysis to slope stability prediction.

5. Conclusions

For rocky slopes containing weak structural surfaces at specified dips, the theory of anchored sections in the slip zone is widely mentioned, where the continuous development and extension of fractures lead to the continuous reduction of the landslide anchorage section (as shown in Figures 7–11), which is considered to be a key factor in the occurrence of landslides. The quantitative examination of the correlation between fracture and shear stress will allow us to infer the evolution of landslide occurrence with greater confidence.

From this perspective, we developed a three-dimensional slip zone soil numerical model in this study, which aims to incorporate the intermediate mechanism of strain softening. However, it is particularly important to note that we propose a mechanical learning method that can be used to predict stability coefficients for slopes where slopes with predetermined shear planes are subjected to cyclic seismic loads under undrained conditions.

Another important insight into the microfracture mechanism comes from the comparison of shear stress and C_n in slip zone soils. When the cyclic loading stress amplitude is large, the slip zone soil will be weakened rapidly, immediately after the shear stress no longer changes significantly, but the number of fractures in the slip zone soil still increase. Meanwhile, the gradual change of confining pressure of the slip zone soil rather gently promotes the growth of microfractures. Nevertheless, there are two possible effects associated with sharp changes in the shear stress of slip zone soils: cumulative progressive damage and significant inertial damage. In the second case, the strain softening of the slip zone soils can be explained by the rearrangement of the particles.

Based on the simulation results, possible field observations are illustrated in the context of landslide occurrence, which evolves from changes in the stability coefficient of the slope. The change of stability coefficients of the slope is compared with the development of fractures in the slip zone soil, and two possible landslide triggering mechanisms have been inferred. The strain of the slip zone soil gradually softens under cyclic loading. The shear stress gradually decreases with increasing confining stress, while cyclic loading stress plays a key role in the crack extension. The continuous development and extension of the fracture lead to the continuous reduction of the stability coefficient of the slope, which is another key driving factor for the occurrence of landslides. When the cyclic loading stress is larger, the shear stress of the slip zone soil tends to decay rapidly with the increase in the number of cyclic loadings.

The mechanical learning model proposed in this paper can be used for the prediction of stability coefficients of slopes, and also provides an accurate and efficient mechanical learning method for stability analysis of slopes under cyclic loading. Furthermore, it has to be highlighted once again that the stability of slopes is influenced by a number of factors, including drainage conditions, slope stress states, other forms of dynamic loading, etc. Therefore, the types of slopes mentioned in this paper are actually limited to the layered rock landslides with predetermined shear planes subjected to seismic loading in undrained conditions, which is only a small part of the complex slope stability phenomena.

Author Contributions: Conceptualization, N.J., T.W. and H.Y.; methodology, N.J., T.W. and H.Y.; software, N.J., C.Z. and X.L.; validation, N.J., X.L. and H.Y.; writing-original draft preparation, N.J. and T.W.; writing-review and editing, N.J. and H.Y.; project administration, N.J. and C.Z.; funding acquisition, N.J. and C.Z. All authors have read and agreed to the published version of the manuscript.

Funding: This research was funded by the National Natural Science Foundation of China (Grant No. 41807265, No. 41972286, No. 42072309) and the Hubei Key Laboratory of Blasting Engineering Foundation (Grant No. HKLBEF202001 and No. HKLBEF202002).

Conflicts of Interest: The authors declare no conflict of interest.

References

1. Werner, E.D.; Friedman, H.P. *Landslides: Causes, Types and Effects;* Nova Science Pub. Incorporated: New York, NY, USA, 2010.
2. Cruden, D.M. A simple definition of a landslide. *Bull. Int. Assoc. Eng. Geol.* **1991**, *43*, 27–29. [CrossRef]
3. Li, H.; Qi, S.; Chen, H.; Liao, H.; Cui, Y.; Zhou, J. Mass movement and formation process analysis of the two sequential landslide dam events in Jinsha River, Southwest China. *Landslides* **2019**, *16*, 2247–2258. [CrossRef]
4. Ma, J.W.; Niu, X.X.; Tang, H.M.; Wang, Y.K.; Wen, T.; Zhang, J.R. Displacement prediction of a complex landslide in the Three Gorges Reservoir Area (China) using a hybrid computational intelligence approach. *Complexity* **2020**, *2020*, 2624547. [CrossRef]
5. Kirschbaum, D.; Stanley, T.; Zhou, Y. Spatial and temporal analysis of a global landslide catalog. *Geomorphology* **2015**, *249*, 4–15. [CrossRef]
6. Kieffer, D.S.; Jibson, R.; Rathje, E.M. Landslides triggered by the 2004 Niigata ken Chuetsu, Japan, earthquake. *Earthq. Spectra* **2006**, *22*, 47–73. [CrossRef]

7. Chigira, M.; Yagi, H. Geological and geomorphological characteristics of landslides triggered by the 2004 Mid Niigta prefecture earthquake in Japan. *Eng. Geol.* **2006**, *82*, 202–221. [CrossRef]
8. Wang, F.; Fan, X.; Yunus, A.P.; Subramanian, S.S.; Alonso-Rodriguez, A.F.; Dai, L.; Xu, Q.; Huang, R. Coseismic landslides triggered by the 2018 Hokkaido, Japan (Mw 6.6), earthquake: Spatial distribution, controlling factors, and possible failure mechanism. *Landslides* **2019**, *16*, 1551–1566. [CrossRef]
9. Uzuoka, R.; Sento, N.; Kazama, M.; Unno, T. Landslides during the earthquakes on May 26 And July 26, 2003 in Miyagi, Japan. *Soils Found.* **2005**, *45*, 149–163. [CrossRef]
10. Ayalew, L.; Yamagishi, H.; Marui, H.; Kanno, T. Landslides in Sado Island of Japan: Part II. GIS-based susceptibility mapping with comparisons of results from two methods and verifications. *Eng. Geol.* **2005**, *81*, 432–445. [CrossRef]
11. Kasai, M.; Ikeda, M.; Asahina, T. LiDAR-derived DEM evaluation of deep-seated landslides in a steep and rocky region of Japan. *Geomorphology* **2009**, *113*, 57–69. [CrossRef]
12. Strait, T. Death Toll from Hokkaido Quake Hits 44, Power SUPPLY and Toyota Output Disrupted. 2018. Available online: https://www.straitstimes.com/asia/east-asia/death-toll-from-hokkaido-quake-hits-44-power-supply-and-toyota-output-disrupted (accessed on 20 January 2022).
13. Evans, S. Japan May Face Billion Dollar Losses from Typhoon Jebi, Hokkaido Quake. 2018. Available online: https://www.artemis.bm/news/japan-may-face-billion-dollar-losses-from-typhoon-jebi-hokkaido-quake/ (accessed on 20 January 2022).
14. Zhang, J.; van Westen, C.J.; Tanyas, H.; Mavrouli, O.; Ge, Y.; Bajrachary, S.; Gurung, D.R.; Dhital, M.R.; Khanal, N.R. How size and trigger matter: Analyzing rainfall- and earthquake-triggered landslide inventories and their causal relation in the Koshi River basin, central Himalaya. *Nat. Hazard Earth Sys. Sci.* **2019**, *19*, 1789–1805. [CrossRef]
15. Collins, M.; Knutti, R.; Arblaster, J.; Dufresne, J.L.; Fichefet, T.; Friedlingstein, P.; Gao, X.; Gutowski, W.; Johns, T.; Krinner, G.; et al. Chapter 12-Long-term climate change: Projections, commitments and irreversibility. In *Climate Change 2013: The Physical Science Basis*; IPCC Working Group I Contribution to AR5, Cambridge University Press(Pub): Cambridge, UK, 2013; pp. 1029–1136.
16. Crozier, M.J. Deciphering the effect of climate change on landslide activity:a review. *Geomorphology* **2010**, *124*, 260–267. [CrossRef]
17. Frei, C.; Schöll, R.; Fukutome, S.; Schmidli, J.; Vidale, P.L. Future change of precipitation extremes in Europe: Intercomparison of scenarios from regional climate models. *J. Geophys. Resatmos.* **2006**, *111*, D4. [CrossRef]
18. Gawriuczenkow, I.; Kaczmarek, Ł.; Kiełbasiński, K.; Kowalczyk, S.; Mieszkowski, R.; Wójcik, E. Đlope stability and failure hazards in the light of complex geological surveys. *Sci. Rev. Eng. Environ. Sci.* **2017**, *26*, 85–98.
19. Kaczmarek, Ł.; Popielski, P. Selected components of geological structures and numerical modelling of slope stability. *Open Geosci.* **2019**, *11*, 208–218. [CrossRef]
20. Yin, Y.P.; Wang, L.Q.; Zhang, W.G.; Dai, Z.W. Research on the collapse process of a thick-layer dangerous rock on the reservoir bank. *Bull. Eng. Geol. Environ.* **2022**, *81*, 109. [CrossRef]
21. Wieczorek, G.F. Landslide triggering mechanisms. In: Turner, Shuster(eds) Landslides: Investigation and Mitigation. *Res. Board Natl. Res. Counc. Spec. Rep.* **1996**, *247*, 76–90.
22. Jirásek, M.; Rolshoven, S. Localization properties of strain-softening gradient plasticity models. Part I: Strain-gradient theories. *Int. J. Solids Struct.* **2009**, *46*, 2225–2238. [CrossRef]
23. Schulz, W.H.; Wang, G. Residual shear strength variability as a primary control on movement of landslides reactivated by earthquake-induced ground motion: Implications for coastal Oregon, U.S. *J. Geophys. Res. Earth Surf.* **2014**, *119*, 1617–1635. [CrossRef]
24. Ghayoomi, M.; Suprunenko, G.; Mirshekari, M. Cyclic Triaxial Test to Measure Strain-Dependent Shear Modulus of Unsaturated Sand. *Int. J. Geomech.* **2017**, *17*, 4017043. [CrossRef]
25. Paul, M.; Sahu, R.; Banerjee, G. Undrained Pore Pressure Prediction in Clayey Soil under Cyclic Loading. *Int. J. Geomech.* **2015**, *15*, 4014082. [CrossRef]
26. Peters, W.H.; Ranson, W.F. Digital imaging techniques in experimental stress analysis. *Opt. Eng.* **1982**, *21*, 213427. [CrossRef]
27. Potyondy, D.O. The bonded-particle model as a tool for rock mechanics research and application: Current trends and future directions. *Geosystem. Eng.* **2015**, *18*, 259–265. [CrossRef]
28. Wu, Y.; Morgan, E. Effect of fabric on the accuracy of computed tomography-based finite element analyses of the vertebra. *Biomech. Model. Mechanobiol.* **2020**, *19*, 505–517. [CrossRef]
29. Alshibli, K.; Alshibli, K.; Reed, A. *Advances in Computed Tomography for Geomaterials GeoX 2010*; John Wiley & Sons: New York, NY, USA, 2010.
30. Wang, B. Geotechnical investigations of an earthquake that triggered disastrous landslides in eastern Canada about 1020 Cal BP. *Geoenviron. Disasters* **2020**, *7*, 21. [CrossRef]
31. Tatard, L. *Statistical Analysis of Triggered Landslides: Implication for Earthquake and Weather Controls*; Université Joseph-Fourier-Grenoble: Saint Martin, France, 2010.
32. Asch, T.W.J.; Buma, J.; Beek, L.P.H. A view on some hydrological triggering systems in landslides. *Geomorphology* **1999**, *30*, 25–32. [CrossRef]
33. Thokchom, S.; Rastogi, B.; Dogra, N.; Pancholi, V.; Sairam, B.; Bhattacharya, F.; Patel, V. Empirical correlation of SPT blow counts versus shear wave velocity for different types of soils in Dholera, Western India. *Nat. Hazards J. Int. Soc.* **2017**, *86*, 1291–1306. [CrossRef]

34. Thai, P.B.; Tien, B.D.; Prakash, I. Landslide susceptibility modelling using different advanced decision trees methods. *Civ. Eng. Environ. Syst.* **2018**, *35*, 139. [CrossRef]
35. Balogun, A.L.; Rezaie, F.; Pham, Q.B. Spatial prediction of landslide susceptibility in western Serbia using hybrid support vector regression (SVR) with GWO, BAT and COA algorithms. *Geosci. Front.* **2021**, *12*, 101104. [CrossRef]
36. Peethambaran, B.; Anbalagan, R.; Kanungo, D.P. A comparative evaluation of supervised machine learning algorithms for township level landslide susceptibility zonation in parts of Indian Himalayas. *Catena* **2020**, *195*, 104751. [CrossRef]
37. Samui, P. Slope stability analysis: A support vector machine approach. *Environ. Geol.* **2008**, *56*, 255. [CrossRef]
38. Jiang, S.H.; Liu, Y.; Zhang, H.L. Quantitatively evaluating the effects of prior probability distribution and likelihood function models on slope reliability assessment. *Rock Soil Mech.* **2020**, *41*, 3087. (In Chinese)
39. Neuland, H. A prediction model of landslips. *Catena* **1976**, *3*, 215. [CrossRef]
40. Pradhan, B.; Lee, S. Landslide susceptibility assessment and factor effect analysis: Backpropagation artificial neural networks and their comparison with frequency ratio and bivariate logistic regression modelling. *Environ. Model. Softw.* **2010**, *25*, 747. [CrossRef]
41. Alavi, A.H.; Gandomi, A.H. A robust data mining approach for formulation of geotechnical engineering systems. *Eng. Comput.* **2011**, *28*, 242. [CrossRef]
42. Martins, F.F.; Miranda, T.F.S. Application of data mining techniques to the safety evaluation of slopes. In *Information Technology in Geo-Engineering: Proceedings of the 1st International Conference (ICITG)*; IOS Press: Amsterdam, The Netherlands, 2010; p. 84.
43. Tang, H.; Zou, Z.; Xiong, C.; Wu, Y.; Hu, X.; Wang, L.; Lu, S.; Criss, R.E.; Li, C. An evolution model of large consequent bedding rockslides, with particular reference to the Jiweishan rockslide in Southwest China. *Eng. Geol.* **2010**, *186*, 17–27. [CrossRef]
44. Zou, Z.; Yan, J.; Tang, H.; Wang, S.; Xiong, C.; Hu, X. A shear constitutive model for describing the full process of the deformation and failure of slip zone soil. *Eng. Geol.* **2020**, *276*, 105766. [CrossRef]
45. Alaei, E.; Mahboubi, A. A discrete model for simulating shear strength and deformation behavior of rockfill material, considering the particle breakage phenomenon. *Granul. Matter.* **2012**, *14*, 707–717. [CrossRef]
46. Potyondy, D.O.; Cundall, P.A. A bonded-particle model for rock. *Int. J. Rock Mech. Min. Ences* **2004**, *41*, 1329–1364. [CrossRef]
47. *Itasca, PFC*, Version 5.0; Itasca Consulting Group Inc.: Minneapolis, MI, USA, 2014.
48. Basha, B.M.; Babu, G.L.S. Seismic rotational displacements of gravity walls by pseudodynamic method with curved rupture surface. *Int. J. Geomech.* **2009**, *10*, 93–105. [CrossRef]
49. Basha, B.M.; Babu, G.L.S. Computation of sliding displacements of bridge abutments by pseudo-dynamic method. *Soil Dyn. Earthq. Eng.* **2009**, *29*, 103–120. [CrossRef]
50. Basha, B.M.; Babu, G.L.S. Reliability assessment of internal stability of reinforced soil structures: A pseudo-dynamic approach. *Soil Dyn. Earthq. Eng.* **2010**, *30*, 336–353. [CrossRef]
51. Zhang, Y.; Wong, L.; Meng, F. Brittle fracturing in low-porosity rock and implications to fault nucleation. *Eng. Geol.* **2021**, *285*, 106025. [CrossRef]
52. Hofmann, H.; Babadagli, T.; Zimmermann, G. A grain based modeling study of fracture branching during compression tests in granites. *Int. J. Rock Mech. Min.* **2015**, *77*, 152–162. [CrossRef]
53. Ma, J.W.; Wang, Y.K.; Niu, X.X.; Jiang, S.; Liu, Z.Y. A comparative study of mutual information-based input variable selection strategies for the displacement prediction of seepage-driven landslides using optimized support vector regression. *Stoch. Environ. Res. Risk Assess.* **2022**, 1–21. [CrossRef]
54. Chen, G.; Guo, T.Y.; Serati, M. Microcracking mechanisms of cyclic freeze–thaw treated red sandstone: Insights from acoustic emission and thin-section analysis. *Constr. Build. Mater.* **2022**, *329*, 127097. [CrossRef]

 sensors

Article

Machine Learning Models for Slope Stability Classification of Circular Mode Failure: An Updated Database and Automated Machine Learning (AutoML) Approach

Junwei Ma [1,2,*], Sheng Jiang [1,2], Zhiyang Liu [1,2], Zhiyuan Ren [1,2], Dongze Lei [1,2], Chunhai Tan [1,2] and Haixiang Guo [3]

[1] Badong National Observation and Research Station of Geohazards (BNORSG), China University of Geosciences, Wuhan 430074, China
[2] Three Gorges Research Center for Geo-Hazards of the Ministry of Education, China University of Geosciences, Wuhan 430074, China
[3] School of Economics and Management, China University of Geosciences, Wuhan 430074, China
* Correspondence: majw@cug.edu.cn

Abstract: Slope failures lead to large casualties and catastrophic societal and economic consequences, thus potentially threatening access to sustainable development. Slope stability assessment, offering potential long-term benefits for sustainable development, remains a challenge for the practitioner and researcher. In this study, for the first time, an automated machine learning (AutoML) approach was proposed for model development and slope stability assessments of circular mode failure. An updated database with 627 cases consisting of the unit weight, cohesion, and friction angle of the slope materials; slope angle and height; pore pressure ratio; and corresponding stability status has been established. The stacked ensemble of the best 1000 models was automatically selected as the top model from 8208 trained models using the H2O-AutoML platform, which requires little expert knowledge or manual tuning. The top-performing model outperformed the traditional manually tuned and metaheuristic-optimized models, with an area under the receiver operating characteristic curve (AUC) of 0.970 and accuracy (ACC) of 0.904 based on the testing dataset and achieving a maximum lift of 2.1. The results clearly indicate that AutoML can provide an effective automated solution for machine learning (ML) model development and slope stability classification of circular mode failure based on extensive combinations of algorithm selection and hyperparameter tuning (CASHs), thereby reducing human efforts in model development. The proposed AutoML approach has the potential for short-term severity mitigation of geohazard and achieving long-term sustainable development goals.

Keywords: automated machine learning (AutoML); slope stability classification; circular mode failure; hyperparameter tuning; stacked ensemble

Citation: Ma, J.; Jiang, S.; Liu, Z.; Ren, Z.; Lei, D.; Tan, C.; Guo, H. Machine Learning Models for Slope Stability Classification of Circular Mode Failure: An Updated Database and Automated Machine Learning (AutoML) Approach. *Sensors* **2022**, *22*, 9166. https://doi.org/10.3390/s22239166

Academic Editor: Giulio Iovine

Received: 25 October 2022
Accepted: 23 November 2022
Published: 25 November 2022

Publisher's Note: MDPI stays neutral with regard to jurisdictional claims in published maps and institutional affiliations.

1. Introduction

Natural hazards like landslide and subsidence have been acknowledged as a major factor disturbing sustainable development in developing countries [1–4]. For example, a catastrophic landfill slope failure occurred on 20 December 2015, in Guangming, Shenzhen, China, took the lives of 69 people [5]. The risk assessment and management of natural hazard will have a short-term benefit for severity mitigation and a long-term benefit for achieving sustainable development goals [1].

The evaluation of slope stability is of primary importance for natural hazard risk assessment and management in mountain areas. Numerous efforts have been made for slope stability assessment [6–9]. However, slope stability assessment for circular mode failure, a typical problem, still remains a challenge for the practitioner and researcher due to inherent complexity and uncertainty [10]. An extensive body of literature exists regarding slope stability assessments of circular failure, and significant progress has been achieved. Three main categories of assessment approaches have emerged: analytical approaches, numerical approaches, and machine learning (ML)-based approaches [11–13]. Limited equilibrium methods, such as the simplified Bishop, Spencer, and Morgenstern-Price methods, are commonly used analytical approaches and have been routinely used in practice. Generally, geometrical data, physical and shear strength parameters (unit weight, cohesion, and friction angle), and the pore pressure ratio are required in limited equilibrium methods [14,15]. However, the results vary across different methods due to different assumptions [9]. Numerical approaches (e.g., finite element methods) have been widely adopted for slope stability assessment. However, due to the requirement of numerous expensive input parameters, these models can be applied only in limited cases [16]. Recently, ML-based approaches have led to giant strides in slope stability assessment. A summary of the slope stability assessments of circular failure using ML approaches is given in Table 1. Among the various ML approaches used, artificial neural networks (ANNs) are widely utilized for slope stability assessment due to their simple structure and acceptable accuracy [11,17,18]. Recently, sophisticated ML algorithms, including but not limited to support vector machine (SVM), decision tree (DT), extreme learning machine (ELM), random forest (RF), and gradient boosting machine (GBM) algorithms, have been utilized for slope stability assessment. Hyperparameter tuning is a fundamental step required for accurate ML modeling [19,20]. As listed in Table 1, grid search (GS) and metaheuristic methods, such as the artificial bee colony (ABC) algorithm, genetic algorithm (GA), and particle swarm optimization (PSO), have been utilized for hyperparameter tuning in ML-based slope stability assessment. For example, Qi and Tang [16] simultaneously trained six firefly algorithm (FA)-optimized ML models, including multilayer perceptron neural network, logistic regression (LR), DT, RF, SVM, and GBM models, based on 148 cases of circular mode failure. The FA-optimized SVM was selected as the final model, with an area under the receiver operating characteristic curve (AUC) of 0.967 for the testing dataset. The performance of eight ensemble learning approaches was compared by [12] based on a dataset with 444 cases of circular mode failure. A stacked model was selected as the final model, with an AUC of 0.9452 for the testing dataset.

Table 1. Summary of the slope stability assessment of circular mode failure using MLs.

Reference	Data Size (Stable/Failure)	Input Features	Data Preprocessing	ML Algorithm Selection	Hyperparameter Tuning	Final Model and Performance
[21]	82 (38/44)	γ, c, φ, β, H, r_u	/	BP	Trial and error GA	GA-optimized BP was selected as the final model, with an AUC of 0.455 for the testing dataset.
[22]	32 (14/18)	γ, c, φ, β, H, r_u	/	ANN	Trial and error	The ANN achieved an ACC of 1.00 for the testing dataset in two cases.
[23]	46 (17/29)	γ, c, φ, β, H, r_u	Data normalization	SVM	PSO	PSO-SVM achieved an ACC of 0.8125 for the testing dataset.
[24]	168 (84/84)	γ, c, φ, β, H, r_u	Data normalization	LSSVM	FA	The FA-optimized LSSVM achieved an AUC of 0.86 for the testing dataset.
[25]	168 (84/84)	γ, c, φ, β, H, r_u	Data normalization	RBF LSSVM ELM	Orthogonal least squares GA Trial and error	The GA-ELM was selected as the final model, with an AUC of 0.8706 for the testing dataset.
[26]	82 (49/33)	γ, c, φ, β, H, r_u	/	NB	/	NB achieved an ACC of 0.846 for the testing dataset.
[27]	107 (48/59)	γ, c, φ, β, H, r_u	/	RF SVM Bayes GSA	Ten-fold CV	The GSA was selected as the final model, with an AUC of 0.889 for the testing dataset.

Table 1. *Cont.*

Reference	Data Size (Stable/Failure)	Input Features	Data Preprocessing	ML Algorithm Selection	Hyperparameter Tuning	Final Model and Performance
[17]	168 (84/84)	γ, c, φ, β, H, r_u	Data normalization	GP QDA SVM ADB-DT ANN KNN Classifier ensemble	GA	The optimum ensemble classifier was selected as the final model, with an AUC of 0.943 for the testing dataset.
[16]	148 (78/70)	γ, c, φ, β, H, r_u	Data normalization	LR DT RF GBM SVM BP	FA GS	The FA-optimized SVM was selected as the final model, with an AUC of 0.967 for the testing dataset.
[18]	221 (115/106)	γ, c, φ, β, H, r_u	Data normalization	ANN SVM RF GBM	Five-fold CV	The GBM-based model was selected as the final model, with an AUC of 0.900 for the testing dataset.
[28]	87 (42/45)	γ, c, φ, β, H, r_u	/	J48	Trial and error	J48 achieved an ACC of 0.9231 for the testing dataset.
[13]	257 (123/134)	γ, c, φ, β, H, r_u	/	XGB RF LR SVM BC LDA KNN DT MLP GNB XRT Stacked ensemble	ABC PSO	The stacked ensemble was selected as the final model, with an AUC of 0.904 for the testing dataset.
[11]	153 (83/70)	γ, c, φ, β, H, r_u	Data normalization and outlier removing	KNN SVM SGD GP QDA GNB DT ANN Bagging ensemble Heterogeneous ensemble	GS	An ensemble classifier based on extreme gradient boosting was selected as the final model, with an AUC of 0.914 for the testing dataset.
[29]	19 (13/6)	γ, c, φ, β, H, r_u	Data normalization	K-means cluster	HS	K-means clustering optimized by HS achieved an ACC of 0.89 for all datasets.
[12]	444 (224/220)	γ, c, φ, β, H, r_u	Data normalization	AdaBoost GBM Bagging XRT RF HGB Voting Stacked	GS	A stacked model was selected as the final model, with an AUC of 0.9452 for the testing dataset.
[30]	422 (226/196)	γ, c, φ, β, H, r_u	Data normalization	MDMSE	GS	The MDMSE model achieved an AUC of 0.8810 for the testing dataset.

Note: Abbreviations in this table are explained in Abbreviations.

Although ML-based models have been widely applied, some studies have been based on a small number of samples, which may affect the generalization ability of the classifier. Moreover, most ML models have been manually developed by researchers with expert knowledge in a trial-and-error approach. In fact, exhaustive steps, including data preprocessing [31], feature engineering [32], ML algorithm selection [33], and hyperparameter tuning, are involved in practical applications of ML. Among them, model selection and hyperparameter tuning remain challenges for successful ML-based modeling [34]. Based on the no-free-lunch theorem [35], there is no algorithm that outperforms all others in all problems. Therefore, at present, according to prior experience, candidate off-the-shelf models are trained with a training dataset and validated by researchers. The ML model that provides the best performance is considered the final model and tested with an out-of-box testing dataset. This traditional workflow makes the model development process knowledge-based and time-consuming [36], and might yield unsatisfactory results [37]. However, most practitioners and researchers lack the knowledge and expertise required to build satisfactory ML models. Hence, an objective workflow with less human effort is needed, providing a basis for the concept of automated ML (AutoML) [38].

From the perspective of automation, AutoML is a systematic framework that automates algorithm selection and hyperparameter tuning and explores different combinations of factors with minimal human intervention [34,39–41]. AutoML has been successfully applied for ML modeling in a variety of fields, including tunnel displacement prediction [36], tunnel boring machine performance prediction [34], and earthquake casualty and economic loss prediction [42]. Thus, the generalization ability of this approach has been confirmed.

In the present study, an updated database with 627 cases consisting of the unit weight, cohesion, and friction angle of the slope materials: slope angle and height, pore pressure ratio, and corresponding stability status of circular mode failure, has been collected. For the first time, an AutoML approach was proposed for slope stability classification. The top model was selected from 8208 trained ML models by exploring numerous combinations of algorithm selection and hyperparameter tuning (CASHs) with minimal human intervention.

The major contribution of this paper is highlighted as follows:

(a) A large database consisting of 627 cases has been collected for slope stability classification.
(b) Based on the updated dataset, an AutoML approach was proposed for slope stability classification without the need for manual trial and error. The proposed AutoML approach outperformed the existing ML models by achieving superior performance.

The rest of this paper is organized as follows: the updated database and methodology are presented in Sections 2 and 3, respectively. Section 4 presents and discusses experimental results. Finally, the conclusions and further work are presented in Section 5.

2. Database

As listed in Table 1, the input features relevant to the slope stability assessment of the circular failure model (schematic illustrated in inset of Figure 1) mainly include the unit weight, cohesion, and friction angle of the slope materials, the slope angle and height, and the pore pressure ratio. Moreover, these features are fundamental input parameters for limit equilibrium methods, such as the simplified Bishop method [15,43]. Based on the previous research listed in Table 1, an updated database consisting of 627 cases was obtained from previous studies [11,12,16,24,30,44] and is listed in Appendix A. The database consists of the unit weight, cohesion, and friction angle of the slope materials, the slope angle and height, the pore pressure ratio, and the corresponding stability status. The numbers of positive (stable) and negative (failure) samples are 311 and 316, respectively. The statistics of the input features are summarized in Table 2. To better visualize the collected dataset, ridgeline plots showing the density distributions of the input features based on kernel density estimation [3] are presented in Figure 1. As shown, the collected dataset was distributed in a wide range of regions, and the distribution was not symmetric.

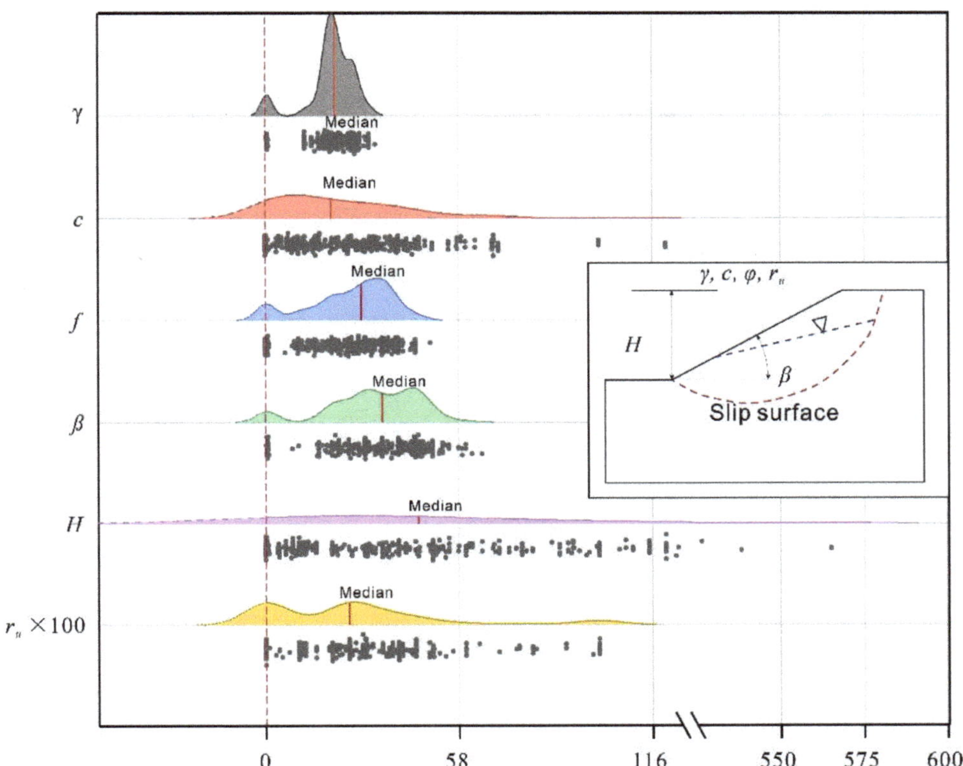

Figure 1. Ridgeline plots showing the density distributions of the input features. The inset shows a schematic diagram of the circular failure model.

Table 2. Summary of the input feature statistics.

Input Feature	Notation	Range	Median	Mean	Std.
Unit weight (kN/m^3)	γ	0.492–33.160	20.959	20.185	7.044
Cohesion (kPa)	c	0–300.00	19.690	25.600	31.036
Friction angle (°)	φ	0–49.500	28.800	25.308	12.331
Slope angle (°)	β	0.302–65.000	34.980	32.605	13.711
Slope height (m)	H	0.018–565.000	45.800	90.289	120.140
Pore pressure ratio	r_u	0–1.000	0.250	0.254	0.260

The Pearson correlation coefficient (R) was adopted to further reveal the linear correlations between input features and the slope stability status and is shown in the lower left half of the panels in Figure 2. As shown, relatively poor linear correlations with correlation coefficients lower than 0.5 were observed between the input features and the slope stability status. Significant linear correlations (R = 0.71, 0.71, and 0.68) were noted for the unit weight, friction angle, and slope angle. Additionally, a moderate correlation (R = 0.51) was found between the unit weight and slope height.

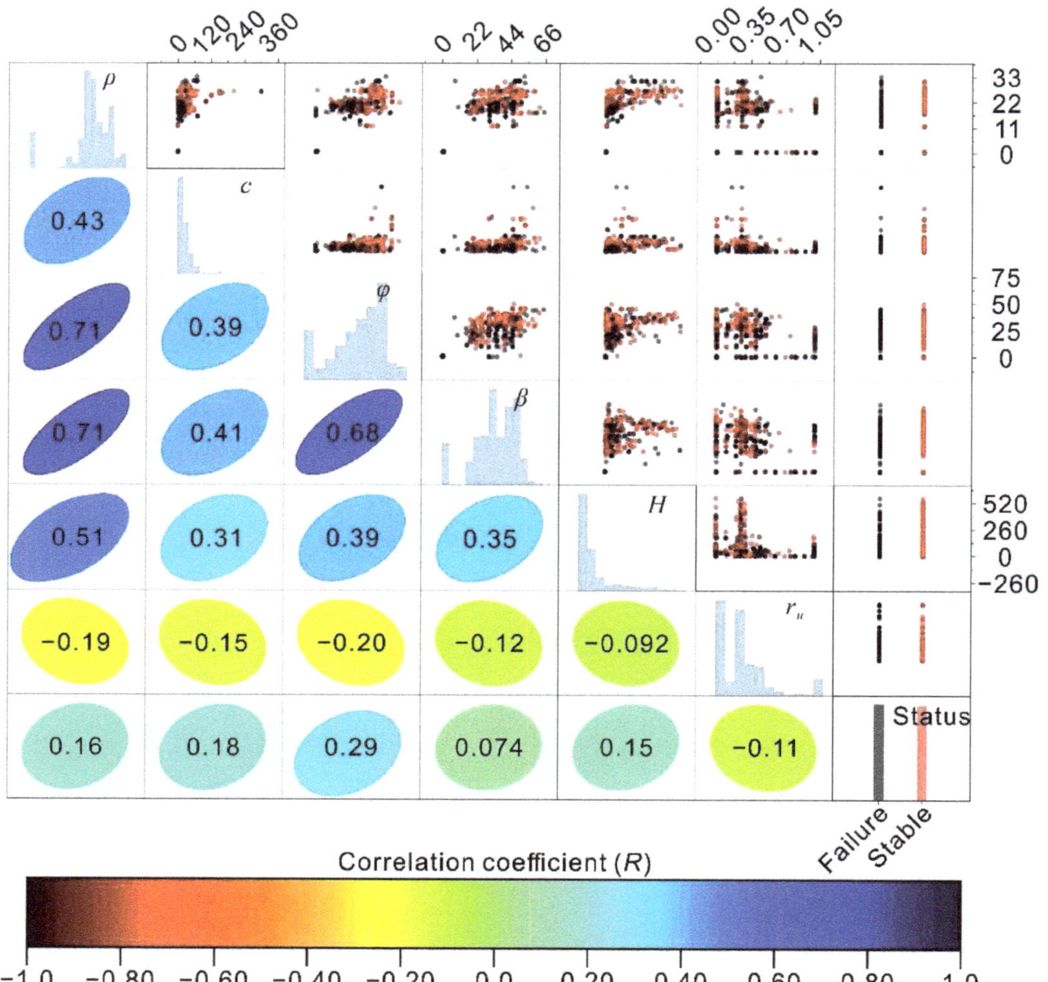

Figure 2. Scatter matrix showing the collected dataset. The panels in the upper right show the data points, and the lower left half of the figure shows the correlation coefficients between the features and the slope stability status.

Furthermore, the multivariate principal component analysis (PCA) technique [45] was applied to enhance the visualization of the statistical relationships among features. The PCA results shown in Figure 3 demonstrate that the first three principal components (PC1-PC3) account for 79.09% of the entire multivariate variance in space. PC1 is mainly associated with the unit weight, friction angle, and slope angle. PC2 corresponds to the pore pressure ratio. Moreover, overlapping among failure and stability classes can be clearly observed. In other words, the decision boundary for separating slope failure and stability is highly nonlinear and complex.

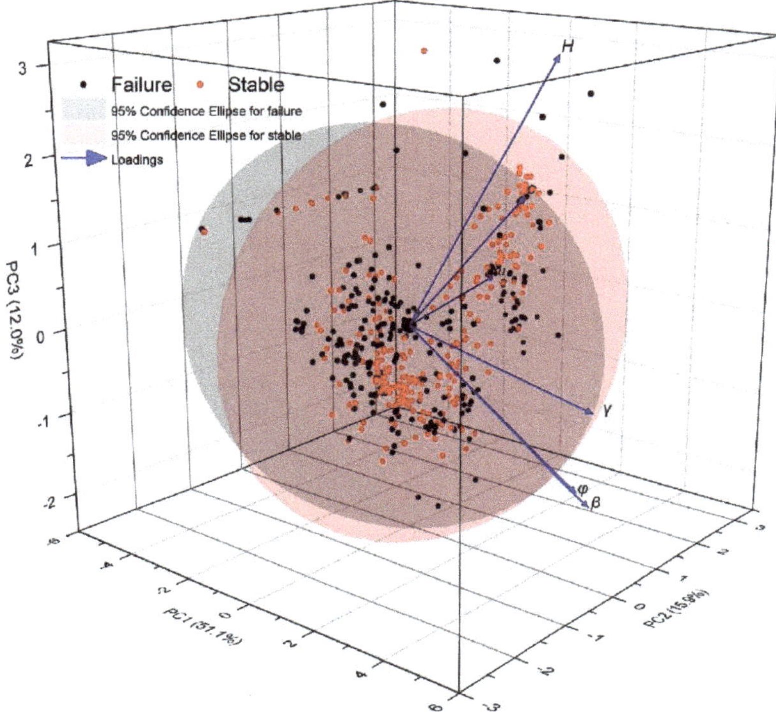

Figure 3. 3D PCA score plot of the input features.

3. Methodology

3.1. AutoML

From the perspective of automation, AutoML is a systematic model that automates the algorithm selection and hyperparameter tuning processes and explores different CASHs with minimal human intervention [34,39,40]. More formally, the CASH problem can be stated as follows. Let $A = \{A^1, A^2, \cdots, A^R\}$ be a set of ML algorithms, $\Lambda = \{\Lambda^1, \Lambda^2, \cdots, \Lambda^R\}$ be the corresponding hyperparameters, and L be the loss function. When adopting k-fold cross validation (CV), the training dataset $D_{training}$ is divided into subsets $\{D^{(1)}_{training}, D^{(2)}_{training}, \cdots, D^{(k)}_{training}\}$ and $\{D^{(1)}_{validation}, D^{(2)}_{validation}, \cdots, D^{(k)}_{validation}\}$. The CASH problem is defined as

$$A^*{}_{\lambda^*} \in \underset{A^{(j)} \in A, \lambda^{(j)} \in \Lambda}{\operatorname{argmin}} \frac{1}{k} \sum_{i=1}^{k} L(A^{(j)}_{\lambda}, D^{(i)}_{training}, D^{(i)}_{validation}) \quad (1)$$

Generally, AutoML consists of the following three key components: a search space, a search strategy, and a performance evaluation strategy [40] (schematically illustrated in Figure 4). The search space refers to a set of hyperparameters and the range of each hyperparameter. The search strategy refers to the strategy of selecting the optimal hyperparameters from the search space. Grid search and Bayesian optimization are commonly used search strategies. The performance evaluation strategy refers to the method used to evaluate the performance of the trained models.

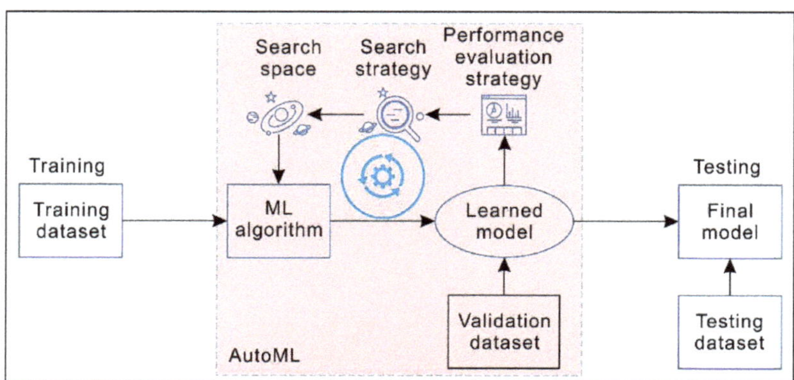

Figure 4. Schematic diagram showing the workflow of AutoML.

Various open-source platforms, such as AutoKeras, AutoPyTorch, AutoSklearn, AutoGluon, and H2O AutoML, have been developed to facilitate the adoption of AutoML [46]. Previous studies [47,48] have demonstrated the strong feature of H2O AutoML for processing large and complicated datasets by quickly searching the optimal model without the need for manual trial and error. Moreover, H2O AutoML provides a user interface for non-experts to import and split datasets, identify the response column, and automatically train and tune models. Therefore, in the present study, the H2O AutoML platform was adopted for the automated assessment of slope.

The H2O AutoML platform includes the following commonly used ML algorithms: generalized linear model (GLM), distributed random forest (DRF), extremely randomized tree (XRT), deep neural network (DNN), and GBM algorithms [49]. The abovementioned ML algorithms in the H2O AutoML platform are briefly described as follows.

GLM is an extended form of a linear model. Given the input variable x, the conditional probability of the output class falling within the class c of observations is defined as follows:

$$\hat{y}_c = Pr(y = c|x) = \frac{ex^T\beta_c + \beta_{c0}}{\sum\limits_{k=1}^{K}(ex^T\beta_k + \beta_{k0})} \quad (2)$$

where β_c is the vector of coefficients for class c.

The DRF is an ensemble learning approach based on decision trees. In the DRF training process, multiple decision trees are built. To reduce the variance, the final prediction was obtained by aggregating the outputs from all decision trees.

Similar to the DRF, XRT is based on multiple decision trees, but randomization is strongly emphasized to reduce the variance with little influence on the bias. The following main innovations are involved in the XRT process: random division of split nodes using cut points and full adoption of the entire training dataset instead of a bootstrap sample for the growth of trees.

The DNN in H2O AutoML is based on a multilayer feedforward artificial neural network with multiple hidden layers. There are a large number of hyperparameters involved in DNN training, which makes it notoriously difficult to manually tune. Cartesian and random grid searches are available in H2O AutoML for DNN hyperparameter optimization.

GBM is an ensemble learning method. The basic idea of GBM is to combine weak base learners (usually decision trees) for the generation of strong learners. The objective is to minimize the error in the objective function through an iterative process using gradient descent.

In addition, stacked ensembles can be built using either the best-performing models or all the trained models.

3.2. Search Space and Search Strategy

In the present study, a random grid search was adopted for hyperparameter tuning in the search space. When adopting k-fold CV, the hyperparameter tuning process can be described as follows (schematically illustrated in Figure 5). First, possible combinations of the tuned parameters are generated. Then, CV is performed using a possible parameter combination. The training dataset is divided into k equal-sized subsets. A single subset is treated as the validation subset, while the remaining subsets are adopted for classification training. The average accuracy from k validation sets is computed and adopted as the performance measure of the k-CV classifier model. The above process is repeated for all possible parameter combinations. A ranking of all trained classifiers by model performance is obtained. The classifier that yields the highest accuracy is selected.

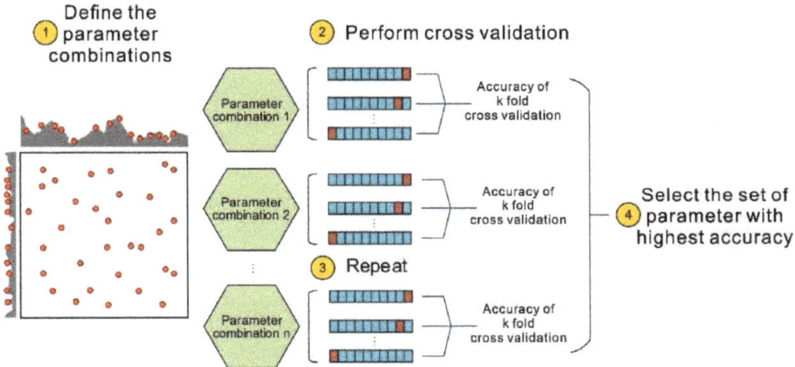

Figure 5. Schematic diagram showing hyperparameter tuning based on the k-fold CV and random grid search methods.

3.3. Performance Evaluation Measures

In the present study, widely applied criteria, including the accuracy (ACC), AUC, sensitivity (SEN), specificity (SPE), positive predictive value (PPV), negative predictive value (NPV), and Matthews correlation coefficient (MCC), were adopted for performance evaluation (Table 3). The AUC can be interpreted as follows: an AUC equal to 1.0 indicates perfect discriminative ability, an AUC value from 0.9 to 1.0 indicates highly accurate discriminative ability, an AUC value from 0.7 to 0.9 indicates moderately accurate discriminative ability, an AUC value from 0.5 to 0.7 demonstrates inaccurate discriminative ability, and an AUC less than 0.5 indicates no discriminative ability.

Table 3. Confusion matrix and performance measures for slope stability assessment.

Predicted / Actual	Stable	Failure	
Stable	True positive (TP)	False negative (FN)	Sensitivity: $SEN = \frac{TP}{TP+FN}$ (The ideal value is 1, whereas the worst is zero.)
Failure	False positive (FP)	True negative (TN)	Specificity $SPE = \frac{TN}{FP+TN}$ (The ideal value is 1, whereas the worst is zero.)
	Positive predictive value (PPV) $PPV = \frac{TP}{TP+FP}$ (The ideal value is 1, whereas the worst is zero.)	Negative predictive value (NPV) $NPV = \frac{TN}{FN+TN}$ (The ideal value is 1, whereas the worst is zero.)	Accuracy $ACC = \frac{TP+TN}{TP+FN+FP+TN}$ (The ideal value is 1, whereas the worst is zero.) Matthews correlation coefficient $MCC = \frac{TP \cdot TN - FP \cdot FN}{\sqrt{(TP+FP) \cdot (TP+FN) \cdot (TN+FP) \cdot (TN+FN)}}$ (The ideal value is 1.)

3.4. Slope Stability Assessment through AutoML

In the present study, the H2O AutoML approach was adopted for ML model development for slope stability classification (schematic illustrated in Figure 6). First, the database listed in Appendix A was randomly divided into training and testing datasets at a ratio of 80% to 20%, respectively. ML models, including GLM, DRF, XRT, DNN, and GBM were automated and developed (schematic illustrated in Figure 6). To enhance the reliability and performance, the common 10-fold CV was performed. A full list of tuned hyperparameters and the corresponding searchable values are given in Table 4. Stacked ensembles were developed based on the best-performing models and all the tuned models. A leaderboard ranking the mode performance accuracy was achieved. The leader models were saved and evaluated on the testing dataset.

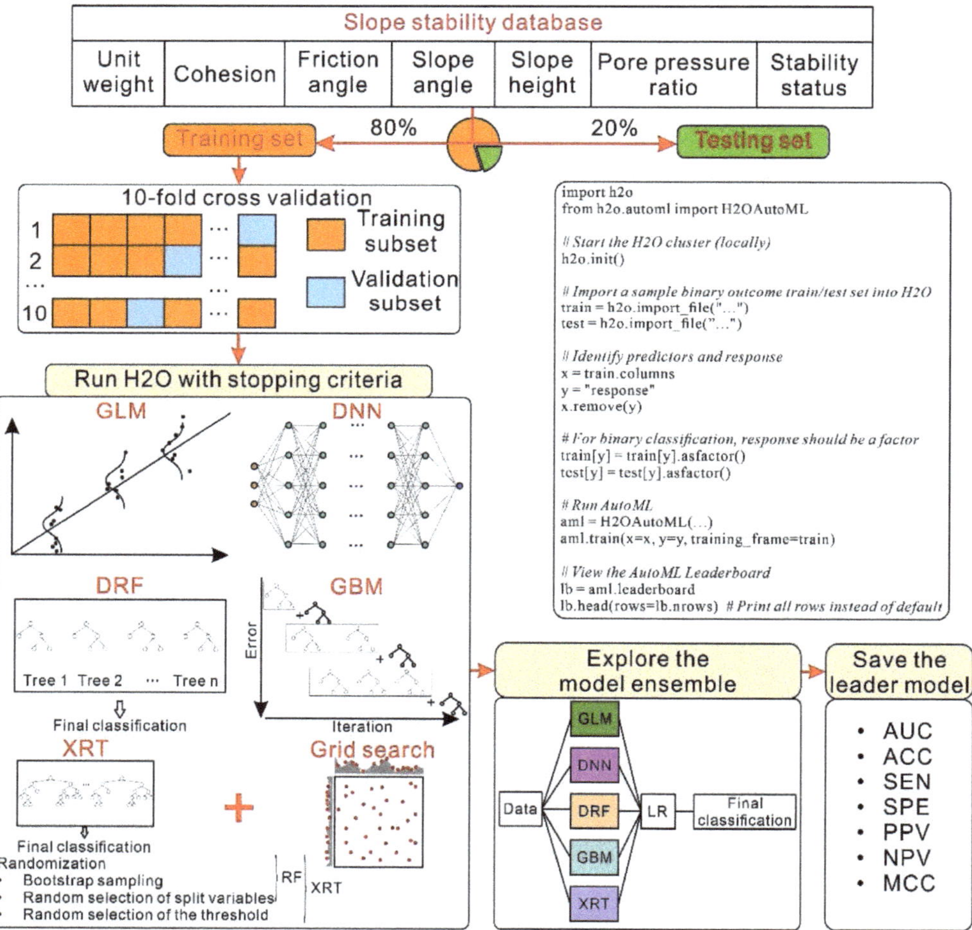

Figure 6. Flowchart of the AutoML-based slope stability classification.

Table 4. The hyperparameter search space for GS optimization for AutoML-based slope stability classification.

Algorithm	Parameter	Searchable values
DNN	Adaptive learning rate time smoothing factor (epsilon)	$\{10^{-6}, 10^{-7}, 10^{-8}, 10^{-9}\}$
	Hidden layer size (hidden)	Grid search 1: {20}, {50}, {100}
		Grid search 2: {20, 20}, {50, 50}, {100, 100}
		Grid search 3: {20, 20, 20}, {50, 50, 50}, {100, 100, 100}
	Hidden_dropout_ratio	Grid search 1: {0.1}, {0.2}, {0.3}, {0.4}, {0.5}
		Grid search 2: {0.1, 0.1}, {0.2, 0.2}, {0.3, 0.3}, {0.4, 0.4}, {0.5, 0.5}
		Grid search 3: {0.1, 0.1, 0.1}, {0.2, 0.2, 0.2} {0.3, 0.3, 0.3}, {0.4, 0.4, 0.4}, {0.5, 0.5, 0.5}
	Input_dropout_ratio	{0.0, 0.05, 0.1, 0.15, 0.2}
	Adaptive learning rate time decay factor (rho)	{0.9, 0.95, 0.99}
GLM	Regularization distribution between L1 and L2 (alpha)	{0.0, 0.2, 0.4, 0.6, 0.8, 1.0}
GBM	Column sampling rate (col_sample_rate)	{0.4, 0.7, 1.0}
	Column sample rate per tree (col_sample_rate_per_tree)	{0.4, 0.7, 1.0}
	Maximum tree depth (max_depth)	{3, 4, 5, 6, 7, 8, 9, 10, 11, 12, 13, 14, 15, 16, 17}
	Minimum number of observations for a leaf (min_rows)	{1, 5, 10, 15, 30, 100}
	Minimum relative improvement in squared error reduction (min_split_improvement)	$\{10^{-4}, 10^{-5}\}$
	Row sampling rate (sample_rate)	{0.50, 0.60, 0.70, 0.80, 0.90, 1.00}

The AutoML process was implemented using H2O AutoML (3.36.1.2) with an Intel(R) Xeon(R) E-2176M @ 2.70 GHz CPU with 64 GB RAM. The maximum time allotted to run generation classifiers, except for the stacked ensembles, was set to 3600 s.

4. Results and Discussions

4.1. Performance Analysis

A total of 8208 ML models, including bypass CV models, were trained with the H2O AutoML platform and saved. The top five models from the leaderboard were selected and listed in Table 4 for testing. The performance evaluation metrics for the top five models on the testing dataset are listed in Table 5.

Table 5. Comparison of the performance of the selected top-five models from AutoML in slope stability assessments of circular mode failure based on the selected test data.

Model ID	Model Type	Hyperparameters	AUC	Confusion Matrix			Performance Measures
$H2O_1$	Stacked ensemble	The base models are the top-1000 trained models, and the metalearner is a GLM. A logit transformation is used for the predicted probabilities.	0.970	Actual \ Predicted	Stable	Failure	SEN = 0.968 SPE = 0.841 PPV = 0.857 NPV = 0.964 ACC = 0.904 MCC = 0.815
				Stable	60	2	
				Failure	10	53	
$H2O_2$	GBM	score_tree_interval = 5; ntrees = 105; max_depth = 7; stopping_metric = logloss; stopping_tolerance = 0.045; learn_rate = 0.1; learn_rate_annealing = 1; sample_rate = 1; col_sample_rate = 0.4; col_sample_rate_change_per_level = 1; col_sample_rate_per_tree = 0.7	0.968	Actual \ Predicted	Stable	Failure	SEN = 0.903 SPE = 0.937 PPV = 0.933 NPV = 0.908 ACC = 0.920 MCC = 0.840
				Stable	56	6	
				Failure	4	59	
$H2O_3$	DRF	Ntrees = 50; max_depth = 20	0.963	Actual \ Predicted	Stable	Failure	SEN = 0.839 SPE = 0.968 PPV = 0.963 NPV = 0.859 ACC = 0.904 MCC = 0.815
				Stable	52	10	
				Failure	2	61	
$H2O_4$	XR	score_tree_interval = 5; max_after_balance_size = 5; max_confusion_matrix_size = 20; ntrees = 50; max_depth = 20; stopping_metric = logloss; stopping_tolerance = 0.045; sample_rate = 0.632	0.963	Actual \ Predicted	Stable	Failure	SEN = 0.871 SPE = 0.937 PPV = 0.931 NPV = 0.881 ACC = 0.904 MCC = 0.810
				Stable	54	8	
				Failure	4	59	
$H2O_5$	GBM	score_tree_interval = 5; ntrees = 97; max_depth = 7; stopping_metric = logloss; stopping_tolerance = 0.045; learn_rate = 0.1; learn_rate_annealing = 1; sample_rate = 0.8; col_sample_rate = 0.8; col_sample_rate_change_per_level = 1; col_sample_rate_per_tree = 0.8	0.960	Actual \ Predicted	Stable	Failure	SEN = 0.968 SPE = 0.810 PPV = 0.833 NPV = 0.962 ACC = 0.888 MCC = 0.786
				Stable	60	2	
				Failure	12	51	

As listed in Table 5, the stacked ensemble of the best 1000 models ($H2O_1$) ranked as the top-performing model. The corresponding ROC curves are shown in Figure 7, which clearly indicates that the top-performing model is capable of providing highly accurate discriminative ability, with AUC of 0.999 and 0.970 for the training and testing dataset, respectively. The model performance was further evaluated using gain and lift charts (Figure 8). A gain chart measures the effectiveness of a classifier based on the percentage of correct classifications obtained with the model versus the percentage of correct classifications obtained by chance (i.e., the baseline). As shown, for the top model, only 30% of the population is required to achieve an accuracy of 60%, compared to 30% for the random model. The top classifier is capable of achieving a maximum lift of 2.1. In other words, when only 10% of the sample was selected, the average accuracy of the top model was approximately two times higher than that of the random model.

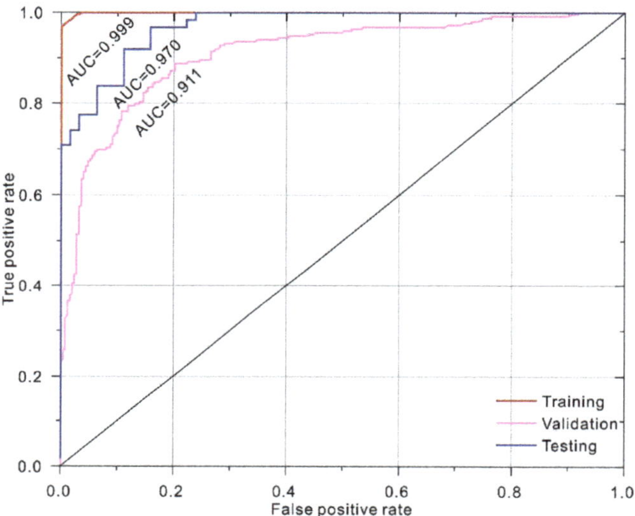

Figure 7. ROC curve of the top-performing model ($H2O_1$) from AutoML.

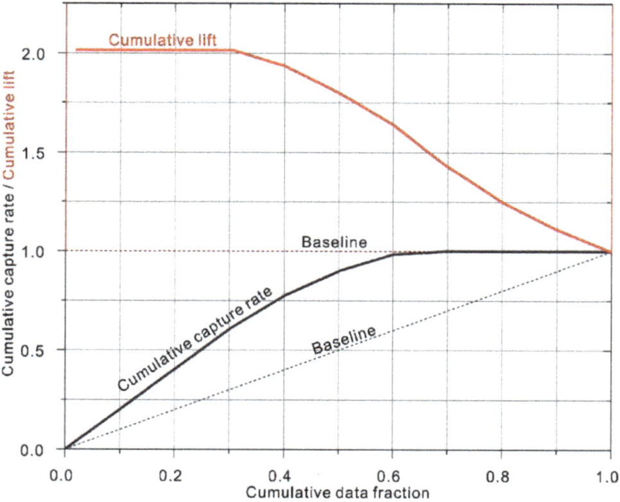

Figure 8. Cumulative gain and lift charts for the top-performing model ($H2O_1$) based on testing data.

Figure 9 demonstrates the correlation between NPV and PPV for the obtained top-five classification models based on the testing dataset. As shown, the top-performing model ($H2O_1$) falls within zone 2, in which the obtained NPV is greater than the PPV. This result indicates that the top-performing model ($H2O_1$) tends to classify slope status as a failure (negative status) more often than stable (positive status). In other words, the top-performing model ($H2O_1$) may overestimate stability.

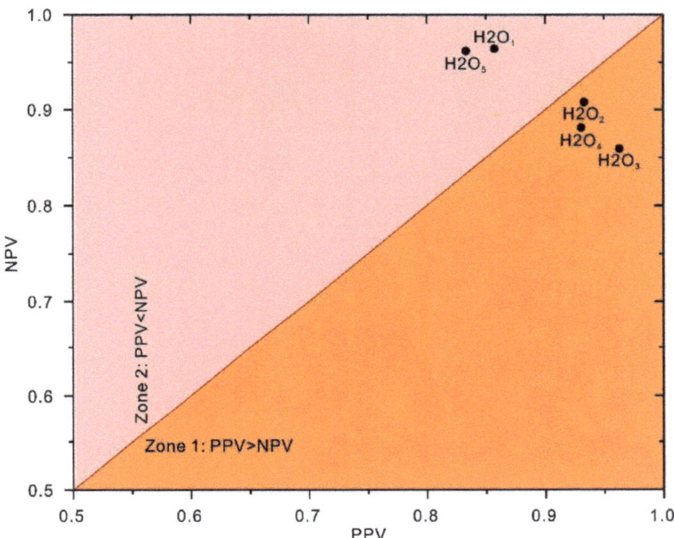

Figure 9. Correlation between the NPV and PPV values of the classification models based on the testing dataset.

4.2. Model Interpretation

In the present study, the partial dependence plot graphically revealing the input–output relationship was adopted for model interpretation. The partial dependence plot has been considered as one of the most popular model agnostic tools due to the advantages of simple definition and easy implementation. The partial dependence relations of the input features in the top-performing model ($H2O_1$) are shown in Figure 10. In partial dependence plots, features with greater variability have more significant effects on the model [18,50]. As shown, the top-performing model ($H2O_1$) is highly influenced by the slope height and friction angle.

4.3. Validation of the AutoML Model in ACADS Example

Furthermore, the predictive capacity of the top-performing model ($H2O_1$) was validated on the Australian Association for Computer-Aided Design (ACADS) referenced slope example EX1, which is a simple homogeneous slope. The slope is 20 m long and 10 m high. The geometry and material properties are shown in Figure 11. With the parameters listed in Figure 11, the example slope was estimated to fail [43]. The top-performing model ($H2O_1$) successfully classified the slope example as a failure case.

4.4. Comparison with Exiting Models

To further assess performance, the top-performing model ($H2O_1$) from the AutoML approach was further compared with a manually derived ML model for slope stability assessment (Table 6). As shown in Table 6, in the previous studies, the firefly algorithm optimized SVM (FA-SVM) provides the best performance with an AUC of 0.967 [16], followed by ensemble classifiers on the extreme gradient boosting (XGB-CM) [11]. Obviously, the top-performing model ($H2O_1$) is of better generalization ability than the existing models shown in Table 6 with the largest AUC and ACC values. These comparative results clearly indicate that the top-performing model ($H2O_1$) from AutoML approach is capable of providing better generalization performance than the manually derived ML and metaheuristics-optimized model.

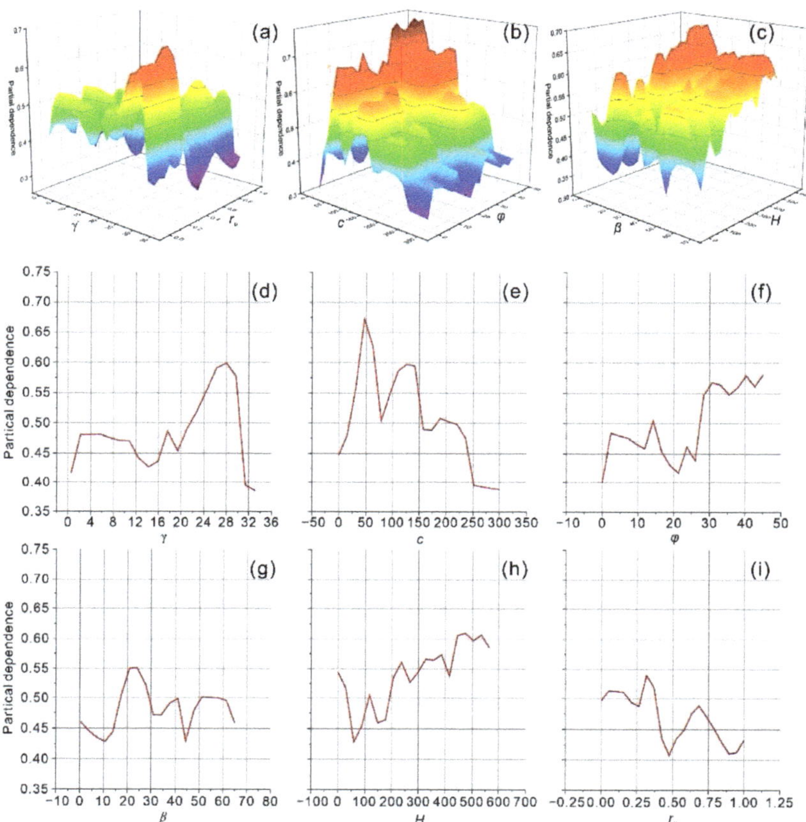

Figure 10. Partial dependence plots of the input features in the top-performing model ($H2O_1$) for the classification of slope stability. (**a**) Unit weight and pore pressure ratio, (**b**) cohesion and friction angle, (**c**) slope angle and slope height, (**d**) unit weight, (**e**) cohesion, (**f**) friction angle, (**g**) slope angle, (**h**) slope height, and (**i**) pore pressure ratio.

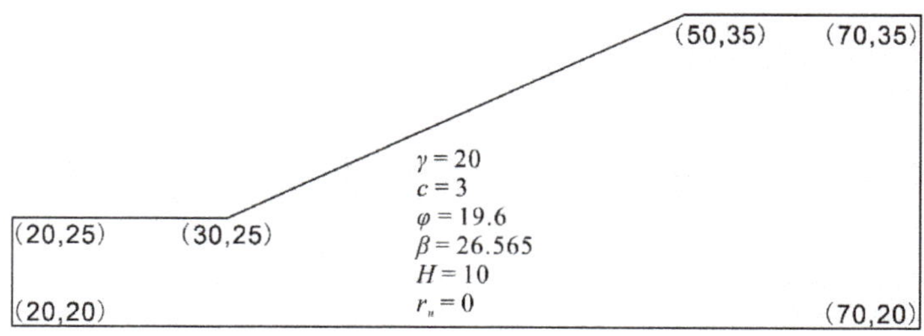

Figure 11. ACADS reference slope example EX1 (Unit: m).

Table 6. Comparison of different ML models for slope stability assessments of circular mode failure.

Reference	Model	AUC	ACC	Reference	Model	AUC	ACC
[24]	BDA LM-ANN SCG-ANN RMV SVM RBP-ANN MO-LSSVM	0.75 0.79 0.81 0.83 0.83 0.84 0.86	/	[25]	RBF LSSVM ELM	/	0.81 0.8706 0.8400
[17]	GA-GP GA-QDA GA-SVM GA-ANN GA-ADB-DT GA-KNN GA-OEC	0.893 0.798 0.908 0.877 0.936 0.908 0.943	/	[27]	RF SVM NB GSA	0.833 0.556 0.667 0.886	/
[16]	FA-LR FA-DT FA-MLP FA-RF FA-GBM FA-SVM	0.822 0.854 0.864 0.957 0.962 0.967	/	[18]	ANN SVM RF GBM	0.888 0.889 0.897 0.900	/
[13]	XGB RF LR SVM BC LDA KNN DT MLP GNB XRT Stacked ensemble	0.77 0.79 0.83 081 0.71 0.80 0.78 0.72 0.83 0.7. 0.74 0.90	/	[11]	KNN SVM SGD GP QDA GNB DT ANN B-KNN B-SVM B-ANN RF AB GBM XGB Heterogeneous ensemble	0.931 0.796 0.688 0.933 0.817 0.775 0.829 0.817 0.938 0.892 0.933 0.904 0.910 0.929 0.950 0.950	0.839 0.806 0.710 0.839 0.774 0.806 0.774 0.806 0.871 0.871 0.839 0.806 0.839 0.774 0.903 0.806
[12]	GBM Bagging Adaboost XRT RF HGB Voting Stacked	0.9199 0.9291 0.9199 0.9519 0.9268 0.8970 0.9588 0.9382	/	[30]	SVM DT LR NB Boosting MDMSE	/	0.8452 0.8333 <0.75 <0.75 0.8214 0.8810
Current study	$H2O_1$ (Stacked Ensem-ble_Best1000)	***0.970***	***0.904***				

Note: The best results are shown in bold italics. The results for relatively small sample sets (less than 100) are not presented or compared.

4.5. Advantages and Limitations of the Proposed Approach

Generally, the traditional ML models require workflows which encompass data preprocessing, feature engineering, ML algorithm selection, and hyperparameter tuning to be constructed, and are often developed based on prior experience. Due to varying levels of knowledge, the traditional ML model may not fully exploit the power of ML, resulting in less optimal results than those obtained with other models. Therefore, it is not objective to claim that one algorithm outperforms another without adjusting the hyperparameters. In contrast, AutoML is capable of automatically implementing the above processes and extensively exploring different workflows with minimal human intervention, resulting in a better model. In fact, previous studies [51,52] have reported that AutoML outperformed traditional ML models that were manually developed by data scientists. Moreover, it takes less computational time to train AutoML, with hundreds of optional pipelines, than it does to train a manually derived ML model, often requiring days to tune. In fact, based on the collected dataset, the computational time of AutoML with 8408 pipelines is one hour. Moreover, various commercial and open-source AutoML platforms have been developed, and many successful implementations have been reported. For example, an AutoML vision model was implemented for production recommendation using Google Cloud AutoML without hiring ML engineers [40]. These results may suggest that AutoML is preferred in some cases. However, due to the complex and involved process required to build an AutoML system from scratch, AutoML is still in an early stage of development. At present, AutoML is not fully automated [37,40]. For example, human efforts are still needed for data collection and data cleaning. For now, clear objectives based on high-quality data must be defined for AutoML. Nevertheless, the AutoML approach holds limitations such as black box, and is computationally expensive for large-scale datasets due to extensive searching of different pipelines.

5. Conclusions

In the present study, an updated database consisting of 627 cases was collected for slope stability classification of circular failure model. For the first time, an AutoML approach was proposed for ML model development. Instead of manually building a pipeline for ML algorithm selection and hyperparameter tuning, AutoML is capable of automatically implementing model development and performing extensive searches of different pipelines with minimal human intervention. The stacked ensemble of the best 1000 models was selected as the top model from 8208 ML trained models. The top-performing model provided highly accurate discriminative ability, with an AUC of 0.970 and an ACC of 0.904 for the testing dataset, achieving a maximum lift of 2.1. The trained AutoML model outperformed traditional manually tuned and metaheuristic-optimized models. AutoML was verified as an effective tool for automated ML model development and slope stability assessments of circular failure.

Given the successful use of AutoML for classification of slope stability for circular mode failure, it seems that such a methodology could be useful for short-term severity mitigation of geohazard and achieving long-term sustainable development goals.

Although the proposed AutoML approach shows promising results, it still has some limitations. Beyond the black box nature, among the major shortcomings of AutoML, a solution is their computational complexity. Future works should focus on developing explainable and interpretable ML models by coupling data-driven models with physical models.

Author Contributions: J.M.: Investigation, Methodology, Data curation, Formal analysis, Writing—original draft, Writing—review & editing, Funding acquisition. S.J.: Visualization, Software. Z.L.: Resources, Investigation. Z.R.: Resources, Investigation. D.L.: Resources, Investigation. C.T.: Resources, Investigation. H.G.: Visualization, Validation. All authors have read and agreed to the published version of the manuscript.

Funding: This research was funded by the Major Program of the National Natural Science Foundation of China (Grant No. 42090055), the National Natural Science Foundation of China (Grant Nos. 42177147 and 71874165), and the Fundamental Research Funds for the Central Universities, China University of Geosciences (Wuhan) (CUG2642022006).

Institutional Review Board Statement: Not applicable.

Informed Consent Statement: Not applicable.

Data Availability Statement: The data used are contained in Appendix A.

Conflicts of Interest: The authors declare no conflict of interest.

Abbreviations

AB: adaptive boost; ABC: artificial bee colony; ACC: accuracy; ACADS: Australian Association for Computer Aided Design; ADB: adaptive boosted decision tree; ANN: artificial neural network; AUC: area under the receiver operating characteristic curve; AutoML: automated machine learning; B-ANN: bagging artificial neural network; BC: bagging classifier; BDA: Bayes discriminant analysis; B-KNN: bagging k-nearest neighbors; BP: back-propagation; B-SVM: bagging support vector machine; CASHs: combinations of algorithm selection and hyperparameter tuning; CV: cross validation; DNN: deep neural network; DRF: distributed random forest; DT: decision tree; ELM: extreme learning machine; FA: firefly algorithm; GA: genetic algorithm; GBM: gradient boosting machine; GLM: generalized linear model; GNB: Gaussian naive bayes; GP: Gaussian process; GS: grid search; GSA: gravitational search algorithm; HGB: hist gradient boosting classifier; HS: harmony search, KNN: k-nearest neighbors; LDA: linear discriminant analysis; LM: Levenberg–Marquardt; LR: logistic regression; LSSVM: least squares support vector machine; MDMSE: margin distance minimization selective ensemble; ML: machine learning; MLP: multilayer perceptron; MO: metaheuristic optimized; NB: naive Bayes; NPV: negative predictive value; OEC: optimum ensemble classifier; PC: principal component; PCA: principal component analysis; PPV: positive predictive value; PSO: particle swarm optimization; QDA: quadratic discriminant analysis; RBF: radial basis function; RBP: resilient back-propagation; RF: random forest; RMV: relevance vector machine; SCG: scaled conjugate gradient; SEN: sensitivity; SGD: stochastic gradient descent; SPE: specificity; Std.: standard deviation; SVM: support vector machine; XRT: extremely randomized tree.

Appendix A. Updated Dataset for Slope Stability Assessments of Circular Mode Failure

No	γ (kN/m^3)	c (kPa)	φ (°)	β (°)	H (m)	r_u	Status
1	17.98	4.95	30.02	19.98	8	0.3	Stable
2	18	5	30	20	8	0.3	Stable
3	21.47	6.9	30.02	31.01	76.8	0.38	Failure
4	21.51	6.94	30	31	76.81	0.38	Failure
5	21.78	8.55	32	27.98	12.8	0.49	Failure
6	21.82	8.62	32	28	12.8	0.49	Failure
7	22.4	10	35	30	10	0	Stable
8	21.4	10	30.34	30	20	0	Stable
9	22.4	10	35	45	10	0.4	Failure
10	27.3	10	39	41	511	0.25	Stable
11	27.3	10	39	40	470	0.25	Stable
12	22.4	10	35	30	10	0.25	Stable
13	21.4	10	30.34	30	20	0.25	Stable
14	27	10	39	41	511	0.25	Stable
15	27	10	39	40	470	0.25	Stable
16	27.3	10	39	40	480	0.25	Stable
17	21.36	10.05	30.33	30	20	0	Stable
18	19.97	10.05	28.98	34.03	6	0.3	Stable
19	22.38	10.05	35.01	30	10	0	Stable
20	22.38	10.05	35.01	45	10	0.4	Failure
21	19.08	10.05	9.99	25.02	50	0.4	Failure
22	19.08	10.05	19.98	30	50	0.4	Failure
23	18.83	10.35	21.29	34.03	37	0.3	Failure
24	16.5	11.49	0	30	3.66	0	Failure
25	16.47	11.55	0	30	3.6	0	Failure
26	19.03	11.7	27.99	34.98	21	0.11	Failure
27	19.06	11.7	28	35	21	0.11	Failure
28	19.06	11.71	28	35	21	0.11	Failure

No	γ (kN/m^3)	c (kPa)	φ (°)	β (°)	H (m)	r_u	Status
29	19.06	11.75	28	35	21	0.11	Failure
30	14	11.97	26	30	88	0	Failure
31	19.63	11.97	20	22	12.19	0.41	Failure
32	14	11.97	26	30	88	0.45	Failure
33	19.63	11.97	20	22	21.19	0.4	Failure
34	18.5	12	0	30	6	0	Failure
35	18.5	12	0	30	6	0.25	Failure
36	19.6	12	19.98	22	12.2	0.41	Failure
37	13.97	12	26.01	30	88	0	Failure
38	18.46	12	0	30	6	0	Failure
39	13.97	12	26.01	30	88	0.45	Failure
40	27.3	14	31	41	110	0.25	Stable
41	27	14	31	41	110	0.25	Stable
42	18.84	14.36	25	20	30.5	0	Stable
43	18.84	14.36	25	20	30.5	0.45	Failure
44	18.84	14.36	25	20.3	50	0.45	Failure
45	18.8	14.4	25.02	19.98	30.6	0	Stable
46	18.8	14.4	25.02	19.98	30.6	0.45	Failure
47	18.8	15.31	30.02	25.02	10.6	0.38	Stable
48	18.84	15.32	30	25	10.67	0.38	Stable
49	20.56	16.21	26.51	30	40	0	Failure
50	20.6	16.28	26.5	30	40	0	Failure
51	27.3	16.8	28	50	90.5	0.25	Stable
52	27	16.8	28	50	90.5	0.25	Stable
53	20.96	19.96	40.01	40.02	12	0	Stable
54	21.98	19.96	36	45	50	0	Failure
55	19.97	19.96	36	45	50	0.25	Failure
56	19.97	19.96	36	45	50	0.5	Failure
57	18.77	19.96	9.99	25.02	50	0.3	Failure
58	18.77	19.96	19.98	30	50	0.3	Failure
59	21.98	19.96	22.01	19.98	180	0	Failure
60	21.98	19.96	22.01	19.98	180	0.1	Failure
61	22	20	36	45	50	0	Failure
62	20	20	36	45	50	0.25	Failure
63	20	20	36	45	50	0.5	Failure
64	18	24	30.15	45	20	0.12	Failure
65	17.98	24.01	30.15	45	20	0.12	Failure
66	18.83	24.76	21.29	29.2	37	0.5	Failure
67	20.41	24.9	13	22	10.67	0.35	Stable
68	20.39	24.91	13.01	22	10.6	0.35	Stable
69	18.5	25	0	30	6	0	Failure
70	18.5	25	0	30	6	0.25	Failure
71	18.46	25.06	0	30	6	0	Failure
72	18.77	25.06	19.98	30	50	0.2	Failure
73	18.77	25.06	9.99	25.02	50	0.2	Failure
74	27.3	26	31	50	92	0.25	Stable
75	27	26	31	50	92	0.25	Stable
76	18.68	26.34	15	35	8.23	0	Failure
77	18.66	26.41	14.99	34.98	8.2	0	Failure
78	28.4	29.41	35.01	34.98	100	0	Stable
79	28.44	29.42	35	35	100	0	Stable
80	18.77	30.01	9.99	25.02	50	0.1	Stable
81	18.77	30.01	19.98	30	50	0.1	Stable
82	20.96	30.01	35.01	40.02	12	0.4	Stable
83	18.97	30.01	35.01	34.98	11	0.2	Stable
84	27.3	31.5	29.7	41	135	0.25	Stable
85	27	31.5	29.7	41	135	0.25	Stable
86	27	32	33	42.6	301	0.25	Failure
87	27	32	33	42.4	289	0.25	Stable
88	27	32	33	42	289	0.25	Stable

No	γ (kN/m^3)	c (kPa)	φ (°)	β (°)	H (m)	r_u	Status
89	20.39	33.46	10.98	16.01	45.8	0.2	Failure
90	20.41	33.52	11	16	45.72	0.2	Failure
91	20.41	33.52	11	16	45.7	0.2	Failure
92	20.96	34.96	27.99	40.02	12	0.5	Stable
93	27	35	35	42	359	0.25	Stable
94	27	37.5	35	37.8	320	0.25	Stable
95	27	37.5	35	38	320	0.25	Stable
96	28.4	39.16	37.98	34.98	100	0	Stable
97	28.44	39.23	38	35	100	0	Stable
98	27	40	35	43	420	0.25	Failure
99	19.97	40.06	30.02	30	15	0.3	Stable
100	19.97	40.06	40.01	40.02	10	0.2	Stable
101	20.96	45.02	25.02	49.03	12	0.3	Stable
102	17.98	45.02	25.02	25.02	14	0.3	Stable
103	25	46	35	47	443	0.25	Stable
104	25	46	35	44	435	0.25	Stable
105	25	46	35	46	432	0.25	Stable
106	25	46	35	46	393	0.25	Stable
107	25	48	40	49	330	0.25	Stable
108	26.43	50	26.6	40	92.2	0.15	Stable
109	26.7	50	26.6	50	170	0.25	Stable
110	27	50	40	42	407	0.25	Stable
111	25	55	36	45.5	299	0.25	Stable
112	25	55	36	44	299	0.25	Stable
113	18.84	57.46	20	20	30.5	0	Stable
114	18.8	57.47	19.98	19.98	30.6	0	Stable
115	26.8	60	28.8	59	108	0.25	Stable
116	31.3	68	37	47	213	0.25	Failure
117	31.3	68	37	46	366	0.25	Stable
118	31.3	68.6	37	47	305	0.25	Failure
119	16	70	20	40	115	0	Failure
120	15.99	70.07	19.98	40.02	115	0	Failure
121	22.38	99.93	45	45	15	0.25	Stable
122	22.4	100	45	45	15	0.25	Stable
123	25	120	45	53	120	0	Stable
124	24.96	120.04	45	53	120	0	Stable
125	26.49	150	33	45	73	0.15	Stable
126	26.7	150	33	50	130	0.25	Stable
127	26.89	150	33	52	120	0.25	Stable
128	26	150	45	30	200	0.25	Stable
129	26	150.05	45	50	200	0	Stable
130	25.96	150.05	45	49.98	200	0	Stable
131	26.81	200	35	58	138	0.25	Stable
132	26.57	300	38.7	45.3	80	0.15	Failure
133	26.78	300	38.7	54	155	0.25	Failure
134	19.9652	19.95665	36	44.997	50	0.25	Failure
135	25.6	38.8	36	25	26	0	Stable
136	22.88	0	31.78	36.86	45.45	0.54	Failure
137	23.5	25	20	49.1	115	0.41	Stable
138	16	7	20	40	115	0	Failure
139	27.3	37.3	31	30	30	0	Stable
140	22	0	36	45	50	0.25	Stable
141	27	31.5	30	41	135	0.25	Stable
142	18.8008	14.4048	25.02	19.981	30.6	0	Stable
143	19.6	17.8	29.2	46.8	201.2	0.37	Stable
144	18.84	15.32	30	25	10.7	0.38	Stable
145	25	46	36	44.5	299	0.25	Stable
146	19.63	11.98	20	22	12.19	0	Failure
147	25	12	45	53	120	0	Stable
148	18.7724	30.01	9.99	25.016	50	0.1	Stable

No	γ (kN/m^3)	c (kPa)	φ (°)	β (°)	H (m)	r_u	Status
149	25	46	35	47	443	0.29	Stable
150	18.7724	25.05835	9.99	25.016	50	0.2	Failure
151	30.95	30.79	27.08	39.77	131.22	0.22	Stable
152	17.4	14.95	21.2	45	15	0.4	Failure
153	23.1	25.2	29.2	36.5	61.9	0.4	Stable
154	21.51	6.94	30	31	76.8	0.38	Failure
155	20.9592	45.015	25.02	49.025	12	0.3	Stable
156	27	32	33	42.6	301	0.29	Failure
157	15.9892	70.07335	19.98	40.015	115	0	Failure
158	12	0	30	45	8	0.29	Failure
159	25	46	35	50	285	0.25	Stable
160	13.9728	12.004	26.01	29.998	88	0.45	Failure
161	18.68	26.34	15	35	8.23	0.25	Failure
162	18.7724	30.01	19.98	29.998	50	0.1	Stable
163	22	0	40	33	8	0.35	Stable
164	20	0	36	45	50	0.25	Failure
165	31.3	68.6	37	47	305	0	Failure
166	22	10	35	45	10	0.403	Failure
167	18	5	26.5	15.52	53	0.4	Failure
168	21.7	32	27	45	60	0	Failure
169	14	11.97	26	30	88	0.25	Failure
170	18.84	14.36	25	20	30.5	0.25	Stable
171	12	0	30	45	4	0.25	Stable
172	18	5	22	15.52	53	0.4	Failure
173	26.2	44.14	32.26	37.71	359.04	0.21	Stable
174	19.9652	19.95665	36	44.997	50	0.5	Failure
175	22	20	36	45	50	0.25	Failure
176	12	0	30	35	4	0	Stable
177	25	120	45	53	120	0.25	Stable
178	31.3	68	37	46	366	0	Failure
179	26.5	36.1	31	35	39	0	Stable
180	20.9592	30.01	35.01	40.015	12	0.4	Stable
181	27.3	10	39	40	470	0.29	Stable
182	27.3	36	1	50	92	0.29	Stable
183	18.84	0	20	20	7.62	0.45	Failure
184	26.2	41.5	36	35	30	0	Stable
185	27.4	38.1	31	25	42	0	Stable
186	26.93	0	41.13	31.68	8.16	0.3	Stable
187	20.8	15.6	20	30	45	0	Failure
188	27	27.3	29.1	34	126.5	0.3	Failure
189	30.33	15.62	24.21	52.5	85.76	0.25	Failure
190	19	11.9	20.4	21.04	54	0.75	Stable
191	18.8	9.8	21	19.29	39	0.25	Failure
192	21.1	34.2	26	30	75	0	Failure
193	20	0.1	36	45	50	0.29	Failure
194	24	0	40	33	8	0.3	Failure
195	24.45	11.34	39.31	44.03	9.79	0.43	Failure
196	18	0	30	33	8	0.303	Stable
197	20.41	24.91	13	22	10.67	0.35	Stable
198	21.8	31.2	25	30	60	0	Failure
199	20	0.1	36	45	50	0.503	Failure
200	24	0	40	33	8	0.303	Stable
201	26.78	26.79	30.66	43.66	249.7	0.25	Stable
202	31.25	25.73	27.97	48.23	91.55	0.21	Failure
203	12	0.03	30	35	4	0.29	Failure
204	22	0	36	45	50	0.25	Failure
205	25	55	36	45	239	0.25	Stable
206	23	24	19.8	23	380	0	Failure
207	21.2	0	35	23.75	150	0.25	Failure
208	20.9592	34.96165	27.99	40.015	12	0.5	Stable

No	γ (kN/m^3)	c (kPa)	φ (°)	β (°)	H (m)	r_u	Status
209	12	0	30	45	8	0.25	Failure
210	27	70	22.8	45	60	0.32	Stable
211	18.7724	19.95665	19.98	29.998	50	0.3	Failure
212	28.44	29.42	35	35	100	0.25	Stable
213	20.8	15.4	21	30	53	0	Failure
214	19.596	12.004	19.98	21.995	12.2	0.405	Failure
215	22.1	24.2	39.7	45.8	49.5	0.21	Stable
216	22.4	29.3	26	50	50	0	Failure
217	20	0	24.5	20	8	0.35	Stable
218	25	55	36	45.5	299	0	Stable
219	17.55	22.08	0	34.99	5.88	0.35	Failure
220	20.52	14.06	26.23	25.38	9.86	0.37	Stable
221	21.9816	19.95665	22.005	19.981	180	0.1	Failure
222	18.46	12.004	0	29.998	6	0	Failure
223	20.45	16	15	30	36	0.25	Stable
224	21.1	33.5	28	40	31	0	Failure
225	22	20	36	45	30	0.29	Failure
226	17.6	10	16	21.8	9	0.4	Stable
227	31.3	68.6	37	47	270	0.25	Failure
228	23.4	15	38.5	30.3	45.2	0.28	Failure
229	16.472	11.55385	0	29.998	3.6	0	Failure
230	23.47	0	32	37	214	0.25	Failure
231	24.86	45.6	39.8	36.31	386.08	0.21	Stable
232	17.2	10	24.25	17.07	38	0.4	Stable
233	14.8	0	17	20	50	0	Failure
234	17.86	0	24.38	22.44	8.23	0.39	Stable
235	18.82	25	14.6	20.32	50	0.4	Failure
236	18.8292	10.35345	21.285	34.026	37	0.3	Failure
237	18.84	57.46	20	20	30.5	0.25	Stable
238	31.3	68.6	37	47	305	0.25	Stable
239	28.01	9.5	37.36	41.86	538.1	0.23	Stable
240	25	63	32	44.5	239	0.25	Stable
241	18.6	0	32	21.8	46	0.25	Stable
242	25.8	38.2	33	27	40	0	Stable
243	31.3	68	37	49	200.5	0.29	Failure
244	16	70	20	40	115	0.25	Failure
245	22	0	40	33	8	0.393	Stable
246	25	46	35	50	284	0.25	Stable
247	20.6	27.8	27	35	70	0	Failure
248	22	40	30	30	196	0	Stable
249	18.9712	30.01	35.01	34.98	11	0.2	Stable
250	26.2	43.8	38	35	68	0	Stable
251	17.9772	4.95165	30.015	19.981	8	0.3	Stable
252	22.4	28.9	24	28	35	0	Failure
253	25.6	39.8	36	30	32	0	Stable
254	19.36	19.8	38.49	43.41	48.88	0.43	Failure
255	20.41	24.9	13	22	10.7	0.35	Stable
256	23.5	10	27	26	190	0	Failure
257	17.4	20	24	18.43	51	0.4	Failure
258	17.6	10	8	21.8	9	0.4	Stable
259	22.3792	10.05335	35.01	29.998	10	0	Stable
260	21.7828	8.55285	31.995	27.984	12.8	0.49	Failure
261	19.63	11.97	20	22	12.19	0.405	Failure
262	25	48	40	45	330	0.25	Stable
263	25.8	39.4	33	25	45	0	Stable
264	27	40	35	47.1	292	0.25	Failure
265	12	0	30	35	8	0.25	Failure
266	22.3792	99.9333	45	44.997	15	0.25	Stable
267	16.5	11.49	0	30	3.66	0.25	Failure
268	25.8	34.7	33	30	50	0	Stable

No	γ (kN/m^3)	c (kPa)	φ (°)	β (°)	H (m)	r_u	Status
269	26.62	0	31.78	42.72	51.48	0.4	Failure
270	24	0	40	33	8	0.3	Stable
271	18.84	0	20	20	7.62	0	Failure
272	18.7724	25.05835	19.98	29.998	50	0.2	Failure
273	22	21	23	30	257	0	Failure
274	23.2	9.5	39.69	39.34	10.49	0.44	Failure
275	21.78	0	34.2	35	7.13	0.32	Stable
276	14.8	0	17	20	50	0.25	Failure
277	31.3	68	37	47	213	0	Failure
278	21.8	32.7	27	50	50	0	Failure
279	21.8	28.8	26	35	99	0	Failure
280	26.2	42.8	37	30	37	0	Stable
281	22	10	35	30	10	0.29	Stable
282	19.6	21.8	29.5	37.8	40.3	0.25	Stable
283	18.6	0	32	26.5	46	0.25	Stable
284	27.3	10	39	41	511	0.29	Stable
285	28.07	35	38.93	44.54	361.51	0.24	Stable
286	19.63	11.97	20	22	12.2	0.41	Failure
287	27	50	40	42	407	0.29	Stable
288	21.73	9.21	30.6	33.06	19.78	0.29	Stable
289	27.3	14	31	41	110	0.29	Stable
290	26.69	50	26.6	50	170	0.25	Stable
291	26.5	35.4	32	30	21	0	Stable
292	26.5	41.8	36	42	54	0	Stable
293	18.7724	19.95665	9.99	25.016	50	0.3	Failure
294	29.7	38.09	32.92	45.48	410.4	0.26	Stable
295	26.2	42.3	36	23	36	0	Stable
296	20.6	16.28	26.5	30	40	0.25	Failure
297	20.9592	19.95665	40.005	40.015	12	0	Stable
298	20.6	28.5	27	40	65	0	Failure
299	17.29	0	37.22	44.55	42.3	0.28	Failure
300	12.34	0	25.92	46.82	8.08	0.43	Failure
301	27	37.5	35	37.8	320	0.29	Stable
302	24.9636	120.04	45	53	120	0	Stable
303	22.1	45.8	49.5	45.8	49.5	0.21	Stable
304	11.94	0	31.75	32.49	3.92	0.11	Stable
305	17.9772	45.015	25.02	25.016	14	0.3	Stable
306	20.6	32.4	26	30	42	0	Failure
307	18.8	8	26	21.8	40	0.4	Failure
308	31.3	68	37	47	360.5	0.25	Failure
309	26.83	13.98	35.46	43.5	96.14	0.23	Stable
310	21.2	0	35	23.75	150	0.25	Stable
311	22	0	36	45	50	0	Failure
312	17.9772	24.008	30.15	44.997	20	0.12	Failure
313	20	20	36	45	30	0.503	Failure
314	24.57	9.98	41.31	35.46	526.13	0.27	Stable
315	21.5	29.8	26	40	70	0	Failure
316	27.1	22	18.6	25.6	100	0.19	Failure
317	22	10	36	45	50	0.29	Failure
318	21.6	6.5	19	40	50	0	Failure
319	20.97	21.8	31.81	38.09	57.75	0.24	Failure
320	26.8	37.5	32	30	26	0	Stable
321	25.9576	150.05	45	49.979	200	0	Stable
322	19.9652	10.05335	28.98	34.026	6	0.3	Stable
323	22.54	29.4	20	24	210	0	Stable
324	26	42.4	37	38	55	0	Stable
325	20.41	24.9	13	22	10.67	0	Stable
326	21	20	24	21	565	0	Stable
327	31.3	68	37	49	200.5	0.25	Failure
328	20.6	32.4	26	35	55	0	Failure

No	γ (kN/m^3)	c (kPa)	φ (°)	β (°)	H (m)	r_u	Status
329	16.05	11.49	0	30	3.66	0	Failure
330	25	46	36	44.5	299	0	Stable
331	19.43	11.16	0	32.34	5.35	0.36	Failure
332	20	30.3	25	45	53	0	Failure
333	21.9816	19.95665	36	44.997	50	0	Failure
334	27.3	31.5	29.703	41	135	0.293	Stable
335	21.5	15	29	41.5	123.6	0.36	Stable
336	20.8	14.8	21	30	40	0	Failure
337	25.8	43.3	37	30	33	0	Stable
338	20.41	33.52	11	16	45.72	0	Failure
339	27	40	35	47.1	292	0	Failure
340	24	40.8	35	35	50	0	Stable
341	22.4	100	45	45	15	0.25	Failure
342	25	63	32	46	300	0.25	Stable
343	18	24	30.2	45	20	0.12	Failure
344	26.81	60	28.8	59	108	0.25	Stable
345	28.35	44.97	33.49	43.16	413.42	0.25	Failure
346	19.0848	10.05335	9.99	25.016	50	0.4	Failure
347	27	27.3	29.1	35	150	0.26	Failure
348	31.3	68	37	8	305.5	0.25	Failure
349	25	48	40	49	330	0	Stable
350	18.8008	57.46915	19.98	19.981	30.6	0	Stable
351	27	32	33	42	301	0.25	Failure
352	25	46	35	46	393	0	Stable
353	18.84	0	20	20	7.6	0.45	Failure
354	20.3912	24.9083	13.005	21.995	10.6	0.35	Stable
355	26	15	45	50	200	0	Stable
356	31.3	58.8	35.5	47.5	438.5	0.25	Failure
357	18.6588	26.4088	14.985	34.98	8.2	0	Failure
358	21.1	10	30.34	30	20	0	Stable
359	25.8	41.2	35	30	40	0	Stable
360	21.4704	6.9023	30.015	31.005	76.8	0.38	Failure
361	23.47	0	32	37	214	0	Failure
362	20	0	20	20	8	0.35	Stable
363	23	20	20.3	46.2	40.3	0.25	Stable
364	31.3	58.8	35.5	47.5	502.7	0.25	Failure
365	26	39.4	36	25	30	0	Stable
366	27.3	10	39	40	480	0	Stable
367	21.8	27.6	25	35	60	0	Failure
368	21.4	28.8	20	50	52	0	Failure
369	19.9652	40.06335	30.015	29.998	15	0.3	Stable
370	20	8	20	10	10	0	Failure
371	23.8	31	38.7	47.5	23.5	0.31	Stable
372	26.6	42.4	37	25	52	0	Stable
373	28.4	39.16305	37.98	34.98	100	0	Stable
374	21.51	17.82	31.75	47.03	49.92	0.52	Failure
375	22	0	40	33	8	0.35	Failure
376	23	0	20	20	100	0.3	Failure
377	21.43	0	20	20	61	0	Failure
378	26.6	40.7	35	35	60	0	Stable
379	27.83	45.01	35.95	47.83	456.38	0.25	Stable
380	25	46	35	44	435	0.29	Stable
381	18.71	4.75	28.12	18.81	8.62	0.31	Stable
382	26.6	44.1	38	35	42	0	Stable
383	28.4	29.4098	35.01	34.98	100	0	Stable
384	19.028	11.7039	27.99	34.98	21	0.11	Failure
385	18.45	0	18.58	17.82	7.55	0.43	Failure
386	27	35	35	42	359	0.29	Stable
387	31.3	68.6	37	47.5	262.5	0.25	Failure
388	31.3	68	37	46	366	0.25	Failure

No	γ (kN/m^3)	c (kPa)	φ (°)	β (°)	H (m)	r_u	Status
389	27	43	35	43	420	0.29	Failure
390	12	0	30	35	4	0.25	Stable
391	26.18	159	44.93	31.5	172.98	0.1	Failure
392	19.32	0	19.44	20.2	68.48	0.45	Failure
393	30	27.38	34.57	43.46	319.21	0.27	Failure
394	12	0	30	45	8	0	Failure
395	28.51	42.34	32.2	43.25	453.6	0.25	Stable
396	11.82	0	33.7	31.26	3.91	0.42	Stable
397	18.84	15.32	30	25	10.67	0	Stable
398	27	35.8	32	30	69	0	Stable
399	18	21	21.33	21.8	40	0.4	Failure
400	17.8	21.2	13.92	18.43	51	0.4	Stable
401	27.3	16.2	28	50	90.5	0.29	Stable
402	22.3	20.1	31	40.2	88	0.19	Stable
403	22.5	20	16	25	220	0	Stable
404	13.9728	12.004	26.01	29.998	88	0	Failure
405	25	46	35	46	432	0.29	Stable
406	20	30	36	45	50	0.29	Failure
407	23.2	31.2	23	30	33	0	Failure
408	25.4	33	33	20	35	0	Failure
409	26	150.05	45	50	200	0.25	Stable
410	19.9652	40.06335	40.005	40.015	10	0.2	Stable
411	20.3912	33.46115	10.98	16.006	45.8	0.2	Failure
412	28.44	39.23	38	35	100	0.25	Stable
413	21	10	30.343	30	30	0.29	Stable
414	22	29	15	18	400	0	Failure
415	27.8	27.8	27	41	236	0.1	Stable
416	26.5	42.9	38	34	36	0	Stable
417	18.8292	24.75825	21.285	29.203	37	0.5	Failure
418	21.9816	19.95665	22.005	19.981	180	0	Failure
419	18.8008	15.3051	30.015	25.016	10.6	0.38	Stable
420	21.83	8.62	32	28	12.8	0	Failure
421	22.85	8.46	38.12	25.67	11.34	0.56	Stable
422	18.5	25	0	30	6.003	0.29	Failure
423	27	38.4	33	25	22	0	Stable
424	24	41.5	36	30	51	0	Stable
425	21.43	0	20	20	61	0.5	Failure
426	26	150	45	30	230	0.29	Stable
427	18.5	12	0	30	6.003	0.29	Failure
428	22.3792	10.05335	35.01	44.997	10	0.4	Failure
429	20.5616	16.2054	26.505	29.998	40	0	Failure
430	31	68	37	46	366	0.25	Failure
431	21.3568	10.05335	30.33	29.998	20	0	Stable
432	25	46	35	50	284	0	Stable
433	27	32	33	42.2	239	0.29	Stable
434	25.6	36.8	34	35	60	0	Stable
435	20	0	36	45	50	0.5	Failure
436	19.0848	10.05335	19.98	29.998	50	0.4	Failure
437	33.16	68.54	41.11	51.98	188.15	0.44	Failure
438	21.2	0	35	18.43	73	0.25	Stable
439	20.6	26.31	22	25	35	0	Failure
440	18.46	25.05835	0	29.998	6	0	Failure
441	22.3	0	40	26.5	78	0.25	Stable
442	12	0	30	35	4	0.29	Stable
443	18.12	10.57	30.84	32.45	21.77	0.11	Failure
444	19.6	29.6	23	40	58	0	Failure
445	27	27.3	29.1	37	184	0.22	Failure
446	25	55	36	45	299	0.25	Stable
447	22.5	18	20	20	290	0	Stable
448	18.8008	14.4048	25.02	19.981	30.6	0.45	Failure

No	γ (kN/m^3)	c (kPa)	φ (°)	β (°)	H (m)	r_u	Status
449	12	0	30	45	4	0	Stable
450	23.47	0	32	37	214	0	Stable
451	20.41	33.52	11	16	10.67	0.35	Stable
452	25.4	33	33	20	35	0	Stable
453	27.3	31.5	30	41	135	0.25	Stable
454	21.4	10	30	30	20	0.25	Stable
455	18.66	8.8	15	35	8.2	0	Failure
456	28.4	9.8	35	35	100	0	Stable
457	25.96	50	45	50	200	0	Stable
458	18.46	8.35	0	30	6	0	Failure
459	21.36	3.35	30	30	20	0	Stable
460	15.99	23.35	20	40	115	0	Failure
461	20.39	8.3	13	22	10.6	0.35	Stable
462	19.6	4	20	22	12.2	0.41	Failure
463	20.39	11.15	11	16	45.8	0.2	Failure
464	19.03	3.9	28	35	21	0.11	Failure
465	17.98	1.65	30	20	8	0.3	Stable
466	20.96	6.65	40	40	12	0	Stable
467	20.96	11.65	28	40	12	0.5	Stable
468	19.97	3.35	29	34	6	0.3	Stable
469	18.77	10	10	25	50	0.1	Stable
470	18.77	10	20	30	50	0.1	Stable
471	18.77	8.35	20	30	50	0.2	Failure
472	20.56	5.4	27	30	40	0	Failure
473	16.47	3.85	0	30	3.6	0	Failure
474	18.8	4.8	25	20	30.6	0	Stable
475	18.8	19.15	20	20	30.6	0	Stable
476	28.4	13.05	38	35	100	0	Stable
477	24.96	40	45	53	120	0	Stable
478	18.46	4	0	30	6	0	Failure
479	22.38	3.35	35	30	10	0	Stable
480	21.98	6.65	36	45	50	0	Failure
481	18.8	5.1	30	25	10.6	0.38	Stable
482	18.8	4.8	25	31	76.8	0.38	Failure
483	21.47	2.3	30	30	88	0.45	Failure
484	13.97	4	26	45	20	0.12	Failure
485	17.98	8	30	45	15	0.25	Failure
486	22.38	33.3	45	45	10	0.4	Stable
487	22.38	3.35	35	45	50	0.25	Failure
488	19.97	6.65	36	45	50	0.25	Failure
489	19.97	6.65	36	45	50	0.5	Failure
490	20.96	15	25	49	12	0.3	Stable
491	20.96	10	35	40	12	0.4	Stable
492	19.97	13.35	30	30	15	0.3	Stable
493	17.98	15	25	25	14	0.3	Stable
494	18.97	10	35	35	11	0.2	Stable
495	19.97	13.35	40	40	10	0.2	Stable
496	18.83	8.25	21	21	37	0.5	Stable
497	18.83	3.45	21	34	37	0.3	Failure
498	18.77	8.35	10	25	50	0.2	Failure
499	18.77	6.65	10	25	50	0.3	Failure
500	19.08	3.35	10	25	50	0.4	Failure
501	18.77	6.65	20	30	50	0.3	Failure
502	19.08	3.35	20	30	50	0.4	Failure
503	21.98	6.65	22	20	180	0	Failure
504	21.98	6.65	22	20	180	0.1	Failure
505	20	20	36	45	50	0	Failure
506	27	27.3	29.1	21	565	0.26	Failure
507	27	27.3	29.1	35	150	0.22	Failure
508	27	27.3	29.1	37	184	0.3	Failure

No	γ (kN/m^3)	c (kPa)	φ (°)	β (°)	H (m)	r_u	Status
509	0.657	0.176	0.333	0.66	0.041	0	Failure
510	1	0.196	0.778	0.66	0.5	0	Stable
511	0.914	1	1	0.943	1	0	Stable
512	0.65	0.167	0	0.566	0.03	0	Failure
513	0.752	0.067	0.674	0.566	0.1	0	Stable
514	0.563	0.467	0.444	0.755	0.575	0	Failure
515	0.718	0.166	0.289	0.415	0.053	0.7	Stable
516	0.69	0.08	0.444	0.415	0.061	0.81	Failure
517	0.767	0.057	0.711	0.528	0.064	0.98	Failure
518	0.718	0.223	0.244	0.302	0.229	0.4	Failure
519	0.67	0.078	0.622	0.66	0.105	0.22	Failure
520	0.633	0.033	0.667	0.377	0.04	0.6	Stable
521	0.738	0.133	0.889	0.755	0.06	0	Stable
522	0.738	0.233	0.622	0.755	0.06	1	Stable
523	0.703	0.067	0.644	0.642	0.03	0.6	Stable
524	0.661	0.2	0.222	0.472	0.25	0.2	Stable
525	0.661	0.2	0.444	0.566	0.25	0.2	Stable
526	0.661	0.167	0.444	0.566	0.25	0.4	Failure
527	0.724	0.108	0.589	0.566	0.2	0	Failure
528	0.58	0.077	0	0.566	0.018	0	Failure
529	0.662	0.096	0.556	0.377	0.153	0	Stable
530	0.662	0.383	0.444	0.377	0.153	0	Stable
531	1	0.261	0.844	0.66	0.5	0	Stable
532	0.492	0.08	0.578	0.566	0.44	0	Failure
533	0.879	0.8	1	1	0.6	0	Stable
534	0.65	0.08	0	0.566	0.03	0	Failure
535	0.788	0.067	0.778	0.566	0.05	0	Stable
536	0.774	0.133	0.8	0.849	0.25	0	Failure
537	0.662	0.102	0.667	0.472	0.053	0.76	Stable
538	0.662	0.096	0.556	0.377	0.153	0.9	Failure
539	0.756	0.046	0.667	0.585	0.384	0.76	Failure
540	0.492	0.08	0.578	0.566	0.44	0.9	Failure
541	0.633	0.16	0.67	0.849	0.1	0.24	Failure
542	0.788	0.666	1	0.849	0.075	0.5	Stable
543	0.788	0.067	0.778	0.849	0.05	0.8	Failure
544	0.703	0.133	0.8	0.849	0.25	0.5	Failure
545	0.703	0.133	0.8	0.849	0.25	1	Failure
546	0.738	0.3	0.556	0.925	0.06	0.6	Stable
547	0.738	0.2	0.778	0.755	0.06	0.8	Stable
548	0.703	0.267	0.667	0.566	0.075	0.6	Stable
549	0.633	0.3	0.556	0.472	0.07	0.6	Stable
550	0.668	0.2	0.778	0.66	0.055	0.4	Stable
551	0.703	0.267	0.889	0.755	0.05	0.4	Stable
552	0.633	0.165	0.473	0.551	0.185	1	Failure
553	0.633	0.069	0.473	0.642	0.185	0.6	Failure
554	0.661	0.167	0.222	0.472	0.25	0.4	Failure
555	0.661	0.133	0.222	0.472	0.25	0.6	Failure
556	0.672	0.067	0.222	0.472	0.25	0.8	Failure
557	0.661	0.133	0.444	0.566	0.25	0.6	Failure
558	0.672	0.067	0.444	0.566	0.25	0.8	Failure
559	0.774	0.133	0.489	0.377	0.9	0	Failure
560	0.774	0.133	0.489	0.377	0.9	0.2	Failure
561	17.6	39.5	30.2	50	38	0.04	Stable
562	17.3	39	30	50	35	0.04	Stable
563	17.8	38.7	30.5	60	26	0	Stable
564	17.9	39	31.2	55	25	0.15	Stable
565	17.3	39	30	50	26	0.2	Stable
566	17.3	37.9	30	45	29	0.37	Stable
567	17.5	38.5	29	50	33	0.2	Stable
568	17.5	39.2	29.7	55	31	0	Stable

No	γ (kN/m^3)	c (kPa)	φ (°)	β (°)	H (m)	r_u	Status
569	17.8	39.8	31.3	45	32	0.34	Stable
570	17.3	39	30	48	30	0.03	Stable
571	18.3	57.2	38.6	38	31	0.64	Stable
572	17.4	5	43.5	58	29	0.05	Failure
573	17.8	14	44.2	65	31	0.07	Failure
574	17.4	0	43.7	60	26	0.4	Failure
575	19.8	57.5	41.3	62	23	0.19	Stable
576	20.5	6.5	12.5	42	70	0	Failure
577	21.4	7.1	16.7	44	70	1	Failure
578	21.5	9.5	11.5	40	75	0	Failure
579	20.6	6.7	9.4	45	30	0	Failure
580	20.9	9.7	18.5	39	38	1	Failure
581	21.4	9.4	21.8	30	106	1	Failure
582	19.9	6.8	19.4	30	80	1	Failure
583	20.2	14.9	18.5	40	70	1	Failure
584	19	9	15.2	45	27	0	Failure
585	19.7	16.4	21.4	30	55	1	Failure
586	21.2	7.8	22.4	45	25	1	Failure
587	19.9	7.4	15.6	44	30	1	Failure
588	19.9	7.1	21.2	30	55	0	Failure
589	22.2	10.7	25.2	35	45	1	Failure
590	21.8	7.2	17.8	40	34	1	Failure
591	21.8	7.2	17.8	42	41	1	Failure
592	21.96	34.77	14.15	28	60	0	Stable
593	21.96	34.77	14.15	24	115	0	Stable
594	22.93	32.33	19.73	30	50	1	Stable
595	22.15	19.47	13.29	28	110	1	Stable
596	23.4	20	9	36.5	50	0	Stable
597	21.8	18.05	9.72	30	40	0	Failure
598	23.98	32.77	17.28	40	100	0	Failure
599	20.57	24.8	15.53	40	50	1	Stable
600	21.2	24.88	17.29	44	52	0	Failure
601	22.15	5	19	45	40	1	Failure
602	21.8	18.05	9.72	35	40	0	Failure
603	23.75	36.78	22.63	42	43	1	Failure
604	20.98	23.59	20	45	65	0	Failure
605	22.6	24.06	14.04	26	190	1	Stable
606	22.29	27.54	10.1	40	70	0	Stable
607	22.1	24.67	16.2	40	70	1	Stable
608	20.25	32.4	11.99	45	36	1	Failure
609	20.8	15.57	8.74	29.7	35	1	Failure
610	21.17	15.44	16	33	32	1	Failure
611	22.94	33.77	23.29	27	170	1	Stable
612	22.95	46.49	25.11	30	42	1	Stable
613	21.92	19.4	15.5	35	80	1	Failure
614	21.42	28.9	16.2	40	30	1	Stable
615	20.8	40.25	19.39	45	123	1	Failure
616	20.1	34.61	24.69	22	94	0	Stable
617	19.19	19.69	17.68	34	43	1	Failure
618	19.18	12.8	9.45	45	20	0	Failure
619	17.8	22.2	6.05	40	51.6	1	Failure
620	19.6	15.53	15.88	35	97	1	Failure
621	19.81	33.75	19.46	20	120	1	Stable
622	19.81	19.97	11.08	35	35	0	Failure
623	19.7	17	9.38	45	20	1	Failure
624	20.2	21.2	19.89	35	62	1	Failure
625	17.96	24.01	28	40	60	1	Failure
626	25	55	36	44.5	299	0.25	Stable
627	21.98	19.96	22.01	19.98	180	0.01	Failure

References

1. Méheux, K.; Dominey-Howes, D.; Lloyd, K. Natural hazard impacts in small island developing states: A review of current knowledge and future research needs. *Nat. Hazards* **2007**, *40*, 429–446. [CrossRef]
2. Iai, S. *Geotechnics and Earthquake Geotechnics towards Global Sustainability*; Springer: Dordrecht, The Netherlands, 2011; Volume 15.
3. Ma, J.W.; Liu, X.; Niu, X.X.; Wang, Y.K.; Wen, T.; Zhang, J.R.; Zou, Z.X. Forecasting of Landslide Displacement Using a Probability-Scheme Combination Ensemble Prediction Technique. *Int. J. Environ. Res. Public Health* **2020**, *17*, 4788. [CrossRef]
4. Niu, X.X.; Ma, J.W.; Wang, Y.K.; Zhang, J.R.; Chen, H.J.; Tang, H.M. A novel decomposition-ensemble learning model based on ensemble empirical mode decomposition and recurrent neural network for landslide displacement prediction. *Appl. Sci.* **2021**, *11*, 4684. [CrossRef]
5. Ouyang, C.J.; Zhou, K.Q.; Xu, Q.; Yin, J.H.; Peng, D.L.; Wang, D.P.; Li, W.L. Dynamic analysis and numerical modeling of the 2015 catastrophic landslide of the construction waste landfill at Guangming, Shenzhen, China. *Landslides* **2017**, *14*, 705–718. [CrossRef]
6. Duncan, J.M. Soil Slope Stability Analysis. In *Landslides: Investigation and Mitigation, Transportation Research Board Special Report 247*; National Academy Press: Washington, DC, USA, 1996; pp. 337–371.
7. Duncan, J.M.; Wright, S.G. The accuracy of equilibrium methods of slope stability analysis. *Eng. Geol.* **1980**, *16*, 5–17. [CrossRef]
8. Zhu, D.Y.; Lee, C.F.; Jiang, H.D. Generalised framework of limit equilibrium methods for slope stability analysis. *Géotechnique* **2003**, *53*, 377–395. [CrossRef]
9. Liu, S.Y.; Shao, L.T.; Li, H.J. Slope stability analysis using the limit equilibrium method and two finite element methods. *Comput. Geotech.* **2015**, *63*, 291–298. [CrossRef]
10. Li, A.J.; Merifield, R.S.; Lyamin, A.V. Limit analysis solutions for three dimensional undrained slopes. *Comput. Geotech.* **2009**, *36*, 1330–1351. [CrossRef]
11. Pham, K.; Kim, D.; Park, S.; Choi, H. Ensemble learning-based classification models for slope stability analysis. *Catena* **2021**, *196*, 104886. [CrossRef]
12. Lin, S.; Zheng, H.; Han, B.; Li, Y.; Han, C.; Li, W. Comparative performance of eight ensemble learning approaches for the development of models of slope stability prediction. *Acta Geotech.* **2022**, *17*, 1477–1502. [CrossRef]
13. Kardani, N.; Zhou, A.; Nazem, M.; Shen, S.-L. Improved prediction of slope stability using a hybrid stacking ensemble method based on finite element analysis and field data. *J. Rock Mech. Geotech. Eng.* **2021**, *13*, 188–201. [CrossRef]
14. Wang, H.B.; Xu, W.Y.; Xu, R.C. Slope stability evaluation using Back Propagation Neural Networks. *Eng. Geol.* **2005**, *80*, 302–315. [CrossRef]
15. Wang, L.; Chen, Z.; Wang, N.; Sun, P.; Yu, S.; Li, S.; Du, X. Modeling lateral enlargement in dam breaches using slope stability analysis based on circular slip mode. *Eng. Geol.* **2016**, *209*, 70–81. [CrossRef]
16. Qi, C.; Tang, X. Slope stability prediction using integrated metaheuristic and machine learning approaches: A comparative study. *Comput. Ind. Eng.* **2018**, *118*, 112–122. [CrossRef]
17. Qi, C.; Tang, X. A hybrid ensemble method for improved prediction of slope stability. *Int. J. Numer. Anal. Methods Geomech.* **2018**, *42*, 1823–1839. [CrossRef]
18. Zhou, J.; Li, E.; Yang, S.; Wang, M.; Shi, X.; Yao, S.; Mitri, H.S. Slope stability prediction for circular mode failure using gradient boosting machine approach based on an updated database of case histories. *Saf. Sci.* **2019**, *118*, 505–518. [CrossRef]
19. Ma, J.; Xia, D.; Wang, Y.; Niu, X.; Jiang, S.; Liu, Z.; Guo, H. A comprehensive comparison among metaheuristics (MHs) for geohazard modeling using machine learning: Insights from a case study of landslide displacement prediction. *Eng. Appl. Artif. Intell.* **2022**, *114*, 105150. [CrossRef]
20. Ma, J.; Xia, D.; Guo, H.; Wang, Y.; Niu, X.; Liu, Z.; Jiang, S. Metaheuristic-based support vector regression for landslide displacement prediction: A comparative study. *Landslides* **2022**, *19*, 2489–2511. [CrossRef]
21. Feng, X.-T. *Introduction of Intelligent Rock Mechanics*; Science Press: Beijing, China, 2000.
22. Lu, P.; Rosenbaum, M.S. Artificial Neural Networks and Grey Systems for the Prediction of Slope Stability. *Nat. Hazards* **2003**, *30*, 383–398. [CrossRef]
23. Xue, X.; Yang, X.; Chen, X. Application of a support vector machine for prediction of slope stability. *Sci. China Technol. Sci.* **2014**, *57*, 2379–2386. [CrossRef]
24. Hoang, N.-D.; Pham, A.-D. Hybrid artificial intelligence approach based on metaheuristic and machine learning for slope stability assessment: A multinational data analysis. *Expert Syst. Appl.* **2016**, *46*, 60–68. [CrossRef]
25. Hoang, N.-D.; Tien Bui, D. Chapter 18—Slope Stability Evaluation Using Radial Basis Function Neural Network, Least Squares Support Vector Machines, and Extreme Learning Machine Slope Stability Evaluation Using Radial Basis Function Neural Network, Least Squares Support Vector Machines, and Extreme Learning Machine. In *Handbook of Neural Computation*; Samui, P., Sekhar, S., Balas, V.E., Eds.; Academic Press: Washington, DC, USA, 2017; pp. 333–344. [CrossRef]
26. Feng, X.; Li, S.; Yuan, C.; Zeng, P.; Sun, Y. Prediction of Slope Stability using Naive Bayes Classifier. *KSCE J. Civ. Eng.* **2018**, *22*, 941–950. [CrossRef]
27. Lin, Y.; Zhou, K.; Li, J. Prediction of Slope Stability Using Four Supervised Learning Methods. *IEEE Access* **2018**, *6*, 31169–31179. [CrossRef]
28. Amirkiyaei, V.; Ghasemi, E. Stability assessment of slopes subjected to circular-type failure using tree-based models. *Int. J. Geotech. Eng.* **2020**, *16*, 301–311. [CrossRef]

29. Haghshenas, S.S.; Haghshenas, S.S.; Geem, Z.W.; Kim, T.-H.; Mikaeil, R.; Pugliese, L.; Troncone, A. Application of Harmony Search Algorithm to Slope Stability Analysis. *Land* **2021**, *10*, 1250. [CrossRef]
30. Zhang, H.; Wu, S.; Zhang, X.; Han, L.; Zhang, Z. Slope stability prediction method based on the margin distance minimization selective ensemble. *CATENA* **2022**, *212*, 106055. [CrossRef]
31. Zou, Z.; Yang, Y.; Fan, Z.; Tang, H.; Zou, M.; Hu, X.; Xiong, C.; Ma, J. Suitability of data preprocessing methods for landslide displacement forecasting. *Stoch. Environ. Res. Risk Assess.* **2020**, *34*, 1105–1119. [CrossRef]
32. Ma, J.W.; Wang, Y.K.; Niu, X.X.; Jiang, S.; Liu, Z.Y. A comparative study of mutual information-based input variable selection strategies for the displacement prediction of seepage-driven landslides using optimized support vector regression. *Stoch. Environ. Res. Risk Assess.* **2022**, *36*, 3109–3129. [CrossRef]
33. Wang, Y.K.; Tang, H.M.; Huang, J.S.; Wen, T.; Ma, J.W.; Zhang, J.R. A comparative study of different machine learning methods for reservoir landslide displacement prediction. *Eng. Geol.* **2022**, *298*, 106544. [CrossRef]
34. Zhang, Q.; Hu, W.; Liu, Z.; Tan, J. TBM performance prediction with Bayesian optimization and automated machine learning. *Tunn. Undergr. Space Technol.* **2020**, *103*, 103493. [CrossRef]
35. Wolpert, D.H.; Macready, W.G. No free lunch theorems for optimization. *IEEE Trans. Evol. Comput.* **1997**, *1*, 67–82. [CrossRef]
36. Zhang, D.; Shen, Y.; Huang, Z.; Xie, X. Auto machine learning-based modelling and prediction of excavation-induced tunnel displacement. *J. Rock Mech. Geotech. Eng.* **2022**, *14*, 1100–1114. [CrossRef]
37. Sun, Z.; Sandoval, L.; Crystal-Ornelas, R.; Mousavi, S.M.; Wang, J.; Lin, C.; Cristea, N.; Tong, D.; Carande, W.H.; Ma, X.; et al. A review of Earth Artificial Intelligence. *Comput. Geosci.* **2022**, *159*, 105034. [CrossRef]
38. Jiang, S.; Ma, J.W.; Liu, Z.Y.; Guo, H.X. Scientometric Analysis of Artificial Intelligence (AI) for Geohazard Research. *Sensors* **2022**, *22*, 7814. [CrossRef]
39. Fallatah, O.; Ahmed, M.; Gyawali, B.; Alhawsawi, A. Factors controlling groundwater radioactivity in arid environments: An automated machine learning approach. *Sci. Total Environ.* **2022**, *830*, 154707. [CrossRef]
40. Quan, S.Q.; Feng, J.H.; Xia, H. Automated Machine Learning in Action. Manning Publications, Co.: New York, NY, USA, 2022.
41. Mahjoubi, S.; Barhemat, R.; Guo, P.; Meng, W.; Bao, Y. Prediction and multi-objective optimization of mechanical, economical, and environmental properties for strain-hardening cementitious composites (SHCC) based on automated machine learning and metaheuristic algorithms. *J. Clean. Prod.* **2021**, *329*, 129665. [CrossRef]
42. Chen, W.; Zhang, L. An automated machine learning approach for earthquake casualty rate and economic loss prediction. *Reliab. Eng. Syst. Saf.* **2022**, *225*, 108645. [CrossRef]
43. Erzin, Y.; Cetin, T. The prediction of the critical factor of safety of homogeneous finite slopes using neural networks and multiple regressions. *Comput. Geosci.* **2013**, *51*, 305–313. [CrossRef]
44. Sakellariou, M.G.; Ferentinou, M.D. A study of slope stability prediction using neural networks. *Geotech. Geol. Eng.* **2005**, *23*, 419–445. [CrossRef]
45. Hoang, N.-D.; Bui, D.T. Spatial prediction of rainfall-induced shallow landslides using gene expression programming integrated with GIS: A case study in Vietnam. *Nat. Hazards* **2018**, *92*, 1871–1887. [CrossRef]
46. Ferreira, L.; Pilastri, A.; Martins, C.M.; Pires, P.M.; Cortez, P. A Comparison of AutoML Tools for Machine Learning, Deep Learning and XGBoost. In Proceedings of the 2021 International Joint Conference on Neural Networks (IJCNN), Shenzhen, China, 18–22 July 2021; pp. 1–8.
47. Sun, A.Y.; Scanlon, B.R.; Save, H.; Rateb, A. Reconstruction of GRACE Total Water Storage Through Automated Machine Learning. *Water Resour. Res.* **2021**, *57*, e2020WR028666. [CrossRef]
48. Babaeian, E.; Paheding, S.; Siddique, N.; Devabhaktuni, V.K.; Tuller, M. Estimation of root zone soil moisture from ground and remotely sensed soil information with multisensor data fusion and automated machine learning. *Remote Sens. Environ.* **2021**, *260*, 112434. [CrossRef]
49. Cook, D. *Practical Machine Learning with H2O: Powerful, Scalable Techniques for Deep Learning and AI*; O'Reilly Media, Inc.: Sebastopol, CA, USA, 2016.
50. Laakso, T.; Kokkonen, T.; Mellin, I.; Vahala, R. Sewer Condition Prediction and Analysis of Explanatory Factors. *Water* **2018**, *10*, 1239. [CrossRef]
51. Padmanabhan, M.; Yuan, P.; Chada, G.; Nguyen, H.V. Physician-Friendly Machine Learning: A Case Study with Cardiovascular Disease Risk Prediction. *J. Clin. Med.* **2019**, *8*, 1050. [CrossRef]
52. Ou, C.; Liu, J.; Qian, Y.; Chong, W.; Liu, D.; He, X.; Zhang, X.; Duan, C.-Z. Automated Machine Learning Model Development for Intracranial Aneurysm Treatment Outcome Prediction: A Feasibility Study. *Front. Neurol.* **2021**, *12*, 735142. [CrossRef]

 sensors

Article

Study on Machine Learning Models for Building Resilience Evaluation in Mountainous Area: A Case Study of Banan District, Chongqing, China

Chi Zhang [1,2], Haijia Wen [1,2,*], Mingyong Liao [1,2], Yu Lin [1,2], Yang Wu [3] and Hui Zhang [4]

1 Key Laboratory of New Technology for Construction of Cities in Mountain Area, School of Civil Engineering, Ministry of Education, Chongqing 400044, China; zc158400@163.com (C.Z.); 201916131116@cqu.edu.cn (M.L.); FLyulin@163.com (Y.L.)
2 National Joint Engineering Research Center of Geohazards Prevention in the Reservoir Areas, Chongqing University, Chongqing 400045, China
3 China Railway Guizhou Tourism and Culture Development Co., Ltd., Guiyang 550000, China; cquwy55555@163.com
4 Investment Management Company of China Construction Fifth Bureau, Changsha 410007, China; 201816131097@cqu.edu.cn
* Correspondence: jhw@cqu.edu.cn; Tel.: +86-132-5132-1327

Abstract: 'Resilience' is a new concept in the research and application of urban construction. From the perspective of building adaptability in a mountainous environment and maintaining safety performance over time, this paper innovatively proposes machine learning methods for evaluating the resilience of buildings in a mountainous area. Firstly, after considering the comprehensive effects of geographical and geological conditions, meteorological and hydrological factors, environmental factors and building factors, the database of building resilience evaluation models in a mountainous area is constructed. Then, machine learning methods such as random forest and support vector machine are used to complete model training and optimization. Finally, the test data are substituted into models, and the models' effects are verified by the confusion matrix. The results show the following: (1) Twelve dominant impact factors are screened. (2) Through the screening of dominant factors, the models are comprehensively optimized. (3) The accuracy of the optimization models based on random forest and support vector machine are both 97.4%, and the F1 scores are greater than 94.4%. Resilience has important implications for risk prevention and the control of buildings in a mountainous environment.

Keywords: building resilience; machine learning; evaluation model; factor screening; model optimization

Citation: Zhang, C.; Wen, H.; Liao, M.; Lin, Y.; Wu, Y.; Zhang, H. Study on Machine Learning Models for Building Resilience Evaluation in Mountainous Area: A Case Study of Banan District, Chongqing, China. *Sensors* **2022**, *22*, 1163. https://doi.org/10.3390/s22031163

Academic Editors: Junwei Ma, Jie Dou and Ali Khenchaf

Received: 29 November 2021
Accepted: 1 February 2022
Published: 3 February 2022

1. Introduction

'Resilience', derived from the Latin word 'resilio' [1], was first introduced into the field of ecology by Holling [2] in the 1970s. Subsequently, scholars have broadened the definition of 'resilience' to various research fields [3–7]. Different research and application fields have different definitions [8,9], corresponding to different evaluation methods. In the fields of engineering and construction, resilience is the ability to absorb or avoid damage without suffering complete failure and is an objective of design, maintenance and restoration for buildings and infrastructure, as well as communities [10,11]. At present, there are different research methods regarding resilient cities and resilient communities [12], but most of them consider the assets (economy, society, environment and infrastructure) and functions (social capital, community function, transportation and communication links and planning) of the community. Buildings, an important part of infrastructure, are inevitably damaged to varying degrees in the actual use and operation process, resulting in property and even life loss. The building resilience in a mountainous area [13] can be understood as the ability of buildings that are under the conditions of the mountainous environment to still maintain their normal function and resist damage or to recover from the comprehensive effects of its various attributes, natural environment and the passage of time.

Many scholars have carried out a series of studies from different perspectives on the issue of building resilience. A resilience-based performance evaluation [14] is employed within a multiobjective optimization methodology for the design optimization of 4, 7, 10 and 15-story buildings under seismic hazard using both life span and conditional analyses. Himoto et al. [15] developed a computational framework using a multi-layer zone model to evaluate the fire resilience of buildings. Dong et al. [16] proposed a method to evaluate the seismic resilience of a steel structure considering economic, social and environmental aspects. For key infrastructure such as hospitals, Bruneau [17] explored the operational and physical resilience of acute care facilities.

In recent years, machine learning algorithms have attracted increasing attention in the field of risk assessment management [18–21]. Riedel et al. [22] carried out seismic vulnerability assessment of urban environments in moderate-to-low seismic hazard regions using association rule learning and support vector machine methods. Xie et al. [23] reviewed the promise of implementing machine learning in earthquake engineering. Some scholars have also used machine learning methods to study the building classification problem [24,25]. Several works [26] have described a hybrid information fusion approach to quantitatively evaluate the seismic resilience of Nepal by formulating nine indicators at the geological, building and social dimensions. Mangalathu et al. [27] used discriminant analysis, k-nearest neighbors, decision trees and random forests to study the damage degree of houses after an earthquake. Zhang et al. [28] used the support vector machine to study the physical resilience evaluation of landslide disasters in cities.

At present, the research on building resilience mainly considers the single factor effect represented by earthquakes. Moreover, it mainly considers the building structure, ignoring the complexity of the interaction between time and the internal and external factors of buildings combined. Studies on resilience in a mountainous environment are limited. Barua et al. [29] studied the resilience of rural mountain communities in relation to climate change and poverty in a mountainous region of India. Mountains are among the regions most affected by climate change [30,31], and climatic factors have an impact on building resilience. Meanwhile, the geographical and geological conditions in a mountainous environment are complex, and natural disasters such as collapse and landslide are likely to occur. Therefore, it is necessary to study the building resilience in mountainous environments specifically. With reference to the provisions of the technical guidelines for rural housing safety appraisal [32] and standards for dangerous building appraisal [33] in China, this paper classifies buildings as Grade I, II and III with regards to building resilience in mountainous areas (Table 1).

With the development of spatial and information technologies, a large amount of temporal and spatial data can be collected, processed and presented [34]. The objective of this study is to develop models for evaluating the resilience of mountainous buildings that take into account the combined effects of the various internal building properties, the natural environment and the passage of time. Firstly, the evaluation index system of building resilience in a mountainous area is constructed, and the dominant factors are screened using the feature recursive elimination method. Secondly, the building resilience models are completed by machine learning methods, including random forest and support vector machine, and the model evaluations are performed by confusion matrix. Finally, the predicted data are substituted into the model to obtain the classification evaluation of building resilience in the area to be studied. The original determination of the resilience grade requires a personal visit by professionals, which is labor-intensive. Through the machine learning method, the building resilience rating of the area to be studied can be determined quickly without visiting the site and without spending considerable time and manpower. This method provides additional value and reference significance in risk prevention and the control of buildings in a mountainous environment.

Table 1. Resilience grades of typical buildings.

Building Resilience Grades	Grade I	Grade II	Grade III
Grading criteria	Buildings whose structure is basically safe for use	Local dangerous buildings in which a part of the load-bearing structure cannot meet the requirements of safe use.	Whole dangerous buildings in which the load-bearing structure cannot meet the requirements of safe use.
Pictures from the scene			

2. Study Area

Banan District is located in the south of Chongqing central city, with an area of 1825 square kilometers and a built-up area of 84.5 square kilometers. It is a typical mountainous county. The gap between urban and rural areas is large, and there are huge differences in the quality of buildings and their ability to withstand natural environmental disasters. The selection of Banan District as the research area of building resilience in a mountainous area has high theoretical value and practical significance.

Our research team and Chongqing Municipal Public Housing Administration Office collected and analyzed data through field research. They obtained data from 1387 buildings in Banan District, including 122 buildings with resilience grade I, 352 buildings with resilience grade II and 913 buildings with resilience grade III. Figure 1 shows the geographical location of Banan District and the distribution of surveyed buildings.

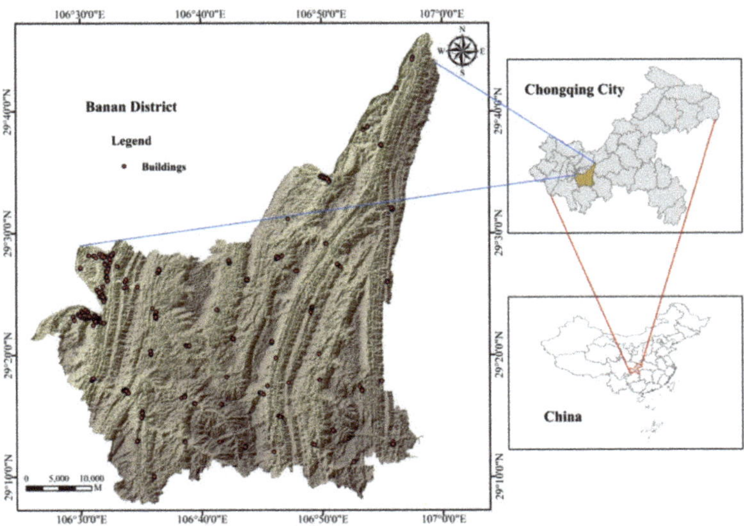

Figure 1. Location and buildings' distribution of Banan District.

3. Data and Methods

3.1. Data

3.1.1. Data Selection

The resilience of buildings in a mountainous area is affected by a combination of various internal and external factors, such as geographical and geological factors, meteorological and hydrological factors, environmental factors and building factors [35,36]. Based on the above four dimensions, 21 factors were select to establish the factor database of building resilience evaluation in a mountainous area. They are as follows: elevation, slope, slope aspect, slope position, curvature, plan curvature, profile curvature, micro-landform [37], terrain humidity index (TWI), terrain roughness index (TRI), lithology, average annual rainfall (AAR), aridity, temperature, distance from fault, distance from roads, distance from rivers, building structure, construction time, building storey and building category.

Geographical and geological factors fully consider the particularity of mountain building topography. Elevation affects climate and human activities. Slope affects the stress distribution of rocks and soil. Slope aspect and slope position influence hydrogeology. Curvature affects soil erosion through water flow on the slope. Plan curvature refers to the change rate of surface aspect at any point on the ground. Section curvature refers to the change rate of surface slope at any point on the ground. Micro-landform is a small terrain fluctuation with the surface complexity of large geomorphology, which affects the strength and weathering degree of rock and soil. TWI considers comprehensively the influence of terrain and soil characteristics on water distribution. TRI refers to the degree of concavity of the soil surface, reflecting the effects of wind and water erosion on the soil. Due to the different formation times and weathering degrees, the bearing capacity of different lithologies is also different. Meteorological and hydrological factors take into account the effect of time, average annual rainfall, aridity and temperature. They affect the durability of buildings. Environmental factors affect the original rock stress and slope stability of buildings through natural (fault, rivers) and human engineering activities (roads). Building factors are internal factors that lead to differences in housing quality and ability to resist natural disasters. Different building structures, categories, storeys and construction times lead to different building materials, weights and aging degrees.

3.1.2. Data Source

Data were obtained from 1387 buildings in Banan District, including the building structure, construction time, building storey and building category, through field investigation by the School of Civil Engineering of Chongqing University and Chongqing Municipal Public Housing Administration Office. DEM of ArcGIS was used to extract and process the data of slope, slope aspect, slope position, curvature, plan curvature, profile curvature, micro-landform, TWI and TRI. Other data sources, types and scale are shown in Table 2.

Table 2. Statistics of data sources.

Category	Data	Data Source	Scale
Geographical and geological factors	Elevation	ASTER	30 m
	Lithology	National Geological Archives of China	1:200000
Meteorological and hydrological factors	Average annual rainfall	China Meteorological Data Service Centre-Resource and Environment Science and Data Center	30 m
	Aridity	Resource and Environment Science and Data Center	500 m
	Temperature	Resource and Environment Science and Data Center	1000 m
Environmental factors	Fault	National Geological Archives of China	1:200,000
	Roads	Google remote sensing images	1:250,000
	Rivers	Google remote sensing images	1:250,000

3.1.3. Data Processing

The factors were quantified and reclassified. The continuous factors such as elevation, slope, curvature, plan curvature, profile curvature, TWI, TRI, average annual rainfall, aridity and temperature were classified by ArcGIS natural breaks method (Jenks). The 360° was divided into eight regions on average, and the flat was assigned separately, so the slope aspect was divided into nine categories. The distances from fault, roads and rivers were obtained by multiple ring buffer of fault, roads and rivers, respectively, through ArcGIS. The qualitative factors such as slope position, micro-landform, lithology, building structure and building category were classified according to their respective characteristics. In this paper, building structure categories were distinguished mainly based on building materials. The structures, which include timber structure, adobe–timber structure, brick–timber structure, brick–concrete structure, as well as steel and reinforced concrete structure, were named directly using the names of materials. The simple structure referred to the building with simple materials such as brick or wood panels. In addition, only a few buildings built of stone–timber and stone–concrete materials were situated in the study area, which were collectively referred to as mixed structures. The construction time was grouped by a minimum of ten years based on data distribution. The building storey adopted the original data. According to their different uses, the buildings in this paper were divided into several categories, including residential building, commercial building, teaching building, auxiliary building and other building. Auxiliary buildings refer to buildings with auxiliary functions as their main purpose. For rural areas, they include buildings such as toilets and those used for storage of agricultural production tools, breeding of farm animals, drying and storage of food crops, etc. For urban areas, they comprise buildings such as public toilets, gatehouses, those used for auxiliary housing and public services, etc. Buildings that did not meet the above criteria were classified as other buildings. The reclassification of impact factors is shown in Table 3.

Table 3. Reclassification of impact Factors.

Category	Impact Factors	Number of Categories	Classification Criteria
Geographical and geological factors	Elevation (m)	9	(1) ≤244; (2) 244~312; (3) 312~377; (4) 377~448; (5) 448~525; (6) 525~605; (7) 605~691; (8) 691~802; (9) ≥802
	Slope (°)	9	(1) ≤5.03°; (2) 5.03°~8.70°; (3) 8.70°~12.33°; (4) 12.33°~16.07°; (5) 16.07°~20.08°; (6) 20.08~24.57; (7) 24.57~29.88; (8) 2 9.88~36.94; (9) ≥36.94
	Slope aspect	9	(1) Flat; (2) N; (3) NE; (4) E; (5) SE; (6) S; (7) SW; (8) W; (9) NW
	Slope position	6	(1) Valleys; (2) Lowslope; (3) Flat; (4) Midslope; (5) Uppslope; (6) Ridge
	Curvature	9	(1) ≤−4.09; (2) −4.09~−2.46; (3) −2.46~−1.29; (4) −1.29~−0.47; (5) −0.47~0.35; (6) 0.35~1.17; (7) 1.17~2.24; (8) 2.24~4.09; (9) ≥4.09
	Plan curvature	9	(1) ≤−1.97; (2) −1.97~−1.21; (3) −1.21~−0.65; (4) −0.65~−0.23; (5) −0.23~0.19; (6) 0.19~0.61; (7) 0.61~1.17; (8) 1.17~2.00; (9) ≥2.00
	Profile curvature	9	(1) ≤−2.88; (2) −2.88~−1.70; (3) −1.70~−0.95; (4) −0.95–0.41; (5) −0.41~0.12; (6) 0.12~0.66; (7) 0.66~1.41; (8) 1.41~2.59; (9) ≥2.59
	Micro-landform	10	(1) Canyons, deeply incised streams; (2) Midslope drainages, shallow valleys; (3) Upland drainages, headwaters; (4) U-shape valleys; (5) Plains; (6) Open slopes; (7) Upper slopes, mesas; (8) Local ridges hills in valleys; (9) Midslope ridges, small hills in plains; (10) Mountain tops, high ridges

Table 3. *Cont.*

Category	Impact Factors	Number of Categories	Classification Criteria
Geographical and geological factors	TWI	9	(1) ≤4.68; (2) 4.68~5.87; (3) 5.87~7.16; (4) 7.16~8.56; (5) 8.56~10.18; (6) 10.18~12.12; (7) 12.12~14.71; (8) 14.71~17.95; (9) ≥17.95
	TRI	9	(1) ≤1.018; (2) 1.018~1.041; (3) 1.041~1.071; (4) 1.071~1.108; (5) 1.108~1.155; (6) 1.155~1.217; (7) 1.217~1.304; (8) 1.304~1.450; (9) ≥1.450
	Lithology	7	(1) Lower Triassic; (2) Middle Triassic; (3) Upper Triassic; (4) Triassic; (5) Middle-Lower Jurassic; (6) Middle Jurassic; (7) Upper Jurassic
Meteorological and hydrological factors	Average annual rainfall (mm)	9	(1) ≤117.0; (2) 117.0~119.2; (3) 119.2~120.7; (4) 120.7~122.3; (5) 122.3~124.0; (6) 124.0~125.8; (7) 125.8~127.7; (8) 127.7~129.9; (9) ≥129.9
	Aridity	9	(1) ≤0.808; (2) 0.808~0.828; (3) 0.828~0.852; (4) 0.852~0.881; (5) 0.881~0.907; (6) 0.907~0.927; (7) 0.927~0.948; (8) 0.948~0.971; (9) ≥0.971
Environmental factors	Temperature (°)	9	(1) ≤16.214; (2) 16.214~16.889; (3) 16.889~17.401; (4) 17.401~17.807; (5) 17.807~18.139; (6) 18.139~18.431; (7) 18.431~18.715; (8) 18.715~19.048; (9) ≥19.048
	Distance from fault (m)	6	(1) ≤1000; (2) 1000~2000; (3) 2000~3000; (4) 3000~4000; (5) 4000~5000; (6) ≥ 5000
	Distance from roads (m)	6	(1) ≤10; (2) 10~20; (3) 20~30; (4) 30~40; (5) 40~50; (6) ≥ 50
	Distance from rivers (m)	6	(1) ≤100; (2) 100~200; (3) 200~300; (4) 300~400; (5) 400~500; (6) ≥500
Building factors	Building structure	7	(1) Timber structure; (2) Simple structure; (3) Adobe–timber structure; (4) Brick–timber structure; (5) Brick–concrete structure; (6) Hybrid structure; (7) Steel and reinforced concrete structure
	Construction time	7	(1) before 1939; (2) 1940~1949; (3) 1950~1959; (4) 1960~1969; (5) 1970~1979; (6) 1980~1999; (7) after 2000;
	Building storey	8	(1) 1; (2) 2; (3) 3; (4) 4; (5) 5; (6) 6; (7) 7; (8) ≥8;
	Building category	5	(1) Residential building; (2) Commercial building; (3) Teaching building; (4) Auxiliary building; (5) Other building

After reclassification, the impact factors' data were normalized. All values were normalized to the distribution between (0,1). All factors were in the same order of magnitude in order to facilitate correct and rapid modelling. The normalization formula is denoted as follows

$$X^* = (X - X_{min})/(X_{max} - X_{min}) \quad (1)$$

In the formula, X^* is the normalized data, X is the original data, X_{max} and X_{min} are the maximum and minimum of the data, respectively.

For better data management and visual representation, the corresponding thematic layers were constructed by ArcGIS, as shown in Figure 2. The specific distribution of the impact factors of geographical and geological factors, meteorological and hydrological factors, environmental factors and building factors can be displayed visually. However, due to the small building area and the large study area, the buildings were only shown as points under the full view of the study area. Figure 2r shows the construction of building factor layers of ArcGIS with the building storey as an example. The attribute table corresponding to the building recorded all the information of each building, including building structure, construction time, building storey and building category. Changing its fields in properties switches it to other building factor layers.

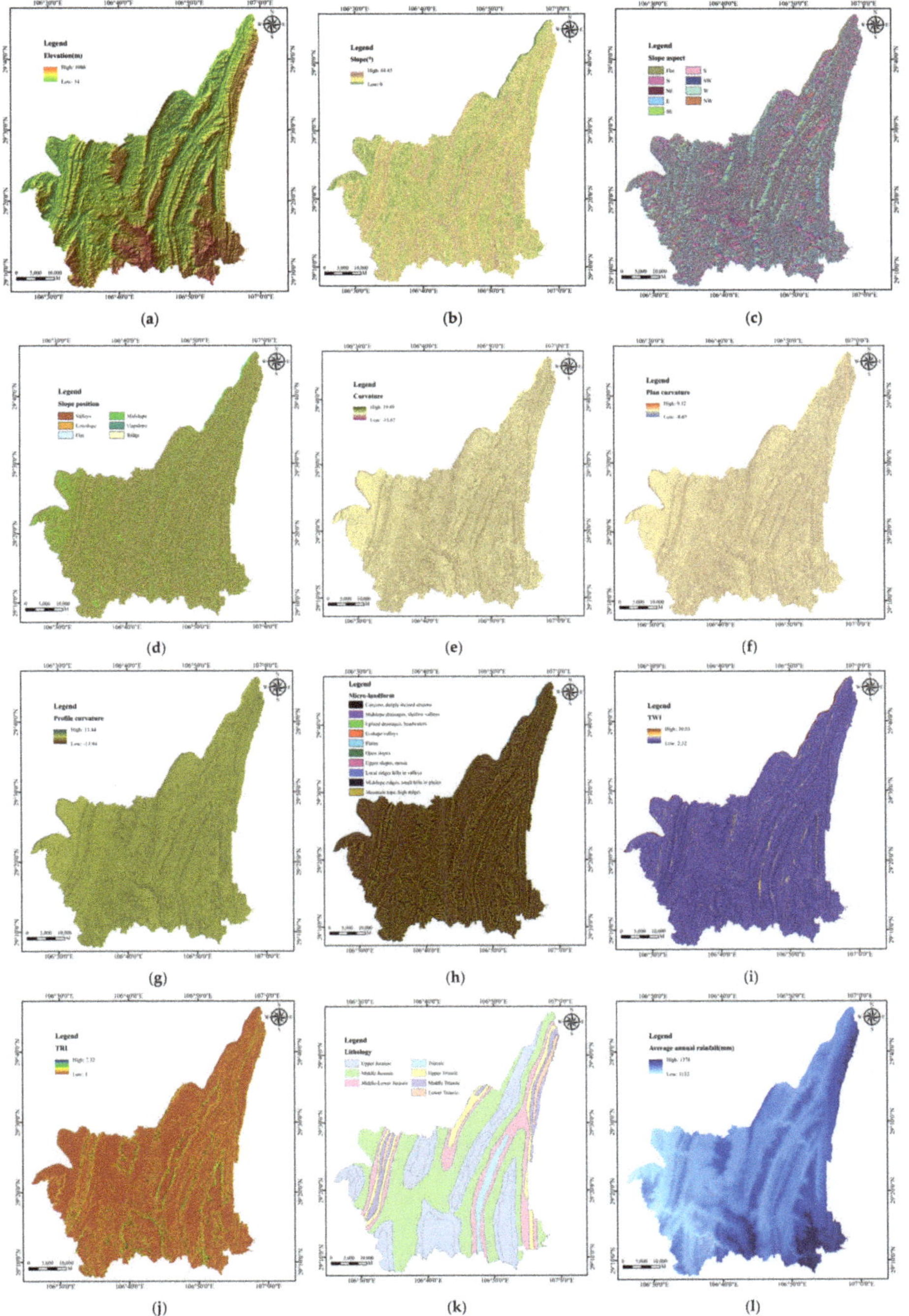

Figure 2. *Cont.*

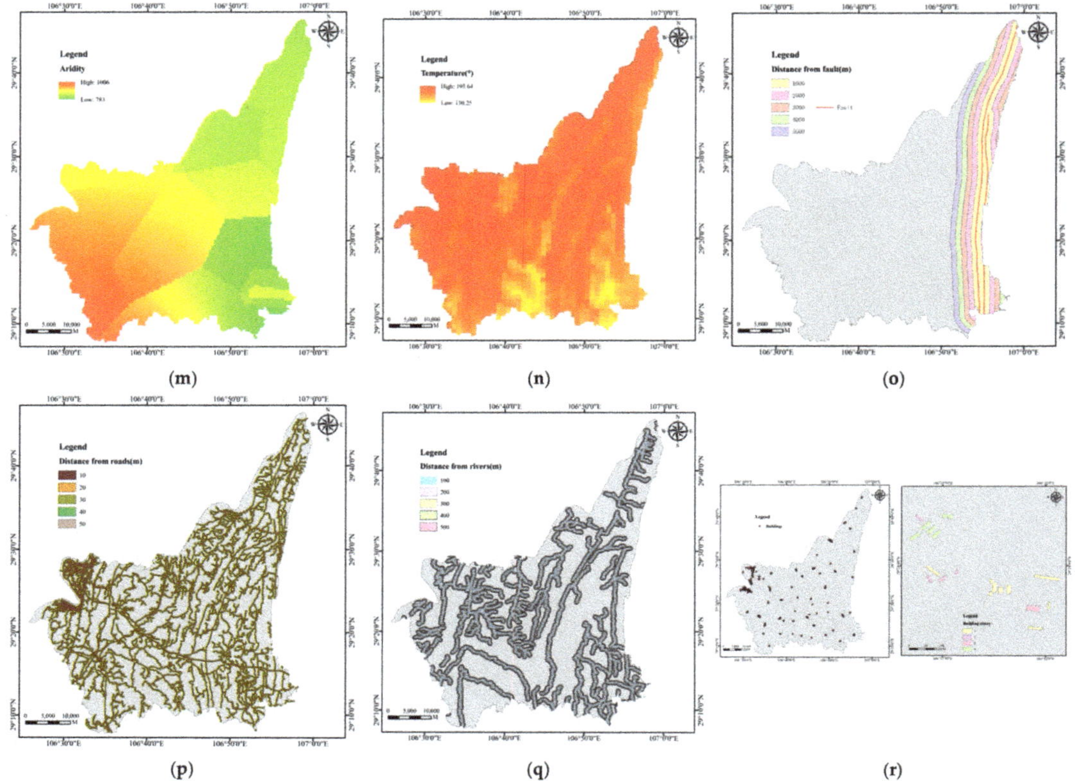

Figure 2. Thematic layers of impact factors: (**a**) Elevation; (**b**) Slope; (**c**) Slope aspect; (**d**) Slope position; (**e**) Curvature; (**f**) Plan curvature; (**g**) Profile curvature; (**h**) Micro-landform; (**i**) TWI; (**j**) TRI; (**k**) Lithology; (**l**) Average annual rainfall; (**m**) Aridity; (**n**) Temperature; (**o**) Distance from fault; (**p**) Distance from roads; (**q**) Distance from rivers; (**r**) Building factors.

3.2. Methodology

3.2.1. Random Forest

Random forest (RF) is a data mining algorithm that contains multiple decision trees. Based on each decision tree, the final classification result is obtained by voting [38]. Random forest model has strong robustness and accuracy in data processing. This study selected random forest as one of the processing algorithms of the model.

By calling the random forest program package through R language, the data obtained from the 1387 buildings containing all the information of influencing factors in the study area were regarded as the total samples, which were randomly divided into 971 training samples and 416 test samples according to the ratio of 7:3. The ratio of 7:3 is an empirical value that has been used by many researchers. The optimal parameter *mtry* was selected by cyclic iteration, and it was substituted into the code to view the error stability of the model and find the optimal *ntree*. *Mtry* refers to the number of variables used for binary trees in nodes, and *ntree* refers to the number of decision trees contained in random forests.

3.2.2. Support Vector Machine

In recent years, many scholars have carried out in-depth research on disaster risk assessment using the support vector machine (SVM) algorithm [39–41]. The basic idea is to use kernel function to project nonlinear separable samples into high-dimensional space to construct linear separable samples. According to the spatial distribution of sample features, the optimal hyperplane solution with the farthest distance between the two groups of classifications was found, so as to correctly divide the data set. This project used the ksvm function of kernlab software package [42]. For the three-classification problem, ksvm used 'one-to-one' method to construct three secondary classifiers by permutation and combination, and judged the resilience grade of buildings in mountainous area by voting. In this study, the SVM model was selected as another prediction model to measure the reliability of the RF model.

In the support vector machine model, the parameters were also optimized first. The kernlab package was called by R language, and the optimal parameter combination *sigma* and *C* value were selected in the for-loop iteration through the tenfold cross validation. *Sigma* determines the width of the kernel function, and *C* refers to the tolerance of allowing classification errors. Then, the above optimal parameter combination was substituted to establish the model.

3.2.3. Feature Recursive Elimination

In machine learning, not all the results of variable prediction are related. Some irrelevant variables may have a negative impact on the model prediction accuracy. Through feature selection, the results of model effect optimization can be achieved. The main idea of feature recursive elimination method is to eliminate the factor with the smallest ranking criterion score at each time on the basis of all the initial influencing factors and to construct the model repeatedly until the final feature set is obtained [43]; the ranking of features is obtained at the same time.

3.2.4. Model Evaluation Methods

In this paper, the resilience of buildings in mountainous area is divided into grades I, II and III. The prediction effect is analyzed by confusion matrix analysis model. The confusion matrix is an error matrix that measures the predicted and actual values, which can be used to evaluate the accuracy and stability of machine learning algorithms. In order to simplify the expression, the data are referred to by the combination of the real value before and the predicted value after (Table 4). N_{ij} (I = 1,2,3; j = 1,2, 3) represents the number of samples that actually belong to i but are predicted to be j [44].

Table 4. Three-classification confusion matrix.

		Predicted Grade		
		I	**II**	**III**
Actual grade	I	N_{11}	N_{12}	N_{13}
	II	N_{21}	N_{22}	N_{23}
	III	N_{31}	N_{32}	N_{33}

Accuracy rate refers to the proportion of samples with correct prediction, considering the total samples. It is the most basic, intuitive and simple method to measure the evaluation effect of classification model. Precision refers to the proportion of the true values of a grade, considering all the samples predicted as a certain grade, reflecting the precision of the model prediction. Recall rate represents the proportion that is predicted accurately in the actual sample of a certain grade. In order to take both precision and recall into account, the harmonic mean F1 score was used as another reference index. The calculation formulas are as follows

$$Accuracy = \sum_{i=1}^{3} N_{ii} / \sum_{i=1}^{3} \sum_{j=1}^{3} N_{ij} \tag{2}$$

$$Precision_i = N_{ii} / \sum_{k=1}^{3} N_{ki} \tag{3}$$

$$Recall_i = N_{ii} / \sum_{k=1}^{3} N_{ik} \tag{4}$$

$$F_1score = 2 \times Precision_i \times Recall_i / (Precision_i + Recall_i) \tag{5}$$

4. Results and Discussion

4.1. Optimization Models of Building Resilience Based on Dominant Factors

4.1.1. Screening of Dominant Factors

This paper selected the feature recursive elimination (FRE) method to filter the dominant factors for model optimization. Based on the R language call code, when the number of impact factors was 12, the model worked best (Figure 3). The dominant factors screened were elevation, lithology, TRI, aridity, temperature, average annual rainfall, distance from roads, distance from rivers, building structure, building category, construction time and building storey.

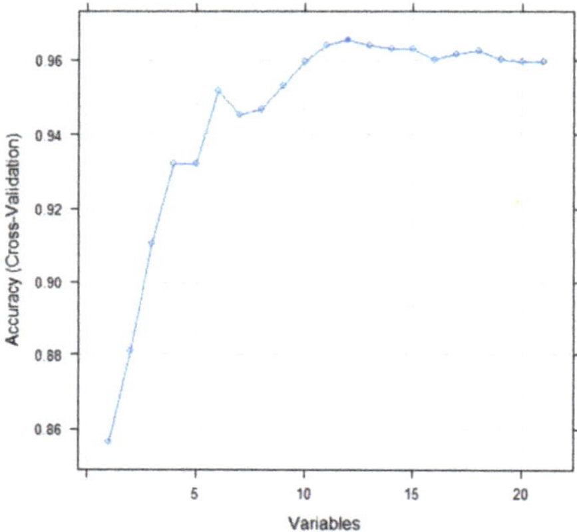

Figure 3. Screening diagram of dominant factors by using FRE.

4.1.2. Optimization models' results of building resilience based on dominant factors

The dominant factor was used as input layer, while mountain building resilience grade was used as output layer. After debugging, in the random forest model, the optimal parameters *mtry* = 8 and *ntree* = 1000 were selected. In the support vector machine model, the optimal parameter combination kpar = list (*sigma* = 0.21) and C = 5 were selected. Thus, the confusion matrix of the prediction results of the training samples, test samples and total samples based on the random forest and support vector machine algorithm was obtained (Figure 4). The nine data in the matrix center are the direct output results of the confusion matrix. The three data on the left side of the last line are the precision of building resilience grades I, II and III. The three data above the last column are the recall of their respective grades. The data in the bottom-right corner are the model accuracy.

Based on random forest and support vector machine, the accuracies of the building resilience optimization models in mountainous area are calculated using training samples, test samples and total samples, respectively. For training samples, the model accuracies based on random forest and support vector machine are 99.7% and 98.7%, respectively. For test samples, both are 97.4%; for the total samples, they are 99.0% and 98.3%, respectively. Accuracy is a metric in confusion matrix for evaluating the mountainous building resilience model, and a larger accuracy rate indicates a better model. Observing the precision of the model in the test samples, RF and SVM are very good in the prediction of grade I buildings. In the prediction of grade II buildings, the random forest model is better than the support vector machine model. The support vector machine model is better in the prediction of grade III buildings. All precisions are above 94.9%. Observing the recall of the model in the test samples, RF and SVM are very good in the prediction of grade I buildings. In the prediction of grade II buildings, the SVM model is better than the RF model, while the RF model is better in the prediction of grade III buildings. All recalls are above 93.0%. The F1 score comprehensively considers the precision and recall. The two models have good prediction effect on grade I buildings. There are occasional misjudgments in grade II and grade III buildings, but all values are greater than 94.4%.

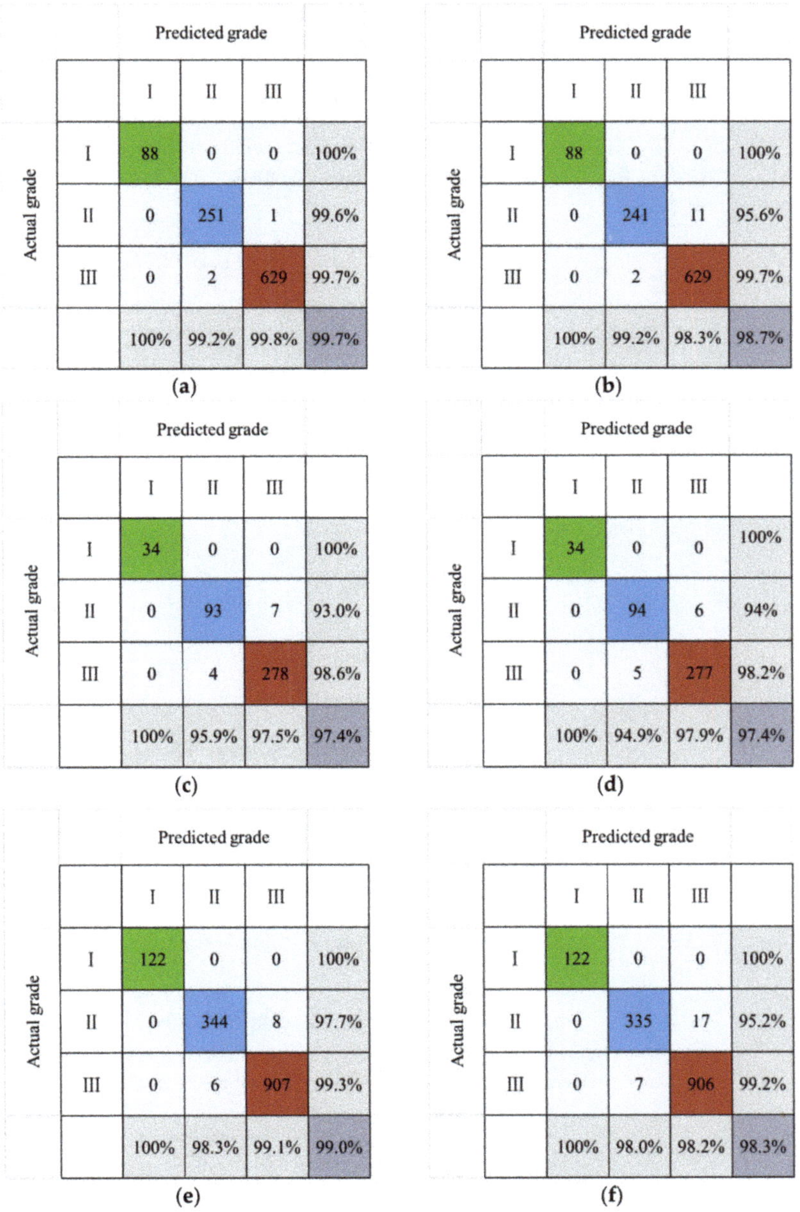

Figure 4. Confusion matrices of optimization models based on machine learning: (**a**) Training samples-RF; (**b**) Training samples-SVM; (**c**) Test samples-RF; (**d**) Test samples-SVM; (**e**) Total samples-RF; (**f**) Total samples-SVM.

In summary, the prediction accuracy, precision, recall and F1 scores of random forest and support vector machine are high, which proves that the machine learning method is reliable for resilience evaluation of buildings in mountainous area.

4.2. Optimization Effect Comparison

The training samples were used to construct the model, and all the evaluation indexes in the confusion matrix are the maximum values. The total samples cover part of the modelling data, and the values are between the training samples and the test samples. The test samples do not participate in model building, but can better detect model performance. The effects of model optimization are analyzed for the test samples.

Accuracy is the most basic evaluation index of the model. After optimization, the accuracy of the random forest model was improved from 95.7% to 97.4%, and the accuracy of the support vector machine model was improved from 95.4% to 97.4% (Figure 5).

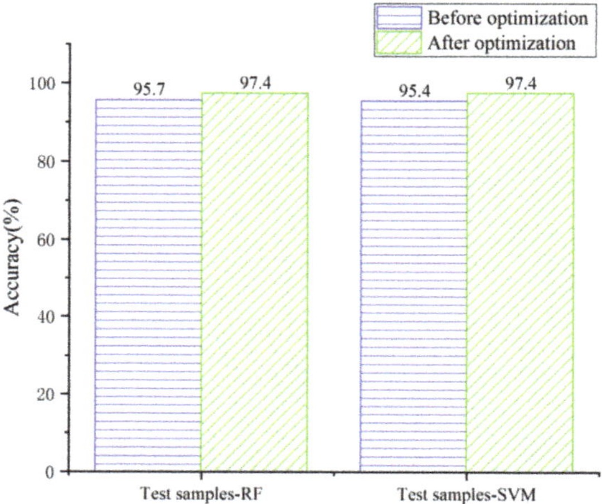

Figure 5. Comparison of test samples' accuracy before and after optimization.

As shown in Figure 6, compared with the pre-optimization state, the minimum value of each index of the model based on the dominant factors' screening improved from 88% to 93%. The model effect was comprehensively improved. The best optimization effect of SVM was that the precision of grade II increased by 5.6%, and the best optimization effect of RF was that the recall of grade II increased by 5%. The range of variation of indicators for each building's resilience grade was inconsistent, which may be due to the quantity and quality of the data themselves. The two machine learning algorithms have different emphases on model optimization but the effects are remarkable.

4.3. Discussion

4.3.1. Comparison of Two Machine Learning Models

In the test samples, the evaluation indexes of RF and SVM optimization models were compared (Table 5). It was observed that the two machine learning methods have the same evaluation results for accuracy rate, recall, F1 score of grade I buildings and F1 score of grade III buildings. The RF model is superior to the SVM model in the evaluation of the precision of grade II buildings and the recall of grade III buildings. The SVM model is better than the RF model in the evaluation of grade III buildings' precision, grade II buildings' recall and F1 score. Both methods have advantages and disadvantages in each evaluation index, but the absolute value of the difference does not exceed 1%. It was proved that RF and SVM are reliable in the evaluation of building resilience in a mountainous area.

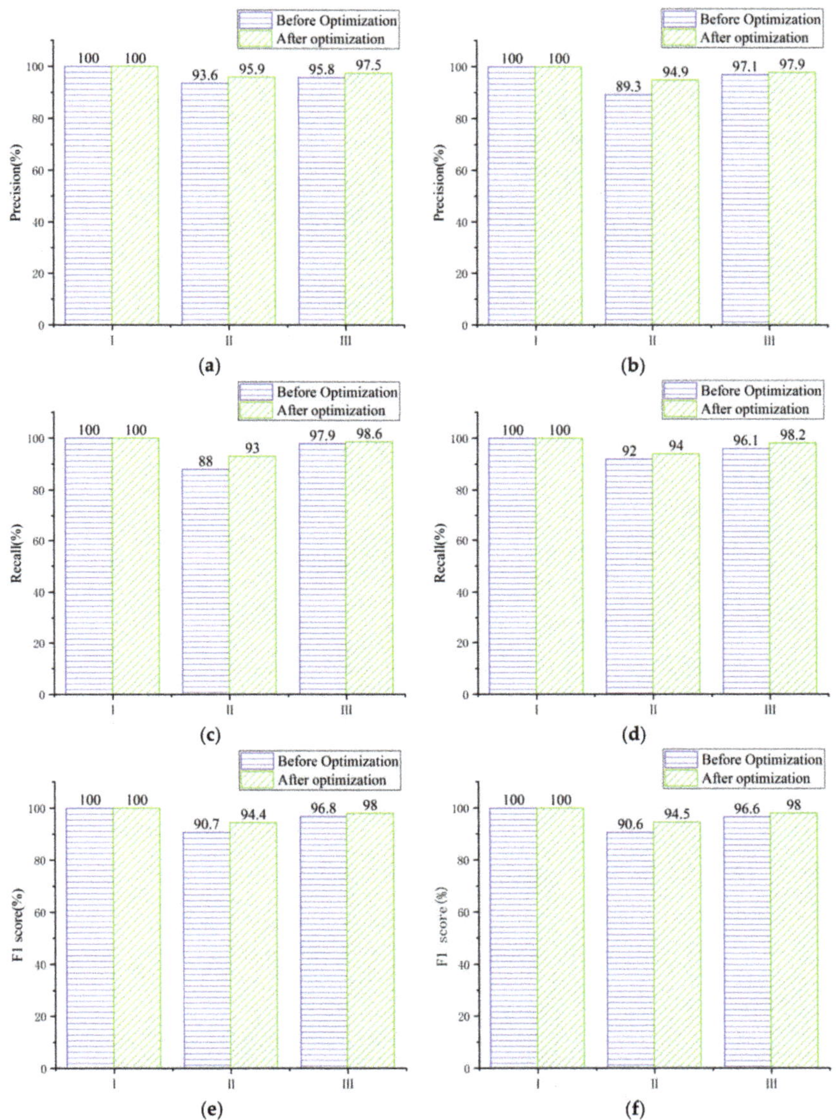

Figure 6. Comparison of test samples' evaluation indexes before and after optimization: (**a**) Precision-RF; (**b**) Precision-SVM; (**c**) Recall-RF; (**d**) Recall–SVM; (**e**) F_1 score-RF; (**f**) F_1 score-SVM.

Table 5. RF and SVM optimization model evaluation indexes for the test samples.

	Accuracy	Precision			Recall			F_1 score		
		I	II	III	I	II	III	I	II	III
RF	97.4%	100%	95.9%	97.5%	100%	93%	98.6%	100%	94.4%	98.0%
SVM	97.4%	100%	94.9%	97.9%	100%	94%	98.2%	100%	94.5%	98.0%
Difference	0	0	1%	0.4%	0	1%	0.4%	0	0.1%	0

4.3.2. Importance of Resilience Impact Factors

The importance ranking of impact factors reflects the contribution of variables to the resilience evaluation model of buildings in a mountainous area. Random forest provides two methods for ranking the importance of features: Mean Decrease Accuracy (MDA) and Mean Decrease Gini (MDG) [45]. MDA is the change in the error rate of model results caused by disrupting the value of an impact factor in the test set. MDG is the sum of all decreases in Gini impurity due to a given variable. Based on the study by Han et al. [46], this paper combined MDA and MDG for a comprehensive measure. The 12 variables were assigned scores of 12, 11, . . . , 2, and 1 based on the values of MDA and MDG from highest to lowest, respectively. The scores obtained from both were then added and re-ranked to obtain the combined ranking results of the importance of the influencing factors (Table 6).

Table 6. Ranking the importance of impact factors.

Category	Impact Factors	Value of MDA	Score of MDA	Value of MDG	Score of MDG	Score of MDA and MDG	Comprehensive Ranking
Geographical and geological factors	Elevation	27.15	4	6.81	2	6	10
	TRI	86.78	11	91.35	11	22	2
	Lithology	37.78	7	11.50	3	10	8
Meteorological and hydrological factors	Average annual rainfall	18.16	3	5.24	1	4	12
	Aridity	44.21	9	16.84	7	16	4
	Temperature	42.91	8	13.10	6	14	5
Environmental factors	Distance from roads	11.72	1	12.02	4	5	11
	Distance from rivers	33.80	6	13.00	5	11	7
Building factors	Building structure	91.26	12	170.82	12	24	1
	Construction time	33.02	5	50.37	9	14	5
	Building storey	16.66	2	23.27	8	10	8
	Building category	56.82	10	68.79	10	20	3

The results are in the following order: building structure, TRI, building category, aridity, construction time, temperature, distance from rivers, lithology, building storey, elevation, distance from roads and average annual rainfall. For the optimized dominant factor index, all building factors, all meteorological and hydrological factors, three geographical and geological factors, and two environmental factors are selected, which are comprehensive and representative. The alternate arrangement of internal and external factors fully illustrates the necessity of exploring the combined effect of various factors on buildings in a mountainous area. Figure 7 shows the degree of importance of each impact factor clearly.

4.3.3. Model Improvement Options

The building resilience models in a mountainous area work well, but there is still room for improvement. Regarding improvement from the perspective of impact factors, more impact factors such as extreme temperature should be considered in the preliminary selection stage. Moreover, regarding improvement from the perspective of machine learning methods, data imbalance should be the focus in subsequent research. The classification algorithm will produce a certain bias when processing the data set according to the amount of data in different categories. For unbalanced data sets, assigning different weights for processing should be considered.

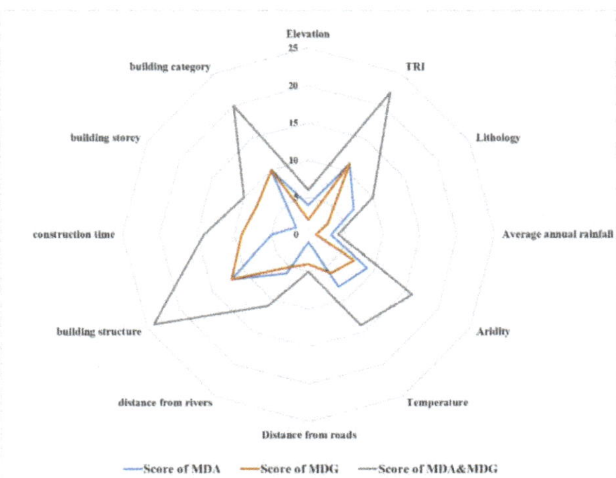

Figure 7. Impact factors' assignment score chart.

5. Conclusions

Based on machine learning, this paper proposed a resilience evaluation method for buildings in a mountainous area. Considering the multi-dimensional effects of geographical and geological conditions, meteorological and hydrological factors, environmental factors and building factors, the database of impact factors was constructed. The models were trained and optimized by machine learning methods, including random forest and support vector machine, and the resilience evaluation models of buildings in a mountainous area were established. Then, the predicted data were substituted into the model to obtain the classification evaluation of building resilience in the area to be studied.

(1) By combining MDA and MDG to form a comprehensive measure, the impact factors of the optimization models were ranked in order of importance: building structure, TRI, building category, aridity, construction time, temperature, distance from rivers, lithology, building storey, elevation, distance from roads and average annual rainfall. In the respective rankings of MDA and MDG, the impact factors in the top three rankings are the same, and the remaining impact factors tend to differ between the two. The alternate arrangement of internal and external factors fully illustrates the necessity of exploring the combined effect of various factors on buildings in a mountainous area.

(2) Through the screening of dominant factors, the minimum value of each index in the model test sets was increased from 88% to 93%, the models were comprehensively optimized, demonstrating the need for factor screening. The two machine learning algorithms have different emphases on model optimization, but the effects were remarkable.

(3) The accuracy of the optimization models based on random forest and support vector machine were both 97.4%, and the F1 scores were greater than 94.4%, which proves that the machine learning method is reliable for resilience evaluation of buildings in a mountainous area. This study has the advantages of accuracy, efficiency and visualization. It provides additional value and reference significance in risk prevention and the control of mountainous environment building construction.

Author Contributions: Conceptualization, H.W.; methodology, C.Z.; formal analysis, C.Z.; investigation, M.L., C.Z., Y.L., Y.W. and H.Z.; resources, H.W.; data curation, H.W. and C.Z.; writing—original draft preparation, C.Z.; supervision, H.W.; funding acquisition, H.W. All authors have read and agreed to the published version of the manuscript.

Funding: This research was funded by Key Technologies Research and Development Program, grant number 2018YFC1505501 and Chongqing Science and Technology Commission, grant number cstc2018jscx-msybX0310.

Institutional Review Board Statement: Not applicable.

Informed Consent Statement: Not applicable.

Data Availability Statement: Data sharing not applicable.

Acknowledgments: Thanks to Chongqing Municipal Public Housing Administration Office for providing some of the building data.

Conflicts of Interest: The authors declare no conflict of interest.

References

1. Cimellaro, G.P.; Reinhorn, A.M.; Bruneau, M. Framework for analytical quantification of disaster resilience. *Eng. Struct.* **2010**, *32*, 3639–3649. [CrossRef]
2. Holling, C.S. Resilience and stability of ecological systems. *Annu. Rev. Ecol. Syst.* **1973**, *4*, 1–23. [CrossRef]
3. Wildavsky, A.B. Searching for Safety. *J. Risk Insur.* **1988**, *57*, 564. [CrossRef]
4. ZHOU, Y.; LU, X. State-of-the-art on rocking and self-centering structures. *J. Build. Struct.* **2011**, *32*, 1–10. [CrossRef]
5. Adger, W.N. Social and ecological resilience: Are they related? *Prog. Hum. Geog.* **2000**, *24*, 347–364. [CrossRef]
6. Folke, C. Resilience: The emergence of a perspective for social–ecological systems analyses. *Glob. Environ. Change* **2006**, *16*, 253–267. [CrossRef]
7. Rasulo, A.; Pelle, A.; Briseghella, B.; Nuti, C. A Resilience-Based Model for the Seismic Assessment of the Functionality of Road Networks Affected by Bridge Damage and Restoration. *Infrastructures* **2021**, *6*, 112. [CrossRef]
8. Park, J.; Seager, T.P.; Rao, P.S.C.; Convertino, M.; Linkov, I. Integrating Risk and Resilience Approaches to Catastrophe Management in Engineering Systems. *Risk Anal* **2013**, *33*, 356–367. [CrossRef] [PubMed]
9. Rose, A. Economic resilience to natural and man-made disasters: Multidisciplinary origins and contextual dimensions. *Environ. Hazards* **2007**, *7*, 383–398. [CrossRef]
10. Jennings, B.J.; Vugrin, E.D.; Belasich, D.K. Resilience certification for commercial buildings: A study of stakeholder perspectives. *Environ. Syst. Decis.* **2013**, *33*, 184–194. [CrossRef]
11. Herrera, M.; Abraham, E.; Stoianov, I. A Graph-Theoretic Framework for Assessing the Resilience of Sectorised Water Distribution Networks. *Water Resour. Manag.* **2016**, *30*, 1685–1699. [CrossRef]
12. Marasco, S.; Cardoni, A.; Zamani Noori, A.; Kammouh, O.; Domaneschi, M.; Cimellaro, G.P. Integrated platform to assess seismic resilience at the community level. *Sustain. Cities Soc.* **2021**, *64*, 102506. [CrossRef]
13. Wen, H.; Xie, P.; Xie, Q.; Hu, J.; Wu, S. An Evaluation Method Based on Big Data for Building Resilience in Mountainous Area. Chinese Patent CN2018107915103 [P/OL], 8 November 2021. Available online: https://kns.cnki.net/kcms/detail/detail.aspx?FileName=CN109214643B&DbName=SCPD2021 (accessed on 10 October 2021).
14. Joyner, M.D.; Gardner, C.; Puentes, B.; Sasani, M. Resilience-Based seismic design of buildings through multiobjective optimization. *Eng. Struct.* **2021**, *246*, 113024. [CrossRef]
15. Himoto, K.; Suzuki, K. Computational framework for assessing the fire resilience of buildings using the multi-layer zone model. *Reliab. Eng. Syst. Safe* **2021**, *216*, 108023. [CrossRef]
16. Dong, Y.; Frangopol, D.M. Performance-based seismic assessment of conventional and base-isolated steel buildings including environmental impact and resilience. *Earthq. Eng. Struct. D* **2016**, *45*, 739–756. [CrossRef]
17. Bruneau, M.; Reinhorn, A. Exploring the concept of seismic resilience for acute care facilities. *Earthq. Spectra* **2007**, *23*, 41–62. [CrossRef]
18. Liu, D.; Fan, Z.; Fu, Q.; Li, M.; Faiz, M.A.; Ali, S.; Li, T.; Zhang, L.; Khan, M.I. Random forest regression evaluation model of regional flood disaster resilience based on the whale optimization algorithm. *J. Clean. Prod.* **2020**, *250*, 119468. [CrossRef]
19. Deng, W.; Zhou, J. Approach for feature weighted support vector machine and its application in flood disaster evaluation. *Disaster Adv.* **2013**, *6*, 51–58.
20. Motta, M.; de Castro Neto, M.; Sarmento, P. A mixed approach for urban flood prediction using Machine Learning and GIS. *Int. J. Disast. Risk Re.* **2021**, *56*, 102154. [CrossRef]
21. Riedel, I.; Guéguen, P.; Dunand, F.; Cottaz, S. Macroscale vulnerability assessment of cities using association rule learning. *Seismol. Res. Lett.* **2014**, *85*, 295–305. [CrossRef]
22. Riedel, I.; Guéguen, P.; Dalla Mura, M.; Pathier, E.; Leduc, T.; Chanussot, J. Seismic vulnerability assessment of urban environments in moderate-to-low seismic hazard regions using association rule learning and support vector machine methods. *Nat. Hazards* **2015**, *76*, 1111–1141. [CrossRef]
23. Xie, Y.; Ebad Sichani, M.; Padgett, J.E.; DesRoches, R. The promise of implementing machine learning in earthquake engineering: A state-of-the-art review. *Earthq. Spectra* **2020**, *36*, 1769–1801. [CrossRef]

24. Liuzzi, M.; Aravena Pelizari, P.; Geiß, C.; Masi, A.; Tramutoli, V.; Taubenböck, H. A transferable remote sensing approach to classify building structural types for seismic risk analyses: The case of Val d'Agri area (Italy). *B Earthq. Eng.* **2019**, *17*, 4825–4853. [CrossRef]
25. Mangalathu, S.; Burton, H.V. Deep learning-based classification of earthquake-impacted buildings using textual damage descriptions. *Int. J. Disast. Risk Re.* **2019**, *36*, 101111. [CrossRef]
26. Chen, W.; Zhang, L. Resilience assessment of regional areas against earthquakes using multi-source information fusion. *Reliab. Eng. Syst. Safe* **2021**, *215*, 107833. [CrossRef]
27. Mangalathu, S.; Sun, H.; Nweke, C.C.; Yi, Z.; Burton, H.V. Classifying earthquake damage to buildings using machine learning. *Earthq. Spectra* **2020**, *36*, 183–208. [CrossRef]
28. Zhang, X.; Song, J.; Peng, J.; Wu, J. Landslides-oriented urban disaster resilience assessment—A case study in Shen Zhen, China. *Sci. Total Environ.* **2019**, *661*, 95–106. [CrossRef]
29. Barua, A.; Katyaini, S.; Mili, B.; Gooch, P. Climate change and poverty: Building resilience of rural mountain communities in South Sikkim, Eastern Himalaya, India. *Reg. Environ. Change* **2014**, *14*, 267–280. [CrossRef]
30. Kohler, T.; Giger, M.; Hurni, H.; Ott, C.; Wiesmann, U.; Wymann Von Dach, S.; Maselli, D. Mountains and Climate Change: A Global Concern. *Mt. Res. Dev.* **2010**, *30*, 53–55. [CrossRef]
31. Kato, T.; Rambali, M.; Blanco-Gonzalez, V. *Strengthening Climate Resilience in Mountainous Areas*; OECD Development Co-operation Working Papers, No. 104; OECD Publishing: Paris, France, 2021.
32. Ministry of Housing and Urban-Rural Development of the People's Republic of China Web Page. Available online: http://www.mohurd.gov.cn/gongkai/fdzdgknr/tzgg/201912/20191202_242931.html (accessed on 2 October 2019).
33. Ministry of Housing and Urban-Rural Development of the People's Republic of China Web Page. Available online: https://www.mohurd.gov.cn/gongkai/fdzdgknr/tzgg/201608/20160801_228380.html (accessed on 1 August 2016).
34. Xiao, L.; Zhang, Y.; Peng, G. Landslide Susceptibility Assessment Using Integrated Deep Learning Algorithm along the China-Nepal Highway. *Sensors* **2018**, *18*, 4436. [CrossRef]
35. Narjabadifam, P.; Hoseinpour, R.; Noori, M.; Altabey, W. Practical seismic resilience evaluation and crisis management planning through GIS-based vulnerability assessment of buildings. *Earthq. Eng. Eng. Vib.* **2021**, *20*, 25–37. [CrossRef]
36. Xue, M.; Wen, H.; Lin, Y.; Sun, D. Random Forest Evaluation Model for Physical Toughness of Slopes Along Mountain Roads-Taking Maoxian County of Sichuan Province as an example. *Bull. Fo Soil Water Conserv.* **2020**, *40*, 168–175. [CrossRef]
37. Sun, D.; Wen, H.; Zhang, Y.; Xue, M. An optimal sample selection-based logistic regression model of slope physical resistance against rainfall-induced landslide. *Nat. Hazards.* **2020**, *105*, 1255–1279. [CrossRef]
38. Taalab, K.; Cheng, T.; Zhang, Y. Mapping landslide susceptibility and types using Random Forest. *Big Earth Data* **2018**, *2*, 159–178. [CrossRef]
39. Al Rifat, S.A.; Liu, W. Measuring community disaster resilience in the conterminous coastal United States. *Isprs Int J Geo-Inf* **2020**, *9*, 469. [CrossRef]
40. El-Tawil, S.; Ibrahim, A.; Eltawil, A. Stick-Slip Classification Based on Machine Learning Techniques for Building Damage Assessment. *J. Earthq. Eng. JEE* **2021**, 1–18. [CrossRef]
41. Mahmoudi, S.N.; Chouinard, L. Seismic fragility assessment of highway bridges using support vector machines. *B Earthq. Eng.* **2016**, *14*, 1571–1587. [CrossRef]
42. kernlab: Kernel-Based Machine Learning Lab Web Page. Available online: http://cran.r-project.org/web/packages/kernlab/index.html (accessed on 12 November 2019).
43. Zhou, X.; Wen, H.; Zhang, Y.; Xu, J.; Zhang, W. Landslide susceptibility mapping using hybrid random forest with GeoDetector and RFE for factor optimization. *Geosci. Front.* **2021**, *12*, 101211. [CrossRef]
44. Deng, X.; Liu, Q.; Deng, Y.; Mahadevan, S. An improved method to construct basic probability assignment based on the confusion matrix for classification problem. *Inform. Sci.* **2016**, *340–341*, 250–261. [CrossRef]
45. Banks, S.; Millard, K.; Pasher, J.; Richardson, M.; Wang, H.; Duffe, J. Assessing the Potential to Operationalize Shoreline Sensitivity Mapping: Classifying Multiple Wide Fine Quadrature Polarized RADARSAT-2 and Landsat 5 Scenes with a Single Random Forest Model. *Remote Sens.* **2015**, *7*, 13528–13563. [CrossRef]
46. Han, H.; Guo, X.; Yu, H. Variable Selection Using Mean Decrease Accuracy and Mean Decrease Gini Based on Random Forest. In Proceedings of the 2016 7th IEEE International Conference on Software Engineering and Service Science (ICSESS), Beijing, China, 26–28 August 2016; pp. 219–224. [CrossRef]

 *sensors*

Article

Capture and Prediction of Rainfall-Induced Landslide Warning Signals Using an Attention-Based Temporal Convolutional Neural Network and Entropy Weight Methods

Di Zhang, Kai Wei, Yi Yao, Jiacheng Yang, Guolong Zheng and Qing Li *

National and Local Joint Engineering Laboratories for Disaster Monitoring Technologies and Instruments, China Jiliang University, Hangzhou 310018, China
* Correspondence: lq13306532957@163.com

Abstract: The capture and prediction of rainfall-induced landslide warning signals is the premise for the implementation of landslide warning measures. An attention-fusion entropy weight method (En-Attn) for capturing warning features is proposed. An attention-based temporal convolutional neural network (ATCN) is used to predict the warning signals. Specifically, the sensor data are analyzed using Pearson correlation analysis after obtaining data from the sensors on rainfall, moisture content, displacement, and soil stress. The comprehensive evaluation score is obtained offline using multiple entropy weight methods. Then, the attention mechanism is used to weight and sum different entropy values to obtain the final landslide hazard degree (LHD). The LHD realizes the warning signal capture of the sensor data. The prediction process adopts a model built by ATCN and uses a sliding window for online dynamic prediction. The input is the landslide sensor data at the last moment, and the output is the LHD at the future moment. The effectiveness of the method is verified by two datasets obtained from the rainfall-induced landslide simulation experiment.

Keywords: rainfall-induced landslide; attention mechanism; entropy weight methods; an attention-based temporal convolutional neural network; landslide hazard degree

Citation: Zhang, D.; Wei, K.; Yao, Y.; Yang, J.; Zheng, G.; Li, Q. Capture and Prediction of Rainfall-Induced Landslide Warning Signals Using an Attention-Based Temporal Convolutional Neural Network and Entropy Weight Methods. *Sensors* **2022**, *22*, 6240. https://doi.org/10.3390/s22166240

Academic Editors: Junwei Ma and Jie Dou

Received: 21 July 2022
Accepted: 18 August 2022
Published: 19 August 2022

1. Introduction

Rainfall-induced landslides are geological hazards triggered by prolonged rainfall or short-term heavy rainfall. Scholars have conducted in-depth research on landslide susceptibility mapping [1], data modeling [2], and mechanism analysis [3].

Machine learning (ML) and deep learning (DL) are important methods for landslide prediction because of their ability to achieve complex nonlinear modeling. Many ML and DL methods are used for landslide detection and prediction with better performance than traditional methods. Wei et al. proposed an attention-constrained neural network with overall cognition (OC-ACNN) to capture features to predict landslides [4]. Ghorbanzadeh et al. used different deep convolutional neural networks (CNNs) for landslide remote sensing images and achieved better results in landslide mapping [5]. An integrated framework of DL models with rule-based object-based image analysis (OBIA) to detect landslides was explored by Ghorbanzadeh et al. [6]. Wang et al. optimized the Elman neural network with the genetic algorithm and used it to implement the prediction of landslide displacement [7]. Wang et al. compared five machine learning methods for reservoir displacement prediction, and the Hodrick–Prescott filter decomposed the cumulative displacement into trend displacement and periodic displacement [8]. Wang et al. predicted the intrinsic evolution trend of landslide displacement by (double exponential smoothing, DES) DES-VMD-LSTM, based on the Gaussian process regression (GPR) model to assess the uncertainty in the first prediction [9]. Miao et al. applied the fruit fly optimization algorithm back-propagation neural network (FOA-BPNN) for the prediction of random displacements [10]. Gong et al. considered the problem of interval prediction of landslide displacements and proposed a new method of interval prediction of landslide displacements combining dual-output least squares support vector machine (DO-LSSVM) and particle swarm optimization (PSO) algorithms [11]. Time series analysis and long short-term memory neural networks are used in landslide displacement prediction [12,13]. Lin et al. analyzed the internal relationship between rainfall, reservoir water level, and periodic landslide displacement and used the double-bidirectional long short-term memory (Double-BiLSTM) model to predict landslide displacement [14]. Zhang et al. proposed a method based on Gated Recurrent Unit (GRU) and Fully Integrated Empirical Decomposition of Adaptive Noise (CEEMDAN) for the dynamic prediction of landslide displacement [15]. The application of hybrid methods based on metaheuristics (MH) in the field of geohazards is a recent research direction in disaster prediction. Ma et al. conducted a comparative study on MHs and proposed a new hybrid algorithm, namely MH-based support vector machine regression (SVR) [16]. The hybrid method has high performance in terms of accuracy and reliability for landslide displacement prediction. Meanwhile, the hybrid method combined with a multiverse optimization (MVO) for hyperparameter optimization of MHs [17] improves the reliability of disaster prediction modeling.

Rainfall is commonly used for early warning as an important trigger for landslides. Cost-sensitive rainfall thresholds were investigated by Sala et al. and sensitivity analysis was performed [18]. However, rainfall thresholds that are difficult to standardize cannot be used as early warning signals for the occurrence of landslides. Changes in soil moisture are an important factor in landslides. Domínguez-Cuesta et al. focused on the role of rainfall and soil moisture as triggering and evolutionary factors for unstable events [19]. Soil moisture saturation and sudden rainfall are more likely to lead to landslides. Chen et al. analyzed the role of soil moisture index (SWI) in landslides based on 279 mass movements that occurred in Taiwan during 2006–2017 [20].

These data-driven approaches effectively implement the displacement prediction problem for landslides; however, these models do not consider correlations among multiple sensor data and do not capture warning signals in sensor data well. Entropy value, as a physical quantity describing the degree of data chaos, has also been used to analyze landslide risk [21]. However, landslide hazard analysis using the information entropy value method does not take into account the effects of different entropy values on landslide sensor data. A single entropy value method for landslide warning feature analysis failure will result in the possibility of misclassification.

Challenges: First, there are many landslide monitoring sensors, but the methods of effectively capturing warning signals are less studied. Second, there are correlations among different types of landslide sensor data, which need to be analyzed. Third, the accuracy of data-driven rainfall-induced landslide hazard prediction models needs to be improved.

Contributions:

- We combine an attention mechanism with multiple entropy weight methods and propose an attention-fusion entropy weight method (En-Attn) to capture warning signals based on massive landslide sensor data.
- We propose an attention-based temporal convolutional neural network for landslide warning signals prediction based on massive sensor data.
- We carry out the experimental simulation of rainfall-induced landslides, collect sensor data when landslides occur, analyze the precursory warning characteristics of the data, and use a variety of entropy weight methods to analyze the characteristics of warning signals offline.
- Our model is validated on two datasets obtained from rainfall-induced simulation experiments, and our model has high accuracy compared with similar landslide warning capture and prediction methods.

2. Methods

2.1. Capture Models of Landslide Warning Signal

We obtain massive sensor data from landslide simulation experiments, including rainfall, the soil moisture content in shallow layers, the soil moisture content in deep layers, soil stress, and displacement. The evaluation of landslide warning signals is to extract the warning features from these massive sensor data to characterize the landslide warning situation. The entropy weight methods (EWM) can be used to assess the degree of landslide hazard [21].

2.1.1. Entropy Weight Methods

Entropy is a measure of uncertain information. The smaller the entropy value, the greater the amount of information and the greater the weight. The entropy weight method (EWM) [22] is an objective weighting method. The canonical EWM uses information entropy (*InEn*) [23] as the basis for calculation. In fact, there are many entropy methods, namely approximate entropy [24], sample entropy [25], fuzzy entropy [26], and permutation entropy [27]. Therefore, an improved entropy method can be obtained by replacing the information entropy in the canonical entropy weight method with the following four entropy values: approximate entropy (*ApEn*), sample entropy (*SampEn*), fuzzy entropy (*FuzzyEn*), permutation entropy (*PeEn*).

The calculation process of the EWM [28] has five steps.

Step 1: Data normalization using Equation (1).

Step 2: Calculate the entropy value using Equation (2).

Step 3: Calculate the coefficient of variation using Equation (3).

Step 4: Calculate weights using Equation (4).

Step 5: Calculate the entropy weight score using Equation (5).

$$x_{ij} = z_{ij} / \sum_{i=1}^{N} z_{ij} \tag{1}$$

$$e_j = f_{En}(x_{ij}), i \in [1, N], e_j \in [0, 1] \tag{2}$$

$$d_j = 1 - e_j \tag{3}$$

$$\omega_j = d_j / \sum_{j=1}^{N} d_j \tag{4}$$

$$s_i = \sum_{j=1}^{M} \omega_j x_{ij}, i = 1, 2, \cdots, N \tag{5}$$

where

z_{ij} is the raw data at row i and column j in the sensor dataset.

x_{ij} is the data normalized by z_{ij}.

e_j is the entropy value of x_{ij}.

f_{En} is the method for calculating the entropy values using Equations (6)–(26) for the specific formula.

N is the number of rows in the sensor dataset.

d_j is the coefficient of variation of x_{ij}.

ω_j is the corresponding weight of each column of data obtained by the EWM.

s_i is the weight entropy score.

M is the number of columns in the sensor dataset.

Information entropy (*InEn*) [23] can be calculated by Equation (6).

$$f_{InEn_j} = -\frac{1}{\ln N} \sum_{i=1}^{N} x_{ij} \ln x_{ij}, e_j \in [0, 1] \tag{6}$$

where

ln denotes the natural logarithm.

f_{InEn_j} denotes the information entropy value.

The calculation of *ApEn* can also be understood as the degree of self-similarity of a sequence in the pattern. For the change of a signal sequence, the change of the approximate entropy value can be used to achieve the purpose of effective identification. The biggest advantage of the approximate entropy calculation is that it does not require a large amount of data, most of the measured time series can meet the requirements, and the obtained results are robust and reliable [29]. The calculation of approximate entropy (*ApEn*) is as follows:

$$X_i = [x(i), x(i+1), \cdots, x(i+m-1)] \tag{7}$$

$$d[X_i, X_j] = \max|x(i+k) - x(j+k)|, k \in (0, m-1) \tag{8}$$

$$B_i(r) = num\{d[X_i, X_j] < r\} \tag{9}$$

$$\Phi_i^m(r) = \frac{B_i}{N-m+1} \tag{10}$$

$$f_{ApEn} = \Phi^m(r) - \Phi^{m+1}(r) \tag{11}$$

where

$d[X_i, X_j]$ denotes the distance between the vector X_i and X_j.

B_i is the number of items that satisfy the condition $d[X_i, X_j] < r$.

r denotes the similarity tolerance threshold.

Φ_i^m denotes the ratio of the approximate quantity to the total quantity, namely the approximate ratio.

f_{ApEn} denotes the approximate entropy value of sequence X_i.

m is the dimension of X_i, which is an artificially set parameter value.

ApEn characterizes the complexity of a sequence. The value of *ApEn* is less affected by the amount of data and is suitable for non-stationary and nonlinear sequences. *ApEn* preserves the time series information in the original signal sequence and reflects the characteristics of the signal sequence on the structural distribution. The entropy value of the fault signal will be greater for fault data present in a set of continuous data, so *ApEn* is often used to detect the fault signal. The fault signal here refers to the presence of multiple abnormal signals in a set of sequential signals.

SampEn is an improved method based on *ApEn* [29]. The *SampEn* has better consistency. If one time series has a higher *SampEn* value than another time series, then the other r and m values also have higher *SampEn* values. Meanwhile, *SampEn* is not sensitive to missing data [29].

The calculation of sample entropy (*SampEn*) is as follows:

$$B_i^m(r) = \frac{1}{N-m} num\{d[X_i, X_j] < r\} \tag{12}$$

$$B^m(r) = \frac{1}{N-m+1} \sum_{i=1}^{N-m+1} B_i^m(r) \tag{13}$$

$$f_{SampEn} = -\ln(B^{m+1}(r)/B^m(r)) \tag{14}$$

where

B_i^m denotes the ratio of the number of $d[X_i, X_j] < r$ to the total number of vectors *N-m*, for a given threshold r ($r > 0$).

f_{SampEn} denotes the sample entropy value of the sequence X_i.

In the definitions of *ApEn* and *SampEn*, the similarity of vectors is determined by the difference in absolute values of the data. Correct analysis results cannot be obtained when there are slight fluctuations in the data used or baseline drift. *FuzzyEn* removes the influence of baseline drift through mean operation, and the similarity of vectors is no longer determined by the absolute amplitude difference, but determined by the shape of the fuzzy function determined by the exponential function, thereby fuzzifying the similarity measure [26]. The *FuzzyEn* uses an exponential function to fuzzify the similarity measurement formula. The continuity of the exponential function makes the fuzzy entropy change continuously and smoothly with the parameter change.

The calculation of fuzzy entropy (*FuzzyEn*) is as follows:

$$Y_i = [x(i), x(i+1), \cdots, x(i+m-1)] - x_0(i), i = 1, 2, \cdots, N-m+1 \tag{15}$$

$$x_0(i) = \frac{1}{m}\sum_{j=0}^{m-1} x(i+j) \tag{16}$$

$$d_{i,j}^m = d[Y_i, Y_j] = \max_{k\in(0,m-1)} \left| x(i+k) - x_0(i) - x(j+k) - x_0(j) \right| \tag{17}$$

$$D_{i,j}^m = \exp\left[-\frac{(d_{i,j}^m)^n}{r} \right] \tag{18}$$

$$\psi^{m+1}(r) = \frac{1}{N-m+1}\sum_{i=1}^{N-m+1}\left(\frac{1}{N-m}\sum_{j=1,j\neq i}^{N-m+1} D_{i,j}^m \right) \tag{19}$$

$$f_{FuzzyEn} = -\ln(\psi^{m+1}(r)/\psi^m(r)) \tag{20}$$

where

m denotes the embedding dimension.

Y denotes the sequence after the phase space reconstruction of X.

x_0 is the mean of m consecutive $x(i+j)$.

$d_{i,j}^m$ denotes the maximum value of the difference between the corresponding endpoints of Y_i and Y_j.

$D_{i,j}^m$ is the similarity between Y_i and Y_j after using the fuzzy membership function.

ψ^m is a function defined like Φ_i^m and B_i^m.

$f_{FuzzyEn}$ denotes the fuzzy entropy value of sequence X_i.

Permutation entropy (*PeEn*) is a method to detect the randomness and dynamic mutation behavior of time series. The *PeEn* has the characteristics of simple and fast calculation, strong anti-noise ability, and can realize the characteristics of online monitoring of mutation signals. *PeEn* introduces the idea of permutation when calculating the complexity between reconstructed subsequences.

The calculation of permutation entropy (*PeEn*) is as follows:

$$Y_i = [x(i), x(i+\tau), \cdots, x(i+(m-1)\tau)], i = 1, 2, \cdots, N-m+1 \tag{21}$$

$$x(i+(j_1-1)\tau) \leq x(i+(j_2-1)\tau) \leq \cdots \leq x(i+(j_m-1)\tau) \tag{22}$$

$$S(l) = (j_1, j_2, \cdots, j_m), l = 1, 2, \cdots, k, and\ k \leq m! \tag{23}$$

$$P_i = \frac{Number(Y_i)}{N-(m-1)\tau} \tag{24}$$

$$PE(m) = -\sum_{i=1}^{k}(P_i \ln P_i) \tag{25}$$

$$0 \leq f_{PeEn} = PE/\ln(m!) \leq 1 \tag{26}$$

where

m denotes the embedding dimension.

τ denotes the time delay factor.

$k = N-(m-1)\tau, j = 1, 2, \cdots, k$

S is a set of symbol sequences consisting of the index of each element position column after each reconstructed component is rearranged in ascending order.

j_m is the column index of the position of the mth element in the vector.

P_i is the probability of occurrence of each sort.

PE denotes the permutation entropy value of the sequence.

f_{PeEn} denotes the normalized value of the permutation entropy.

The matrix has k reconstruction components in total, and each reconstruction component has m-dimensional embedded elements. Arrange the jth category in the matrix in ascending order according to the size of the array using Equation (22).

$j_1, j_2, \cdots, j_m$ represents the subscript index value of each element in the reconstructed component. Note that the above sequence has a parameter τ, namely the time delay factor, which must be a positive integer. In fact, this parameter can be understood as the downsampling of the sequence. For example, when $\tau = 3$, it is sampling every three data points. When $\tau = 1$, the sequence is the same as the sequence definition of the *ApEn* and *SampEn*.

2.1.2. Attention-Fusion Entropy Method

The attention mechanism can pay attention to important parts of the sequence data [2,30]. Queries and key-value pairs are mapped to outputs. The calculation process of the attention mechanism is shown in Figure 1.

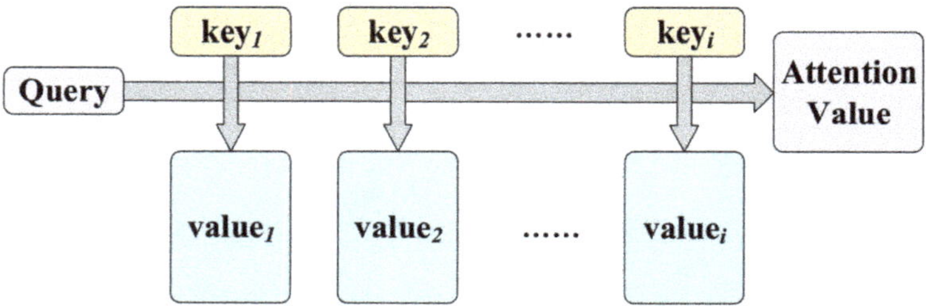

Figure 1. Overview of attention mechanism.

Equation (27) shows the score function, and Equation (28) shows the attention calculation process. The score function is essentially seeking a degree of similarity, and the *Softmax* function is to normalize the weights at all positions so that the sum is equal to one [31].

$$f(Q,K) = \frac{Q^T K}{\sqrt{d}} \tag{27}$$

$$C = Attention(Q,K,V) = Softmax(f(Q,K))V \tag{28}$$

where

Q denotes the queries, and $Q = W^{q_i} X_t$, where W^{q_i} is the weight corresponding to Q.

K denotes the keys $K = W^{k_i} X_t$, where W^{k_i} is the weight corresponding to K.

V denotes the values $V = W^{v_i} X_t$, where W^{v_i} is the weight corresponding to V.

C denotes the result of the weighted summation of weights and variables.

$\frac{1}{\sqrt{d}}$ denotes the scaling factor.

The role of the scaling factor is to keep the dot product of Q and K from becoming too large [31]. Once the dot product is too large, the activation function *Softmax* enters a region with a small gradient. The attention mechanism is used for the calculation to fuse multiple EWMs, and the fused entropy method is obtained, which is named as En-Attn.

Figure 2 shows that the input of the En-Attn model is historical sensor data, including rainfall, shallow moisture content, deep moisture content, displacement, and soil stress. The three types of data are calculated by three EWMs for comprehensive evaluation scores. The difference between these three entropy weight methods is that the entropy is different, namely *InEn, FuzzyEn*, and *PeEn*. The reason why *ApEn* and *SampEn* are not used in the En-Attn model is that *FuzzyEn* is an improvement on *SampEn* and *ApEn*. Meanwhile, in the actual dataset, the difference between these three methods is not obvious. For the same datasets, the result of getting almost the same output needs to be computed three times, which consumes computation time and occupies the memory of the computation space. Therefore, *FuzzyEn* is chosen instead of the three EWMs to reduce the time and space complexity of the En-Attn method. The demonstration of the details of these three EWMs for landslide sensor data processing is presented in Section 4.1.

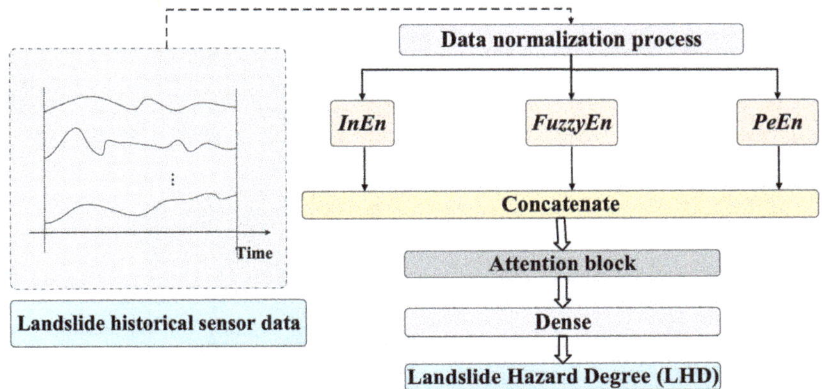

Figure 2. Overview of an attention-fusion entropy weight method (En-Attn).

The attention mechanism is used to fuse the outputs of the three EWMs (*InEn, FuzzyEn*, and *PeEn*) and finally outputs landslide hazard degree (LHD). Algorithm 1 elaborates the specific calculation steps.

Algorithm 1: Attention-fusion entropy weight method (En-Attn).

Initialization: M, m, r, d, W
Input: the raw data z
Entropy weight methods
For j = 1:M
    Data normalization using Equation (1).
    Calculate *InformEn* using Equation (6).
    Calculate *FuzzyEn* using (15)~(20).
    Calculate *PeEn* using (21)~(26).
    Calculate the coefficient of variation using Equation (3).
    Calculate weights using Equation (4).
    Obtain the entropy weight scores using Equation (5).
End if
Output: $S_{InEn}, S_{FuzzyEn}, S_{PeEn}$
Attention calculation
$Q = K = V = W \cdot [S_{InEn}, S_{FuzzyEn}, S_{PeEn}]$
$S_{En-Attn} = Softmax\left(\frac{Q^T K}{\sqrt{d}}\right) V$
LHD $= normalize(S_{En-Attn})$
Output: LHD.

2.2. Prediction Model of Landslide Warning Signal

The prediction model of the hazard degree of rainfall-induced landslides is based on temporal convolutional neural networks (TCNs). TCNs have a good predictive effect on the processing of time series data [32,33]. We add an attention module to the data before TCN input to extract the prediction features of the input data; we also add an attention module to the output data of TCN to extract the features of the output data to improve the performance of TCN.

The TCN incorporating the attention mechanism is shown in Figure 3, including the attention mechanism (I-Attn) in the input stage, the attention mechanism (T-Attn) after the TCN output, and the TCN that plays the main prediction role. The input of I-Attn is sensor data at time t and the hidden layer at time $t-1$, and the output is the attention weight at time t. The input of T-Attn is the hidden layer at time t, and the output is the size of the attention weight at time t and the weight value of the TCN's output, which is the final predicted output value. TCN is composed of multiple residual blocks [32]. The output of the previous residual block is the input of the next residual block. The 1D convolution in TCN enables equal lengths of the input and output sequences [34]. Causal convolution ensures that the prediction process does not suffer from data leakage. TCN enlarges the convolutional field size, which can be obtained from Equation (29). The calculation of the number of residual blocks is obtained from Equation (30).

$$r = 1 + \sum_{i=0}^{n-1} 2(k-1)b^i = 1 + 2(k-1)\frac{b^n - 1}{b-1} \tag{29}$$

$$n = \left\lceil \log_b\left(\frac{(l-1)(b-1)}{2(k-1)} + 1\right)\right\rceil \tag{30}$$

where

k denotes the size of the convolutional kernel.
B denotes the size of the dilated base.
N denotes the number of residual blocks.
L denotes the length of the input tensor.

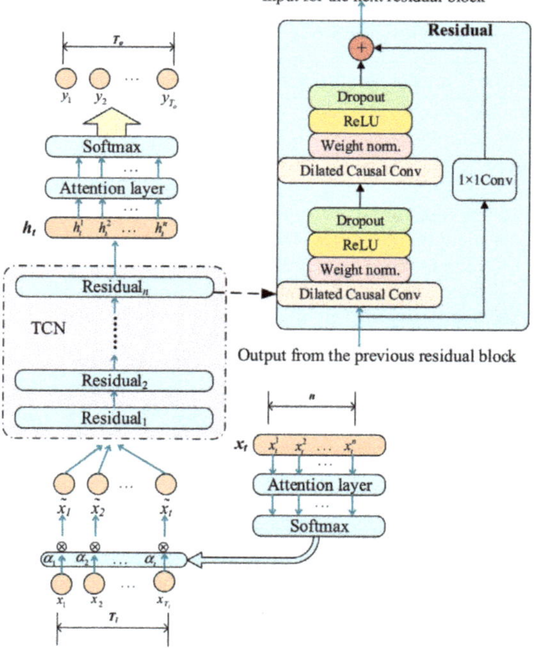

Figure 3. The overall framework of the attention-based temporal convolutional neural network (ATCN).

In the actual landslide experiment, the sensor data are transmitted back to the host computer as a continuous string of arrays. The dynamic sliding prediction of the ATCN model is implemented using a sliding window as a way to process the dynamic data, as shown in Figure 4. The input of the sliding window is the five-dimensional sensor data of T_i length, and the output is the landslide hazard degree (LHD) of T_o length. The sliding window moves forward with the time step while the predicted value is output. Algorithm 2 illustrates the specific steps of the landslide warning signals prediction model (ATCN). The performance of the ATCN is experimentally verified in Section 4.2.

Algorithm 2: Attention-based temporal convolutional neural network (ATCN).

Input: $x_t = \left\{x_t^1, x_t^2, \cdots, x_t^{T_i}\right\}$
Data normalization using Equation (1).
I-Attn calculation:
$Q_i = K_i = V_i = W_i \cdot x_t$
$\widetilde{x}_t = Softmax\left(\frac{Q_i{}^T K_i}{\sqrt{d_i}}\right) V_i$
Predictor:
$h_t = f_{TCN}(\widetilde{x}_t)$
T-Attn calculation:
$Q_o = K_o = V_o = W_o \cdot h_t$
$y_t = Softmax\left(\frac{Q_o{}^T K_o}{\sqrt{d_o}}\right) V_o$
Output: $y_t = \left\{y_t^1, y_t^2, \cdots, y_t^{T_o}\right\}$
Update $x_t \leftarrow x_{t+1}$, and repeat the above steps.

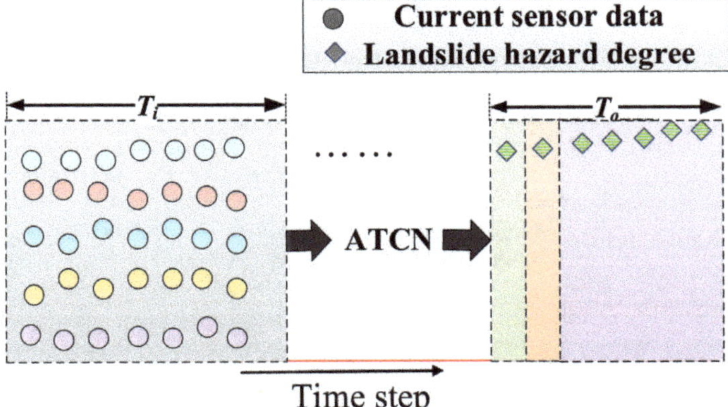

Figure 4. Sliding window for dynamic prediction of sensor data.

3. Data Acquisition and Processing

3.1. Landslide Simulation Platform

The landslide simulation platform (LSP) is built to simulate the occurrence of rainfall-induced landslides. The landslide simulation platform (LSP) simulates a small monitoring area in a mountain rather than a large area such as a natural landslide itself. This is because simulating a mountain in nature is actually very challenging, and all we can do is simulate a certain monitoring area. In nature, multiple monitoring zones work together on a large mountain. The analysis of a monitoring zone is a prerequisite for data analysis and early warning of a large mountain. Figure 5 shows the physical objects of the LSP. The structure of the LSP includes the simulated rainfall system and the sensor measurement system.

Figure 5. Landslide simulation platform (LSP). (**a**) Main view of the LSP; (**b**) Side view of the LSP.

The simulated rainfall system consists of the following components: rainfall sprinklers, soil-carrying box, hydraulic support rods, and lift bars. The rain sprinklers simulate the natural rainfall environment, and controlling the amount of rainfall can simulate the rainstorm. The soil-carrying box contains rock and soil mass to simulate natural slope conditions. The hydraulic support rods and the lifting bars can adjust the angle of the soil-carrying box to simulate the angle of the potential landslide body in nature. Water will seep out of the tube wall as it passes through the porous ceramic tube, simulating underground water in the rock and soil mass.

The experimental process includes five steps:

Step 1: Place the rock and soil mass inside the soil box.

Step 2: Install five types of sensors at the appropriate positions.

Step 3: Use the hydraulic support rod to adjust the soil box to a suitable angle. Here, we chose 30°.

Step 4: Turn on the rain sprinklers for rainfall simulation and use the monitoring software to monitor the sensor data and save it to the database.

Step 5: Analyze and process the sensor data after the experiment is completed.

In the landslide simulation experiment platform, we installed five types of sensors: a tipping bucket rain gauge, a draw-wire displacement sensor, a soil stress gauge, and two moisture content sensors. The installation positions of the sensors are shown in Figure 6.

The locations of the sensors installed in the experiment are as follows:

1. The tipping bucket rain gauge is located in the center of the soil-carrying box, with its opening facing upwards for better rain reception.
2. The position of the draw-wire displacement sensor is in the front third of the soil-carrying box. It monitors the change in soil displacement as the leading edge of the landslide moves.
3. The soil stress gauge is positioned in the front third of the soil-carrying box to monitor the stress changes within the soil at the leading edge of the landslide.
4. The location of the soil moisture sensor for monitoring the shallow moisture content is about 30 cm from the surface, and the location of the soil moisture sensor for monitoring the deep moisture content is about 80 cm from the surface.

Note that the above sensor installation locations are limited by the LSP and are only used as a reference criterion for experiments.

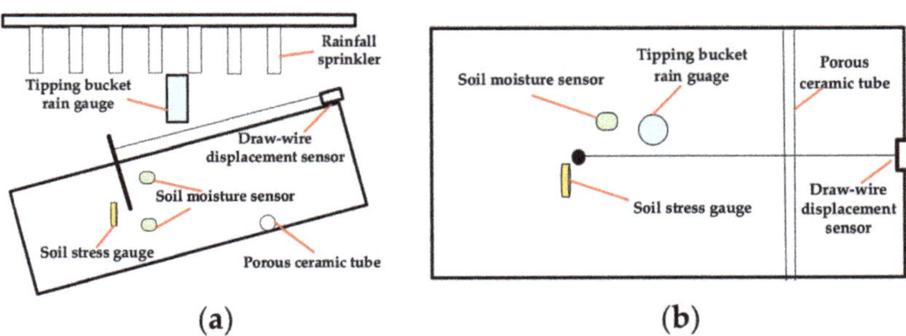

Figure 6. Schematic diagram of sensor installation in the landslide disaster simulation platform. (**a**) Side view of sensor installation schematic; (**b**) Top view of sensor installation schematic.

3.2. Landslide Data Processing

We carry out two experiments on rainfall-induced landslides and obtain datasets for L_1 and L_2. The rainfall, soil stress, and displacement in the datasets are normalized to obtain the sensor data curves in Figure 7.

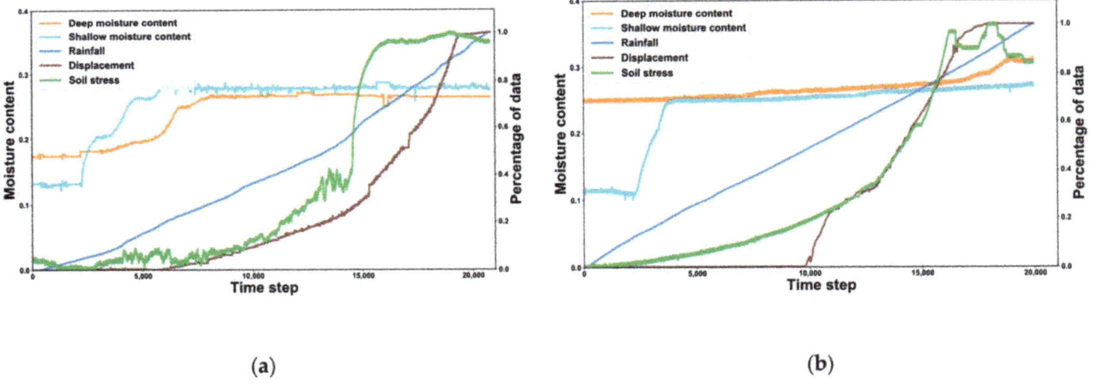

Figure 7. Curve of landslide datasets L_1 and L_2. (**a**) Dataset L_1. (**b**) Dataset L_2.

The ordinate on the left of Figure 7 is moisture content, and the ordinate on the right is the percentage of data. After a period of time, the moisture content of the soil in the shallow layer begins to rise, and the moisture content of the soil in the deep layer rises in response. The reason why the relationship between the two moisture contents in Figure 7b is not significant is that before rainfall, the deep soil moisture content is high and close to saturation.

The Pearson correlation coefficient method is used to analyze the landslide sensor datasets to analyze the correlation between different types of sensor data.

The Pearson correlation coefficient is suitable for two columns of spaced variables (continuous variables) in a normal distribution. The correlation coefficient and the probability of the correlation can be obtained for two columns of data using Equation (31) when they have the same number of data and correspond to each other.

$$r_p = \frac{Cov(X,Y)}{\sigma_X \sigma_Y} = \frac{\sum\limits_{i=1}^{n}(X_i - \overline{X})(Y_i - \overline{Y})}{\sqrt{\sum\limits_{i=1}^{n}(X_i - \overline{X})^2}\sqrt{\sum\limits_{i=1}^{n}(Y_i - \overline{Y})^2}} \tag{31}$$

where

r_p denotes Pearson correlation coefficient.

X represents senor data.

Y represents sensor data other than X.

σ_X denotes the standard deviation of X.

σ_Y denotes the standard deviation of Y.

The Pearson correlation coefficient ranges between -1 and 1. When the Pearson correlation coefficient is 0, the X and Y vectors are not correlated. When its value is greater than 0.8, X and Y are highly correlated.

We let X and Y be one of the five types of sensor data, respectively, and the heatmaps are obtained in Figure 8 after the calculation of Equation (31).

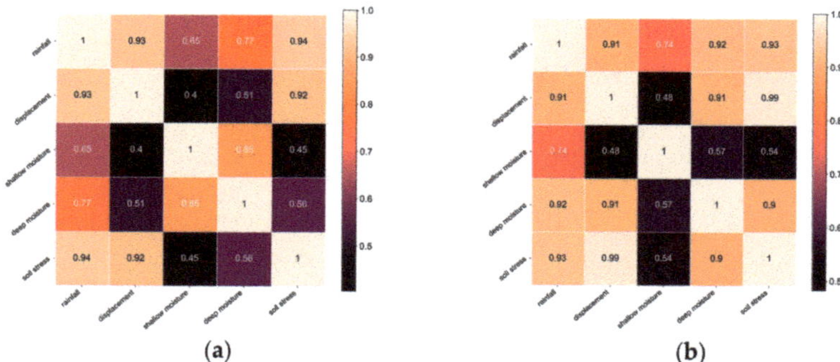

Figure 8. Heatmaps of landslide datasets L_1 and L_2. (**a**) Pearson heatmap of L_1. (**b**) Pearson heatmap of L_2.

In Figure 8a, the rainfall and displacement show a high correlation with the magnitude of soil stress and a moderate correlation with the shallow moisture content and the deep moisture content. The shallow moisture content and the deep moisture content are highly correlated states. The shallow moisture content shows a weak correlation with the displacement amount. Soil stress shows a strong correlation with displacement. In Figure 8b, rainfall displays a strong correlation with displacement, soil stress, and deep moisture content and a moderate correlation with shallow moisture content. The correlation between shallow moisture content and other sensor data is weak. The relationship between the landslide process and different sensor data is analyzed as follows:

1. The amount of rainfall directly affects the moisture content of the shallow soil. Surface water will exist when the surface seepage rate is less than the rainfall.
2. The moisture content of deep soil is significantly higher than that of shallow soil due to groundwater action during the initial stage of rainfall. The moisture content in the deeper layers of the soil would gradually increase as surface water gradually infiltrates into the ground as rainfall continues. However, its moisture content does not exceed the shallow moisture content at this stage. The growth rate of the shallow moisture content would gradually decrease, and the size of the deep moisture content would eventually be approximately equal to the shallow moisture content throughout the entire landslide formation process.
3. The soil stress also varies as the soil layer's moisture content varies. The shear strength of the soil is characterized by soil stress. The soil stress increases quickly for a while when there is no significant displacement of the surface, after which the surface gradually becomes significantly displaced during the sliding phase. As the soil's moisture content rises, the clay in the soil softens and loses some of its slip resistance. It also loses shear strength.
4. The soil moisture content tends to become saturated before the landslide body enters the catastrophic slip phase. When the soil stress increases, the landslide body enters the severe sliding stage. When a landslide reaches the severe slip stage, the surface

displacement dramatically rises, and erosion-created depressions and gullies start to show up near the body's front edge.

5. After entering the stabilization stage, the surface displacement of the landslide body no longer increases, but due to the effects of rainfall and groundwater, the surface and underground runoff still play a role in triggering the secondary landslide.

4. Experiments and Results

In this section, we describe experiments on landslide warning signals and signal prediction. We present the results of two experiments to demonstrate the effectiveness of En-Attn as well as ATCN in landslide warning signal capture and prediction.

4.1. Landslide Hazard Degree and Results

We apply the En-Attn model to process the landslide datasets L_1 and L_2. Figure 9 illustrates the landslide hazard degree (LHD) obtained by En-Attn as well as the three EWMs. The LHD obtained by all six methods shows an increasing trend, indicating a gradual increase in the characteristics of the hazard level during landslide formation. The LHD ranges from 0 to 1. LHD = 0 means no warning feature, and LHD = 1 means the landslide warning feature is significant and enters a very urgent warning situation. For dataset L_1, the LHD increases gradually, and when the time step is greater than 14,000, the LHD increment rate increases. For dataset L_2, the incremental rate of LHD increases when the time step is greater than 10,000, while the volatility of LHD is greater compared to L_1.

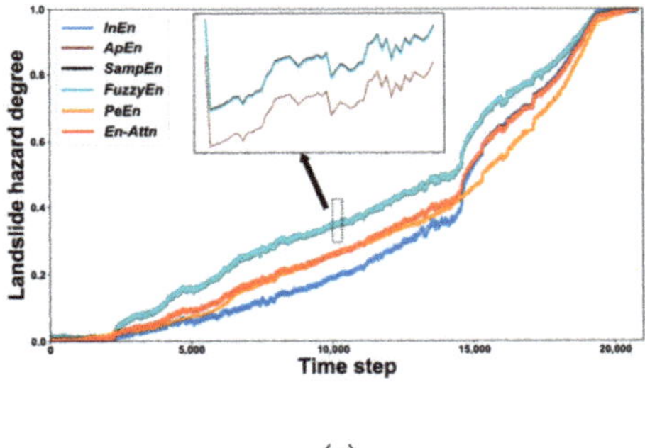

(a)

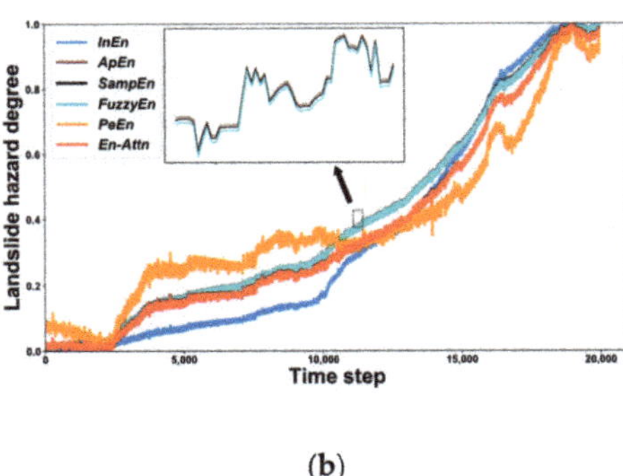

(b)

Figure 9. Landslide hazard degree (LHD) of the landslide datasets L_1 and L_2. (**a**) LHD of L_1. (**b**) LHD of L_2.

Note that the differences in the LHD obtained by *ApEn*, *SampEn*, and *FuzzyEn* are not significant, and the differences exhibited by the local enlarged image are shown in Figure 8a,b. The reason that only *FuzzyEn* is considered in the En-Attn model and not both *ApEn* and *SampEn* is because the differences between the three methods are not significant.

The single entropy value method is prone to fluctuations in the calculation of LHD, as in the case of *PeEn* in Figure 8b. The LHD obtained by the En-Attn model not only demonstrates landslide warning characteristics but also exhibits better stability and robustness. The En-Attn model overcomes the drawbacks of the single EWM and adapts better to the case of multi-sensor data to evaluate landslide warning features.

4.2. Prediction Experiments and Results

We apply the ATCN model to process the landslide datasets L_1 and L_2 and their LHD. The ATCN model is elaborated in Section 2.2. We conducted experiments to test the performance of the ATCN model, comparing long short-term memory neural networks (LSTM) [35], grated recurrent units (GRU) [36], temporal neural networks (TCN) [32,34], convolutional long short-term memory neural networks (ConvLSTM) [37], and dual-stage attention-based recurrent neural networks (DA-RNN) [30]. The metrics [2] for evaluating the performance are root mean square error (RMSE), mean absolute error (MAE) and mean absolute percent error (MAPE), and the specific equations are shown in Equations (32)–(34).

$$\mathrm{MAE} = \frac{1}{N}\sum_{t=1}^{N}|\hat{y}_t - y_t| \tag{32}$$

$$\mathrm{RMSE} = \sqrt{\frac{1}{N}\sum_{t=1}^{N}(\hat{y}_t - y_t)^2} \tag{33}$$

$$\mathrm{MAPE} = \frac{100\%}{N}\sum_{t=1}^{N}\left|\frac{\hat{y}_t - y_t}{y_t}\right| \tag{34}$$

where

N is the total number of test data.

y_t is the true value at the tth time step.

$\hat{y}_t$ is the predicted value at the tth time step.

The model tests are divided into two types of sliding windows, "100-10" and "100-50", which reflect different input data lengths and prediction lengths. The hyperparameters of the TCN and ATCN models are set as follows: filters = 32, batch size = 128, kernel size = 8, where the activation function of the attention mechanism is *Softmax*. The hyperparameters of the LSTM and GRU models are set as follows: the number of units is 16. The activation function is *ReLU*, the optimization algorithm is *Adam*, the initial learning rate is 0.001, and the learning rate can be adjusted according to the loss function subsequently. The hyperparameter experiments of ATCN are shown in Appendix A. All models are run 20 times, and the predicted values are obtained after testing the datasets L_1 and L_2. The average values of RMSE, MAE, and MAPE are shown in Tables 1 and 2.

Table 1. Comparison of LHD prediction effects of different models for dataset L_1.

Model	Metric	Size of Sliding Window	
		100-10	100-50
LSTM	RMSE	0.04973	0.05987
	MAE	0.03483	0.03988
	MAPE (%)	3.45876	4.48301
GRU	RMSE	0.04296	0.11422
	MAE	0.02916	0.10989
	MAPE (%)	3.21155	4.70642
ConvLSTM	RMSE	0.01511	0.02480
	MAE	0.01162	0.02307
	MAPE (%)	1.31189	2.70816
DA-RNN	RMSE	0.02606	0.02044
	MAE	0.01825	0.01590
	MAPE (%)	1.96037	1.68211
TCN	RMSE	0.02009	0.03222
	MAE	0.01500	0.02192
	MAPE (%)	1.68965	2.42844
ATCN	RMSE	0.00892	0.01827
	MAE	0.00718	0.01411
	MAPE (%)	0.82503	1.59699

Table 2. Comparison of LHD prediction effects of different models for dataset L_2.

Model	Metric	Size of Sliding Window	
		100-10	100-50
LSTM	RMSE	0.04465	0.10245
	MAE	0.03571	0.09849
	MAPE (%)	3.74129	6.12409
GRU	RMSE	0.03632	0.06781
	MAE	0.02316	0.05799
	MAPE (%)	2.41790	4.88399
ConvLSTM	RMSE	0.02937	0.05297
	MAE	0.02369	0.03579
	MAPE (%)	2.56583	3.82107
DA-RNN	RMSE	0.01633	0.02966
	MAE	0.01360	0.02266
	MAPE (%)	1.44912	2.38209
TCN	RMSE	0.02540	0.03209
	MAE	0.02059	0.02687
	MAPE (%)	2.16727	2.84709
ATCN	RMSE	0.01082	0.01899
	MAE	0.00950	0.01463
	MAPE (%)	1.02798	1.54598

Tables 1 and 2 demonstrate the RMSE, MAE, and MAPE of ATCN and its counterparts. Table 1 shows that the RMSE, MAE, and MAPE metrics of ATCN are lower for dataset L_1, which implies better performance of ATCN.

The ATCN outperforms other models in the prediction of LHD. Compared with the TCN model, the RMSE, MAE, and MAPE of ATCN decreased by 55.60%, 52.13%, and 51.17%, respectively, with the sliding window set to "100-10". The ATCN can effectively capture the characteristics of landslide prediction. The ATCN also outperforms other models when the sliding window is "100-50". In comparison to the TCN model, the performance of the three metrics is decreased by 43.30%, 35.63%, and 34.24%, respectively. The poor performance is due to the absence of attention mechanisms in the LSTM, GRU, and ConvLSTM, as well as the insignificant features obtained from the complex landslide sensor signals.

Figure 2 displays the metrics for dataset L_2, which is similar to dataset L_1. The classical recurrent neural network models, LSTM and GRU, performed poorly because the predictive properties shown by the sensor data in dataset L_2 are not obvious. The performance of DA-RNN and ATCN with the addition of the attention mechanism is outstanding. The three metrics of ATCN are decreased by 33.74%, 30.15%, and 29.06%, respectively, in comparison to DA-RNN when the sliding window is set to "100-10". The three metrics of ATCN are decreased by 35.97%, 35.44%, and 35.10%, respectively, compared to DA-RNN when the sliding window is set to "100-50".

Comparing the model performance with different prediction lengths, it can be seen that the shorter the prediction length, the smaller the performance metrics, and the better the prediction effect. When the prediction length is long, the attention mechanism captures the long-term dependency characteristics more and more prominently, and the performance of DA-RNN and ATCN with the attention mechanism is better than the other models. Comparing the DA-RNN and ATCN models, ATCN has better prediction results and stable performance when the sliding windows are "100-10" and "100-50". The ATCN model has the lowest error and the best prediction, as seen in Tables 1 and 2. The two sliding windows can be compared to demonstrate that the model's error increases with prediction length. ATCN's prediction accuracy is greater.

5. Discussion and Conclusions

This work adopts the attention mechanism to integrate the multi-entropy values to capture the landslide warning signals and explores the ATCN to realize landslide hazard prediction. Compared with its counterparts, our model has the characteristics of higher accuracy. Compared with current landslide hazard prediction methods, our methods have the following characteristics:

1. Exploring deep learning algorithms combined with big landslide data is an extension of deep learning application scenarios. This model uses a simple attention mechanism combined with a temporal convolutional neural network. Although this model is simple, its prediction effect is better than other complex deep learning models.
2. Effective landslide hazard capture. In the traditional sense, the capture of rainfall-induced landslide hazards is either directly replaced by the landslide displacement or only a single EWM is used to realize the signals capture. The model uses the attention mechanism to integrate a variety of EWMs, and the obtained landslide warning signals are more reliable.
3. Note that our model cannot be adapted for landslide hazard prediction with a small amount of data, as massive data is the basis of our model.

In the future, we intend to design a software system that integrates the algorithms for actual landslide sites. Further, we intend to consider different types of sensor data because more kinds of sensor data represent more comprehensive landslide disaster information. Furthermore, we plan to consider the sensor data of the landslide simulation platform in relation to soil thickness. We use landslide simulation experiments in this study. However, we could not achieve the exact same processes in the laboratory as in nature. For example, simulating different soil layers, which would take millions of years to form in nature. Our future research work will take into account multiple natural environmental factors to improve the experimental setup, including slope angle and dynamics of water extinction.

Author Contributions: Conceptualization, D.Z. and Q.L.; methodology, D.Z.; software, D.Z.; validation, Y.Y. and K.W.; formal analysis, D.Z. and J.Y.; investigation, D.Z. and J.Y.; resources, Q.L.; data curation, D.Z.; writing—original draft preparation, D.Z.; writing—review and editing, K.W., Y.Y. and G.Z.; visualization, D.Z., Y.Y. and K.W.; supervision, D.Z.; project administration, Q.L.; funding acquisition, Q.L. All authors have read and agreed to the published version of the manuscript.

Funding: This research was funded by Key Research and Development Program of Zhejiang Province, China, under grants 2018C03040 and 2021C03016.

Institutional Review Board Statement: Not applicable.

Informed Consent Statement: Not applicable.

Data Availability Statement: Not applicable.

Conflicts of Interest: The authors declare no conflict of interest.

Appendix A. Hyperparameter Experiments of the ATCN

The hyperparameters in ATCN can directly affect the high performance of the landslide prediction model. The kernel size, filters, and training batch size in the model has a large impact on ATCN. With dataset L_2, performance comparison experiments are carried out on the kernel sizes, filters, and batch sizes in the ATCN model. The comparison metrics are RMSE, MAE, and MAPE, and the experiments of each hyperparameter are repeated 20 times, and the mean values of the 20 experiments are counted. The statistical results are shown in Tables A1–A3.

Table A1. Comparison of different batch sizes in the ATCN model.

Batch Size	Metric	Size of Sliding Window	
		100-10	**100-50**
16	RMSE	0.01452	0.01928
	MAE	0.01325	0.01723
	MAPE (%)	1.63992	1.86792
32	RMSE	0.01213	0.01989
	MAE	0.01069	0.01907
	MAPE (%)	1.08400	2.37950
64	RMSE	0.01614	0.01734
	MAE	0.01609	0.01609
	MAPE (%)	1.11208	1.73150
128	RMSE	0.00954	0.01929
	MAE	0.00943	0.01606
	MAPE (%)	1.00213	0.91316
256	RMSE	0.01619	0.01892
	MAE	0.01825	0.01838
	MAPE (%)	2.19243	1.99731

Table A2. Comparison of different filters in the ATCN model.

Filter	Metric	Size of Sliding Window	
		100-10	**100-50**
4	RMSE	0.01674	0.01937
	MAE	0.01531	0.01334
	MAPE (%)	1.64269	1.56591
8	RMSE	0.01016	0.01102
	MAE	0.01158	0.00934
	MAPE (%)	1.31589	1.13547
16	RMSE	0.01023	0.01803
	MAE	0.01709	0.00949
	MAPE (%)	1.82595	1.86010
32	RMSE	0.01953	0.01597
	MAE	0.01897	0.01504
	MAPE (%)	1.07723	1.88453
64	RMSE	0.11779	0.01696
	MAE	0.01085	0.01360
	MAPE (%)	1.42817	1.63355

Table A3. Comparison of different kernel sizes in the ATCN model.

Kernel Size	Metric	Size of Sliding Window	
		100-10	**100-50**
4	RMSE	0.01148	0.01582
	MAE	0.01810	0.01442
	MAPE (%)	1.47336	1.54591
8	RMSE	0.00984	0.01074
	MAE	0.09313	0.00943
	MAPE (%)	1.39457	1.03825

Table A3. *Cont.*

Kernel Size	Metric	Size of Sliding Window	
		100-10	100-50
16	RMSE	0.00949	0.00965
	MAE	0.00809	0.00807
	MAPE (%)	0.89151	0.98417
32	RMSE	0.10553	0.00963
	MAE	0.01805	0.00909
	MAPE (%)	1.37068	1.08417
64	RMSE	0.00959	0.10772
	MAE	0.01168	0.10620
	MAPE(%)	1.21431	1.05872

Table A1 shows the metrics of ATCN for different batch sizes tested with kernel size = 16, filters = 8. The results in Table A1 show that the RMSE, MAE, and MAPE metrics of the model for both sliding window cases are the smallest for batch size = 128. Table A2 provides the metrics of ATCN with different filters tested for batch size = 128 and kernel size = 16. The sliding window "100-50" model exhibits the smallest RMSE, MAE, and MAPE metrics when filter = 8, according to Table A2. Table A3 demonstrates the metrics of ATCN for different kernel sizes with batch size = 128 and filters = 8. The results in Table A3 demonstrate that for the sliding window "100-10" with kernel size = 16, the RMSE, MAE, and MAPE metrics are minimum. The smallest MAE and MAPE metrics are for the sliding window "100-50" with kernel size = 16. The optimal combination of hyperparameters for the ATCN model is batch size = 128, kernel size = 16, and filters = 8.

Note that our model code runs on Windows 10, NVIDIA GeForce GTX 1650 GPU, and the deep learning framework is TensorFlow 2.6.0.

References

1. Kavzoglu, T.; Colkesen, I.; Sahin, E.K. Machine learning techniques in landslide susceptibility mapping: A survey and a case study. *Landslides Theory Pract. Model.* **2019**, *50*, 283–301.
2. Zhang, D.; Yang, J.; Li, F.; Han, S.; Qin, L.; Li, Q. Landslide Risk Prediction Model Using an Attention-Based Temporal Convolutional Network Connected to a Recurrent Neural Network. *IEEE Access* **2022**, *10*, 37635–37645. [CrossRef]
3. Cheng, Q.; Yang, Y.; Du, Y. Failure mechanism and kinematics of the Tonghua landslide based on multidisciplinary pre- and post-failure data. *Landslides* **2021**, *18*, 3857–3874. [CrossRef]
4. Wei, R.; Ye, C.; Ge, Y.; Li, Y. An attention-constrained neural network with overall cognition for landslide spatial prediction. *Landslides* **2022**, *19*, 1087–1099. [CrossRef]
5. Ghorbanzadeh, O.; Blaschke, T.; Gholamnia, K.; Meena, S.R.; Tiede, D.; Aryal, J. Evaluation of different machine learning methods and deep-learning convolutional neural networks for landslide detection. *Remote Sens.* **2019**, *11*, 196. [CrossRef]
6. Ghorbanzadeh, O.; Shahabi, H.; Crivellari, A.; Homayouni, S.; Blaschke, T.; Ghamisi, P. Landslide detection using deep learning and object-based image analysis. *Landslides* **2022**, *19*, 929–939. [CrossRef]
7. Wang, C.; Zhao, Y.; Bai, L.; Guo, W.; Meng, Q. Landslide Displacement Prediction Method Based on GA-Elman Model. *Appl. Sci.* **2021**, *11*, 11030. [CrossRef]
8. Wang, Y.; Tang, H.; Huang, J.; Wen, T.; Ma, J.; Zhang, J. A comparative study of different machine learning methods for reservoir landslide displacement prediction. *Eng. Geol.* **2022**, *298*, 106544. [CrossRef]
9. Wang, H.; Long, G.; Liao, J.; Xu, Y.; Lv, Y. A new hybrid method for establishing point forecasting, interval forecasting, and probabilistic forecasting of landslide displacement. *Nat. Hazards* **2022**, *111*, 1479–1505. [CrossRef]
10. Miao, F.; Xie, X.; Wu, Y.; Zhao, F. Data Mining and Deep Learning for Predicting the Displacement of "Step-like" Landslides. *Sensors* **2022**, *22*, 481. [CrossRef]
11. Gong, W.; Tian, S.; Wang, L.; Li, Z.; Tang, H.; Li, T.; Zhang, L. Interval prediction of landslide displacement with dual-output least squares support vector machine and particle swarm optimization algorithms. *Acta Geotech.* **2022**, *17*, 1–19. [CrossRef]
12. Lin, Z.; Ji, Y.; Liang, W.; Sun, X. Landslide Displacement Prediction Based on Time-Frequency Analysis and LMD-BiLSTM Model. *Mathematics* **2022**, *10*, 2203. [CrossRef]
13. Lin, Z.; Sun, X.; Ji, Y. Landslide Displacement Prediction Model Using Time Series Analysis Method and Modified LSTM Model. *Electronics* **2022**, *11*, 1519. [CrossRef]

14. Lin, Z.; Sun, X.; Ji, Y. Landslide Displacement Prediction Based on Time Series Analysis and Double-BiLSTM Model. *Int. J. Environ. Res. Public Health* **2022**, *19*, 2077. [CrossRef] [PubMed]
15. Zhang, Y.; Tang, J.; Cheng, Y.; Huang, L.; Guo, F.; Yin, X.; Li, N. Prediction of landslide displacement with dynamic features using intelligent approaches. *Int. J. Min. Sci. Technol.* **2022**, *32*, 539–549. [CrossRef]
16. Ma, J.; Xia, D.; Wang, Y.; Niu, X.; Jiang, S.; Liu, Z.; Guo, H. A comprehensive comparison among metaheuristics (MHs) for geohazard modeling using machine learning: Insights from a case study of landslide displacement prediction. *Eng. Appl. Artif. Intell.* **2022**, *114*, 105150. [CrossRef]
17. Ma, J.; Xia, D.; Guo, H.; Wang, Y.; Niu, X.; Liu, Z.; Jiang, S. Metaheuristic-based support vector regression for landslide displacement prediction: A comparative study. *Landslides* **2022**, 1–23. [CrossRef]
18. Sala, G.; Lanfranconi, C.; Frattini, P.; Rusconi, G.; Crosta, G.B. Cost-sensitive rainfall thresholds for shallow landslides. *Landslides* **2021**, *18*, 2979–2992. [CrossRef]
19. Domínguez-Cuesta, M.J.; Quintana, L.; Valenzuela, P.; Cuervas-Mons, J.; Alonso, J.L.; Cortés, S.G. Evolution of a human-induced mass movement under the influence of rainfall and soil moisture. *Landslides* **2021**, *18*, 3685–3693. [CrossRef]
20. Chen, C.-W.; Hung, C.; Lin, G.-W.; Liou, J.-J.; Lin, S.-Y.; Li, H.-C.; Chen, Y.-M.; Chen, H. Preliminary establishment of a mass movement warning system for Taiwan using the soil water index. *Landslides* **2022**, *19*, 1779–1789. [CrossRef]
21. Zhang, N.; Li, Q.; Li, C.; He, Y. Landslide Early Warning Model Based on the Coupling of Limit Learning Machine and Entropy Method. *J. Phys. Conf. Ser.* **2019**, *1325*, 012076. [CrossRef]
22. Fagbote, E.; Olanipekun, E.; Uyi, H. Water quality index of the ground water of bitumen deposit impacted farm settlements using entropy weighted method. *Int. J. Environ. Sci. Technol.* **2014**, *11*, 127–138. [CrossRef]
23. Omar, Y.M.; Plapper, P. A survey of information entropy metrics for complex networks. *Entropy* **2020**, *22*, 1417. [CrossRef] [PubMed]
24. Pincus, S.M. Approximate entropy as a measure of system complexity. *Proc. Natl. Acad. Sci. USA* **1991**, *88*, 2297–2301. [CrossRef] [PubMed]
25. Song, K.-S. Limit theorems for nonparametric sample entropy estimators. *Stat. Probab. Lett.* **2000**, *49*, 9–18. [CrossRef]
26. Chen, W.; Wang, Z.; Xie, H.; Yu, W. Characterization of surface EMG signal based on fuzzy entropy. *IEEE Trans. Neural Syst. Rehabil. Eng.* **2007**, *15*, 266–272. [CrossRef]
27. Bandt, C.; Pompe, B. Permutation entropy: A natural complexity measure for time series. *Phys. Rev. Lett.* **2002**, *88*, 174102. [CrossRef]
28. Park, E.; Ahn, J.; Yoo, S. Weighted-Entropy-Based Quantization for Deep Neural Networks. In Proceedings of the IEEE Conference on Computer Vision and Pattern Recognition, Honolulu, HI, USA, 21–26 July 2017; pp. 5456–5464.
29. Delgado-Bonal, A.; Marshak, A. Approximate entropy and sample entropy: A comprehensive tutorial. *Entropy* **2019**, *21*, 541. [CrossRef]
30. Huang, B.; Zheng, H.; Guo, X.; Yang, Y.; Liu, X. A Novel Model Based on DA-RNN Network and Skip Gated Recurrent Neural Network for Periodic Time Series Forecasting. *Sustainability* **2021**, *14*, 326. [CrossRef]
31. Vaswani, A.; Shazeer, N.; Parmar, N.; Uszkoreit, J.; Jones, L.; Gomez, A.N.; Kaiser, Ł.; Polosukhin, I. Attention is all you need. *Adv. Neural Inf. Process. Syst.* **2017**, *30*, 5998–6008.
32. Chen, Y.; Kang, Y.; Chen, Y.; Wang, Z. Probabilistic forecasting with temporal convolutional neural network. *Neurocomputing* **2020**, *399*, 491–501. [CrossRef]
33. Pelletier, C.; Webb, G.I.; Petitjean, F. Temporal convolutional neural network for the classification of satellite image time series. *Remote Sens.* **2019**, *11*, 523. [CrossRef]
34. Xu, Y.; Hu, C.; Wu, Q.; Li, Z.; Jian, S.; Chen, Y. Application of temporal convolutional network for flood forecasting. *Hydrol. Res.* **2021**, *52*, 1455–1468. [CrossRef]
35. Hochreiter, S.; Schmidhuber, J. Long short-term memory. *Neural Comput.* **1997**, *9*, 1735–1780. [CrossRef] [PubMed]
36. Zhang, Y.-G.; Tang, J.; He, Z.-Y.; Tan, J.; Li, C. A novel displacement prediction method using gated recurrent unit model with time series analysis in the Erdaohe landslide. *Nat. Hazards* **2021**, *105*, 783–813. [CrossRef]
37. Petersen, N.C.; Rodrigues, F.; Pereira, F.C. Multi-output bus travel time prediction with convolutional LSTM neural network. *Expert Syst. Appl.* **2019**, *120*, 426–435. [CrossRef]

MDPI
St. Alban-Anlage 66
4052 Basel
Switzerland
www.mdpi.com

Sensors Editorial Office
E-mail: sensors@mdpi.com
www.mdpi.com/journal/sensors

www.ingramcontent.com/pod-product-compliance
Lightning Source LLC
LaVergne TN
LVHW070149170726
843515LV00007B/1877

* 9 7 8 3 0 3 6 5 9 7 8 6 7 *